Sugarcane Bioenergy for Sustainable Development

In recent years, there has been a rapid expansion of the growing of crops for use in bioenergy production rather than for food. This has been particularly the case for sugarcane in Latin America and Africa. This book examines the further potential in the context of the food versus fuel debate, and as a strategy for sustainable development.

Detailed case studies of two countries, Colombia and Mozambique, are presented. These address the key issues such as the balance between food security and energy security, rural and land development policies, and feasibility and production models for expanding bioenergy. The authors then assess these issues in the context of broader sustainable development strategies, including implications for economics, employment generation, and the environment. The book will be of great interest to researchers and professionals in energy and agricultural development.

Luís A. B. Cortez is a former Professor in the Faculty of Agricultural Engineering (FEAGRI) and Interdisciplinary Center of Energy Planning (NIPE) at the State University of Campinas (UNICAMP), Brazil. Presently, he is Institutional Relations Advisor of the Brazilian Center for Research in Energy and Materials (CNPEM), Brazil.

Manoel Regis L. V. Leal is a Researcher for the LACAf Project at the Interdisciplinary Center of Energy Planning (NIPE) of the State University of Campinas (UNICAMP), Brazil, and Researcher for the Brazilian Bioethanol Science and Technology National Laboratory (CTBE), Brazil.

Luiz A. Horta Nogueira is consultant to the United Nations Economic Commission for Latin America and the Caribbean, and Associate Researcher at the Interdisciplinary Center of Energy Planning (NIPE) of the State University of Campinas (UNICAMP), Brazil.

Routledge Studies in Bioenergy

Miscanthus
For Energy and Fibre
Edited by Michael Jones and Mary Walsh

Bioenergy Production by Anaerobic Digestion
Using Agricultural Biomass and Organic Wastes
Edited by Nicholas Korres, Padraig O'Kiely, John A.H. Benzie and Jonathan S. West

The Biomass Assessment Handbook
Second Edition
Edited by Frank Rosillo-Calle, Peter de Groot, Sarah L. Hemstock and Jeremy Woods

Handbook of Bioenergy Crops
A Complete Reference to Species, Development and Applications
N. El Bassam

Biofuels, Food Security, and Developing Economies
Nazia Mintz-Habib

Bioenergy Crops for Ecosystem Health and Sustainability
Alex Baumber

Biofuels and Rural Poverty
Joy Clancy

Sugarcane Bioenergy for Sustainable Development
Expanding Production in Latin America and Africa
Edited by Luís A. B. Cortez, Manoel Regis L. V. Leal and Luiz A. Horta Nogueira

https://www.routledge.com/Routledge-Studies-in-Bioenergy/book-series/RSBIOEN

Sugarcane Bioenergy for Sustainable Development

Expanding Production in Latin America and Africa

Edited by
Luís A. B. Cortez,
Manoel Regis L. V. Leal
and Luiz A. Horta Nogueira

First published 2019
by Routledge
2 Park Square, Milton Park, Abingdon, Oxon OX14 4RN

and by Routledge
711 Third Avenue, New York, NY 10017

Routledge is an imprint of the Taylor & Francis Group, an informa business

© 2019 selection and editorial matter, Luís A. B. Cortez, Manoel Regis L. V. Leal and Luiz A. Horta Nogueira; individual chapters, the contributors

The right of Luís A. B. Cortez, Manoel Regis L. V. Leal and Luiz A. Horta Nogueira to be identified as the authors of the editorial material, and of the authors for their individual chapters, has been asserted in accordance with sections 77 and 78 of the Copyright, Designs and Patents Act 1988.

All rights reserved. No part of this book may be reprinted or reproduced or utilised in any form or by any electronic, mechanical, or other means, now known or hereafter invented, including photocopying and recording, or in any information storage or retrieval system, without permission in writing from the publishers.

Trademark notice: Product or corporate names may be trademarks or registered trademarks, and are used only for identification and explanation without intent to infringe.

British Library Cataloguing-in-Publication Data
A catalogue record for this book is available from the British Library

Library of Congress Cataloging-in-Publication Data
Names: Cortez, Luís Augusto Barbosa, editor. | Leal, Manoel Regis L. V. (Manoel Regis Lima Verde), editor. | Nogueira, Luiz Augusto Horta, editor. | LACAf Project, author. | Universidade Estadual de Campinas.
Title: Sugarcane bioenergy for sustainable development: expanding production in Latin America and Africa / edited by Luis A.B. Cortez, Manoel Regis L.V. Leal and Luiz A. Horta Nogueira.
Description: Abingdon, Oxon; New York, NY: Routledge, 2018. | Series: Routledge studies in bioenergy | Includes case reports from results from the LACAF Project in which the Colombian and Mozambican cases are presented. Includes work from the State University of Campinas, UNICAMP, Brazil. |
Includes bibliographical references and index.
Identifiers: LCCN 2018020863 | ISBN 9781138312944 (hardback) | ISBN 9780429457920 (ebk).
Subjects: LCSH: Sugarcane industry—Latin America. | Sugarcane industry—Africa. | Biomass energy—Economic aspects. | Sustainable development.
Classification: LCC HD9502.5.B543 L297 2018 | DDC 338.4/766288096—dc23
LC record available at https://lccn.loc.gov/2018020863

ISBN: 978-1-138-31294-4 (hbk)
ISBN: 978-0-429-45792-0 (ebk)

VN: 9 October 2018

Typeset in Bembo
by codeMantra

Printed in Canada

Contents

List of figures	x
List of tables	xv
List of authors	xix
Acknowledgements	xxiii
Foreword by Lee R. Lynd	xxiv
Foreword by Carlos Henrique de Brito Cruz	xxv

Introduction: Production of sugarcane bioenergy in Latin America and Africa — 1
LUÍS A. B. CORTEZ, LUIZ A. HORTA NOGUEIRA
AND MANOEL REGIS L. V. LEAL

PART I
Bioenergy overview: The context — 7

1 The future of fuel ethanol — 9
ROGÉRIO CEZAR DE CERQUEIRA LEITE, LUÍS A. B. CORTEZ,
CARLOS EDUARDO DRIEMEIER, MANOEL REGIS L. V. LEAL,
AND ANTONIO BONOMI

2 Bioenergy in the world: Present status and future — 27
JOSÉ GOLDEMBERG

3 Is there really a food versus fuel dilemma? — 35
FRANK ROSILLO-CALLE

4 Integrating pasture intensification and bioenergy crop expansion — 46
ELEANOR E. CAMPBELL, JOHN JOSEPH SHEEHAN, DEEPAK
JAISWAL, JOHNNY R. SOARES, JULIANNE DE CASTRO OLIVEIRA,
LEONARDO AMARAL MONTEIRO, ANDREW M. ALLEE, RUBENS
AUGUSTO C. LAMPARELLI, GLEYCE K.D. ARAÚJO FIGUEIREDO
AND LEE R. LYND

vi Contents

5 The role of the sugarcane bioeconomy in supporting energy access and rural development 60
FRANCIS JOHNSON AND ROCIO A. DIAZ-CHAVEZ

6 Is the sugarcane capital goods sector prepared to respond to large-scale bioenergy supply in developing countries? 71
JOSÉ LUIZ OLIVÉRIO AND PAULO AUGUSTO SOARES

7 Energy medium- and long-term perspectives: Are we moving toward an "all-electric model"? 87
CYLON GONÇALVES DA SILVA

8 Comparative assessment of non-ethanol biofuel production from sugarcane lignocelluloses in Africa, including synfuels, butanol and jet fuels 97
JOHANN F. GÖRGENS AND MOHSEN MANDEGARI

9 2G 2.0 109
LEE R. LYND

10 Medium- and long-term prospects for bioenergy trade 121
SERGIO C. TRINDADE AND DOUGLAS NEWMAN

PART II
Why bioenergy? 133

11 Why promote sugarcane bioenergy production and use in Latin America, the Caribbean and Southern Africa? 135
LUIZ A. HORTA NOGUEIRA

12 The potential of bioenergy from sugarcane in Latin America, the Caribbean and Southern Africa 141
LUIS CUTZ AND LUIZ A. HORTA NOGUEIRA

13 Reconciling food security, environmental preservation and biofuel production: Lessons from Brazil 154
JOÃO GUILHERME DAL BELO LEITE, MANOEL REGIS L. V. LEAL, LUÍS A. B. CORTEZ, LEE R. LYND AND FRANK ROSILLO-CALLE

14 Energy security and energy poverty in Latin
America and the Caribbean and sub-Saharan Africa 172
MANOEL REGIS L. V. LEAL, JOÃO GUILHERME DAL
BELO LEITE AND LUIZ A. HORTA NOGUEIRA

15 Sustainable bioenergy production in Mozambique as a
vector for economic development 181
MARCELO PEREIRA DA CUNHA, LUIZ GUSTAVO ANTÔNIO DE SOUZA,
ANDRÉ DE TEIVE E ARGOLLO, JOÃO VINÍCIUS NAPOLITANO DA
CUNHA AND LUIZ A. HORTA NOGUEIRA

16 Integrated analysis of bioenergy systems: The path
from initial prospects to feasibility and acceptability 198
LUIZ A. HORTA NOGUEIRA

PART III
Availability of land and sugarcane potential production 213

17 Initial considerations to estimate sugarcane potential
in Mozambique and Colombia 215
MARCELO MELO RAMALHO MOREIRA AND FELIPE
HAENEL GOMES

18 Sustainability aspects: Restrictions and potential
production 217
MARCELO MELO RAMALHO MOREIRA AND FELIPE
HAENEL GOMES

19 Case study: Potential of sugarcane production for
Mozambique 227
MARCELO MELO RAMALHO MOREIRA, FELIPE HAENEL GOMES
AND KARINE MACHADO COSTA

20 Case study: Potential of sugarcane production
for Colombia 243
FELIPE HAENEL GOMES, MARCELO MELO RAMALHO MOREIRA,
MARIANE ROMEIRO AND RAFAEL ALDIGHIERI MORAES

21 Final comments 258
MARCELO MOREIRA

PART IV
Impacts and feasibility of sugarcane production   261

22 Sugarcane production model and sustainability indicators   263
MANOEL REGIS L. V. LEAL AND JOÃO
GUILHERME DAL BELO LEITE

23 Sugarcane and energy poverty alleviation   273
JOÃO GUILHERME DAL BELO LEITE AND MANOEL REGIS L. V. LEAL

24 Why modern bioenergy and not traditional production and use can benefit developing countries?   284
LUÍS A. B. CORTEZ, MANOEL REGIS L. V. LEAL,
LUIZ A. HORTA NOGUEIRA, ARIELLE MUNIZ KUBOTA
AND RICARDO BALDASSIN JUNIOR

25 Sugarcane bioenergy production systems: Technical, socioeconomic and environmental assessment   299
TEREZINHA DE FÁTIMA CARDOSO, MARCOS DJUN BARBOSA
WATANABE, ALEXANDRE SOUZA, MATEUS FERREIRA CHAGAS,
OTÁVIO CAVALETT, EDVALDO RODRIGO DE MORAIS, LUIZ
A. HORTA NOGUEIRA, MANOEL REGIS L. V. LEAL, OSCAR
ANTONIO BRAUNBECK, LUÍS A. B. CORTEZ
AND ANTONIO BONOMI

26 Sustainability of scale in sugarcane bioenergy production   317
MATEUS FERREIRA CHAGAS, OTAVIO CAVALETT, CHARLES
DAYAN FARIA DE JESUS, MARCOS DJUN BARBOSA WATANABE,
TEREZINHA DE FÁTIMA CARDOSO, JOÃO GUILHERME DAL
BELO LEITE, MANOEL REGIS L. V. LEAL, LUÍS A. B. CORTEZ
AND ANTONIO BONOMI

27 Vinasse, a new effluent from sugarcane ethanol production   331
CARLOS EDUARDO VAZ ROSSELL

PART V
Assessing the LACAf Project findings — 345

28 **Final remarks on production model alternatives** — 347
MANOEL REGIS L. V. LEAL AND JOÃO GUILHERME DAL BELO LEITE

29 **Alcohol and LACAf project in Colombia** — 357
JOSÉ MARIA RINCÓN MARTÍNEZ, JESSICA A. AGRESSOT RAMIREZ AND DIANA M. DURÁN HERNÁNDEZ

30 **Sugarcane ethanol in Guatemala** — 370
AIDA LORENZO, MARIO MELGAR AND LUIZ A. HORTA NOGUEIRA

31 **Methodology of assessment and mitigation of risks facing bioenergy investments in sub-Saharan Africa** — 377
RUI CARLOS DA MAIA

32 **Can sugarcane bring about a bioenergy transformation in sub-Saharan Africa?** — 388
WILLEM HEBER VAN ZYL

33 **Sugarcane bioenergy, an asset for Latin America, the Caribbean and Africa?** — 400
PATRICIA OSSEWEIJER

34 **Closing remarks** — 407
LUÍS A. B. CORTEZ, MANOEL REGIS L. V. LEAL AND LUIZ A. HORTA NOGUEIRA

Index — *411*

Figures

I.1	Available land in selected countries in Africa and Latin America	2
I.2	Evolution of Brazilian energy dependence 1970–2015	2
1.1	View of the CTBE (in the center) at CNPEM in Campinas, Brazil, where CTBE is located	13
1.2	Global technology penetrations in LDV stock by scenario (2DS and RTS), 2020–2060	15
1.3	Projections of fuel consumed (in PJ) in different types of transportation in different types of energy (2DS)	17
1.4	The role of liquid biofuels – 2DS, 2020–2060	18
1.5	Participation of different energy sources in the overall energy consumption in the world, considering 2DS, 4DS, and 6DS	19
2.1	Traditional and modern biomass use in 2015 (left) and modern bioenergy growth by sector (2008–2015)	28
2.2	Growth rate of bioenergy to the energy supply in the period 2008–2015	28
2.3	Modern bioenergy in 2050 (EJ)	29
2.4	Biomass potential (2050)	30
2.5	Evolution of ethanol production in Brazil and the United States (1970–2015)	31
2.6	Productivity and sugar content (ATR) from 165 sugar/ethanol mills in the southeast of Brazil	33
3.1	Land use in the production of biofuels in 2016 (Mha)	38
3.2	Worldwide production of total grains, and major crops, from 1971/72 to 2017/18 harvest	39
3.3	Worldwide consumption of grains and major uses, from 2010/11 to 2017/18 harvest	40
3.4	World ethanol fuel production and FAO Food Price Index, from 2008 to 2017	41
4.1	The distribution of land use by humans, as calculated from estimates by Lambin and Meyfroidt (2011), excluding natural forest and unmanaged productive lands from "Global Land Use". Also, the relative contributions of	

Figures xi

	vegetal and livestock products to global protein and Calories (FAO, 2012). *Globally, grazing has been estimated to contribute 3% of dietary protein and 1% of dietary energy (Woods et al., 2015)	47
4.2	Brazilian pasture area and livestock performance from 1950 to 2006 (A). Brazilian land use change for sugarcane expansion from 2000 to 2009 (B). Adapted from Adami et al. (2012); Martha Jr. et al. (2012)	49
4.3	Global spatial distribution of pasturelands from Ramankutty et al. (2008) (A); and global spatial distribution of climate space defined in two dimensions: total annual precipitation and growing degree-days, as reported by WorldClim (B)	50
4.4	Global spatial distribution of livestock production systems under arid and semiarid climates in a spatial resolution of 5 × 5 arcmin, totaling 745 million hectares (A), and associated theoretical ethanol yields from varying species of Agave (B)	53
5.1	Sugarcane production by region (1994–2016)	66
5.2	Global area and production of sugarcane over time (1994–2016)	66
5.3	Top 10 sugarcane producers (1994–2016)	67
5.4	Production and area in Africa over time (1994–2016)	67
6.1	"Blending octane" ratings of various gasoline additives/components	72
6.2	Scenarios to achieve 10% ethanol	74
6.3	Brazilian sugarcane/sugar/ethanol production and the three great leaps	75
6.4	DSM can be implemented in stages and Barralcool Mill is a first-generation four BIOs Mill	81
6.5	DSM: the six BIOproducts Mill is designed to meet the optimization/zero concepts reaching maximum GHG mitigation effect	82
6.6	Impacts in building an EPZ	84
8.1	Annexed biorefinery for biofuel production, integrated into an existing sugar mill	99
9.1	Land required as a function of energy supply as impacted by biomass yield	112
9.2	Linkages between bioenergy and other things we care about	113
10.1	FAME world output − 1,000 tons	124
10.2	FAME output − 1,000 m. Tons − 2008–2017 (United States, Brazil, Argentina, Colombia, Canada)	124
10.3	Ethanol import penetration typically is the greatest in the EU market	128
10.4	Ethanol export production share rises in the United States and falls in Brazil	129

xii *Figures*

12.1	Population and petroleum consumption per capita for selected LAC and SA countries for 2014. Population data from World By Map (2016) and Petroleum consumption per capita from World By Map (2014). The countries are shown in decreasing order of their petroleum consumption	142
12.2	Countries adopting biofuels blending mandates	143
12.3	(a) BAU scenario; (b) NF scenario. Potential ethanol supply in LAC and SA. Identifications (Ex) indicate potential gasoline blend that could be achieved in each country	146
12.4	Potential bioelectricity production in the NF scenario. (x%) indicates the percentage of sugarcane bagasse in the total electricity generation	148
12.5	Potential GHG savings due to gasoline and electricity displacement in LAC and SA for the NF scenario	149
13.1	Availability: (A) average dietary energy supply adequacy (%); and (B) average protein supply (g per capita per day)	158
13.2	Access: (A) prevalence of undernourishment (%); (B) depth of the food deficit (kcal per capita per day); (C) prevalence of food inadequacy (%)	159
13.3	Stability: (A) cereal import dependency ratio (%); utilization: (B) children affected by malnutrition (%)	160
13.4	Food production and cultivated area in Brazil (A), food production and cultivated area per capita (B)	161
13.5	Sugarcane area and ethanol production (A), and primary energy production in Brazil (B)	163
13.6	Deforested area in Brazil ((A) Amazon, Cerrado and Caatinga, and (B) Atlantic forest, Pantanal and Pampa)	164
13.7	Pasture area and cattle herd in Brazil	165
13.8	Planted area of maize (first crop, second crop and total) in Brazil from 1990 to 2014	166
15.1	Relationship between HDI and GDP for 178 countries in 2011	195
15.2	Comparison between estimated and observed HDIs for Mozambique	195
16.1	The structure of the SIByl-LACAf approach	200
16.2	Number of papers (published between 1970 and 2016) applying evaluation method in bioenergy systems	203
16.3	The decision flow in the feasibility and acceptability evaluation in the SIByl-LACAf	207
18.1	Summary of the main characteristics used in the CTC potential production classification	222
19.1	Simplified diagram of the mapping process	229
19.2	Map of land use and land cover in Mozambique in 2013	231
19.3	Strong restrictions (physical and legal)	233
19.4	Moderate restrictions (climate, non-mechanizable and high carbon stocks)	233
19.5	Moderate restrictions (small-scale agriculture and cropland)	234

19.6	Potential for sugarcane production in Mozambique without strong restrictions	236
19.7	Potential for sugarcane production in Mozambique without strong and moderate restrictions	240
20.1	Map of the natural regions of Colombia	244
20.2	Sugarcane harvested area by department	245
20.3	Map of sugar mills and sugarcane in Colombia	245
20.4	Scheme of the methodology used in the study	247
20.5	Land use and land cover map of the Colombian territory for the 2012/2014 period	249
20.6	Map of legally protected areas	250
20.7	Climate related restrictions (temperature, precipitation and hydric deficit)	251
20.8	Non-mechanizable areas, high carbon stocks and areas occupied by agriculture	252
20.9	Map of edaphic suitability for sugarcane in Colombia	253
20.10	Potential for sugarcane production in Colombia without strong restrictions	254
20.11	Potential for sugarcane production in Colombia without strong and moderate restrictions	255
24.1	Estimated renewable energy share of global energy consumption (2015)	284
24.2	Developing countries: use of TB vs. poverty, undernourishment, electricity access and quality of life	285
24.3	Types of bioenergy systems models according to agriculture, industrial scales and technological/management levels	288
24.4	Production costs (dots) and productivity (columns) of sugarcane, soybean and corn in Brazil according to agricultural scale (2014)	292
24.5	Evolution of number of formal jobs and average salary in Brazilian agriculture and livestock (2002–2014)	293
24.6	Evolution of expansion areas and deforestation, and agricultural and livestock production in Brazil (1991–2014)	293
25.1	Evaluated scenarios – description based on main agricultural operations	301
25.2	Sugarcane biomass production costs US$ per ton considering risk assessment. Stalks production is represented in white bars, while straw recovery is represented in gray bars. 1: Manual Burned; 2: Manual Green; 3: Manual Bales; 4; Mechanized Burned; 5: Mechanized Green; 6: Mechanized Bales; 7: Mechanized Integral	308
25.3	Industrial yields of considered scenarios	309
25.4	IRRs of vertically integrated systems considering the uncertainties on biomass production costs. 1: Manual Burned; 2: Manual Green; 3: Manual Bales; 4: Mechanized Burned; 5: Mechanized Green; 6: Mechanized Bales; 7: Mechanized Integral	310

xiv *Figures*

25.5	Environmental impacts per unit on mass of ethanol produced in the evaluated scenarios. CC: climate change; ODP: ozone depletion potential; PMF: particulate matter formation; AP: terrestrial acidification potential; FD: fossil depletion	311
25.6	Number of jobs and occupational accidents, per million liters of ethanol	313
26.1	CanaSoft model structure	319
26.2	Agricultural investments (A) and sugarcane production cost (B)	322
26.3	Industrial investments (A) and ethanol production cost (B)	323
26.4	Comparative environmental impacts per liter of ethanol as a function of scale	324
26.5	Sensitivity assessment for sugarcane production cost in microdistilleries	325
26.6	Sensitivity analysis for ethanol production cost in microdistilleries	326
26.7	Sensitivity analysis for climate change impact of ethanol production in microdistilleries	327
28.1	Impacts of plant size on industrial investments and ethanol production costs	350
29.1	Variation of carbon monoxide emissions by adding ethanol to gasoline: speed characteristic test	358
29.2	Variation of carbon dioxide emissions by adding ethanol to gasoline: speed characteristic test	359
29.3	Anhydrous alcohol obtaining process	360
29.4	Process diagram of cogeneration in the sugar industry	362
29.5	Cogeneration capacity in sugarcane mills (MW)	363
29.6	Complementarity between the production of sugarcane and rainfall	363
29.7	Sugarcane suitability	365
29.8	Future processing of sugarcane	368
30.1	Guatemala sugarcane geographical location (Melgar et al., 2012)	371
30.2	Sugarcane and sugar yields in Guatemala (Meneses et al., 2017)	373
32.1	Over 620 million people in sub-Saharan Africa do not have access to electricity	390
32.2	Areas available and suitable for sugarcane cultivation in southern Africa	393

Tables

1.1	Energy used in different modes of transportation as an effort to maximize fuel savings and emissions – 2DS	14
1.2	Global technology penetrations in LDV stock by scenario (2DS and RTS), 2020–2060	15
1.3	Projections of fuel consumed (in Million Tonnes of Oil Equivalent (MTOE)) in different types of transportation in different types of energy	16
1.4	Projections of fuel consumed (in PJ) in different types of transportation in different types of energy (2DS and RTS)	17
1.5	Energy used for different liquid biofuels – 2DS (EJ), 2020–2060	18
2.1	Biomass of bioenergy use (EJ) (2050)	29
2.2	Comparing the projections of the biomass potential in four categories: farm residues, forest residues, energy crops and port-consumer waste (municipal solid waste)	30
2.3	Comparison between agricultural and industrial yields using traditional biomass and "energy cane"	33
6.1	"New-Mills" installed at "third-great-leap"	76
6.2	"Total-Mills"/"New-Mills" classified by products (2012)	76
6.3	Sugarcane mill evolution performance as a function of the available technology and equipment	77
6.4	The three pillars of sustainable development define the DSM characteristics	81
7.1	"Battery buying power" from the savings on fuel expenses for the EV driver	94
8.1	Comparing the conversion efficiencies and energy demands for the conversion of sugarcane lignocellulose to synfuels or n-butanol in a biorefinery integrated into an existing sugar mill	102
8.2	Comparing the economics of the conversion of sugarcane lignocellulose to synfuels or n-butanol in a biorefinery integrated into an existing sugar mill	103
8.3	Comparing the conversion efficiencies and energy demands for the production of aviation biofuels in a stand-alone conversion facility, not integrated into an existing sugar mill	104

8.4	Comparing the required MJSPs for the production of aviation biofuels in a stand-alone conversion facility, not integrated into an existing sugar mill	104
9.1	Features of companies involved in advanced biofuels over the last decade	114
12.1	Current sugarcane production and 1% of the pastureland in LAC and SA	144
12.2	Number of new 1-Mt sugarcane mills and total investment for the NF scenario	150
14.1	Global energy access 2016	174
14.2	Energy access in selected countries in SSA and LA&C	174
14.3	Global access to clean cooking fuels in 2015	175
14.4	Access to clean cooking fuels in selected countries of SSA and Central and South America in 2014	176
14.5	LA&C sugarcane and sugar statistics for selected countries	177
14.6	Global ethanol production in selected countries	178
14.7	Mozambique and South Africa energy profiles: electricity and gasoline consumption	179
15.1	SAM structure	187
15.2	Type-I output multipliers	189
15.3	Type-II output multipliers	190
15.4	Socioeconomic impacts considering direct and indirect effects	192
15.5	Socioeconomic impacts considering direct, indirect and induced effects	193
16.1	Analytic methods used for evaluating bioenergy systems by IA	202
16.2	GBEP sustainability indicators for bioenergy	205
18.1	WRB Reference Soil Group description and correlation with Brazil and US systems	223
19.1	Data used for the construction of the land use and coverage maps of Mozambique	230
19.2	Land use and land cover in Mozambique in the years 2001 and 2013 (1,000 ha and %)	230
19.3	Total area without strong restrictions in each Mozambican province (1,000 ha and %)	235
19.4	Distribution of potential production according to active moderate restriction	237
19.5	Total area without strong or moderated restrictions in each Mozambican province (1,000 ha and %)	238
19.6	Area (ha) of high, medium and low potentials considering levels of restriction in Mozambique	238
20.1	Land use and land cover in Colombia in the years 2000/2002 and 2012/2014 (1,000 ha and %)	248
20.2	Area (ha) of high, medium and low potentials considering levels of restriction in Colombia	255
22.1	Biofuel project types	266

22.2	GBEP sustainability indicators	268
22.3	Brazilian sugar/ethanol sector profile in 2011	270
23.1	Description of baseline and alternative sugarcane-based scenarios	276
23.2	Production of sugar, electricity and ethanol from a standard SSA sugar mill (baseline) and alternative scenarios	278
23.3	Number of households supplied with electricity and cooking ethanol under different sugarcane-oriented scenarios	278
23.4	Potentially avoided deforestation (ha year^{-1}) by substituting cooking firewood for cooking ethanol	279
24.1	Summary of the advantages/benefits and disadvantages of energy models	294
25.1	Main agricultural parameters considered in the scenarios	302
25.2	Main operational characteristics of the industrial phase	302
25.3	Ranges considered for parameters in the risk assessment of agricultural scenarios	305
25.4	Main components of sugarcane stalks and straw production costs according to CanaSoft model	306
25.5	Sugarcane stalk harvest and straw recovery	307
25.6	Main results of the economic analysis of the vertically integrated scenarios	310
25.7	Workers in agricultural phase in the evaluated scenarios, considering a production of 2 million tons of sugarcane stalks per year	312
25.8	Average wage of workers in the different scenarios	313
26.1	Main agricultural and industrial parameters for large-scale scenarios	320
26.2	Sensitivity analysis parameters for microdistillery yields	321
26.3	Technical results – ethanol and electricity yields per ton of sugarcane and ethanol and electricity annual production	322
27.1	Physical and chemical characteristics of vinasse from molasses, cane juice and mixed musts	331
27.2	Physical and chemical characteristics of vinasse obtained from blackstrap molasses from different fermentation processes	332
28.1	Jobs in agriculture operations	348
28.2	Average wages in agricultural and industrial areas	348
28.3	Sugarcane production costs under different harvesting alternatives	348
28.4	Industrial and economic performance	349
28.5	Sugarcane mill investment under different scenarios	354
29.1	Comparison between Colombia and Brazil	366
30.1	Land use in Guatemala in 2016 (ASAZGUA, 2017)	372
30.2	Area cultivated with sugarcane, milled cane and sugar production in Guatemala	372
30.3	Guatemala sugarcane ethanol production capacity	374
31.1	Problem table – listing barriers to development of bioenergy projects in Africa	381

31.2 Interventions table – enabling actions to help success of
    bioenergy projects in Africa 382
32.1 Population of the world regions forecasted until 2100 389
32.2 Lessons from Brazil experience for sub-Saharan Africa 394

Authors

Aida Lorenzo, Senior Researcher, Asociación de Combustibles Renovables de Guatemala

Alexandre Souza, Research Scientist, Brazilian Bioethanol Science and Technology Laboratory (CTBE), Brazil

André de Teive E Argollo, Institute of Economics, UNICAMP, Brazil

Andrew M. Allee, PhD Student, Thayer School of Engineering, Dartmouth College, USA

Antonio Bonomi, Senior Researcher, Brazilian Bioethanol Science and Technology Laboratory (CTBE), Brazil

Arielle Muniz Kubota, Research Scientist, NIPE – UNICAMP, Brazil

Carlos Eduardo Driemeier, Research Scientist, Brazilian Bioethanol Science and Technology Laboratory (CTBE), Brazil

Carlos Eduardo Vaz Rossell, Senior Researcher, NIPE – UNICAMP, Brazil

Carlos Henrique de Brito Cruz, Scientific Director of FAPESP, FAPESP, Brazil

Charles Dayan Faria de Jesus, Research Scientist, Brazilian Bioethanol Science and Technology Laboratory (CTBE), Brazil

Cylon Gonçalves da Silva, Emeritus Professor, Institute of Physics Gleb Wataghin, UNICAMP, Brazil

Deepak Jaiswal, Postdoctoral Researcher, School of Agricultural Engineering – UNICAMP, Brazil

Diana M. Durán Hernández, Professor, University of Colombia, Colombia

Douglas Newman, Former International Trade Analyst, U.S. International Trade Commission

Edvaldo Rodrigo de Morais, Research Scientist, Brazilian Bioethanol Science and Technology Laboratory (CTBE), Brazil

Eleanor E. Campbell, Research Scientist, Earth Systems Research Center, University of New Hampshire, USA

Felipe Haenel Gomes, Agronomist, Pedologica, Brazil

Francis Johnson, Senior Researcher, Stockholm Environment Institute, Sweden

Frank Rosillo-Calle, Honorary Senior Research Fellow, CEP, Imperial College London, UK

Gleyce K.D. Araújo Figueiredo, Professor, School of Agricultural Engineering – UNICAMP, Brazil

Jessica A. Agressot Ramirez, Professor, University of Colombia, Colombia

João Guilherme Dal Belo Leite, Professor, Universidade Federal da Fronteira Sul (UFFS), Brazil

João Vinícius Napolitano da Cunha, School of Economics, Business and Accounting, University of São Paulo, Brazil

Johann F. Görgens, Professor, Department of Process Engineering, Stellenbosch University, South Africa

John Joseph Sheehan, Senior Researcher, Department of Soil and Crop Sciences, Colorado State University, Fort Collins, Colorado, USA

Johnny R. Soares, Postdoctoral Researcher, School of Agricultural Engineering – FEAGRI, UNICAMP, Brazil

José Goldemberg, Emeritus Professor, University of São Paulo – USP, Brazil

José Luiz Olivério, Former CEO DEDINI, DEDINI, Brazil

José Maria Rincón Martínez, Professor, University of Colombia, Colombia

Julianne de Castro Oliveira, Postdoctoral Researcher, School of Agricultural Engineering – UNICAMP, Brazil

Karine Machado Costa, Researcher, AGROICONE, Brazil

Lee R. Lynd, Professor, Dartmouth College, USA

Leonardo Amaral Monteiro, Postdoctoral Researcher, School of Agricultural Engineering – UNICAMP, Brazil

Luís A. B. Cortez, Professor, School of Agricultural Engineering – UNICAMP, Brazil

Luis Cutz, Research Scientist, Chalmers University of Technology, Sweden

Luiz A. Horta Nogueira, Professor, NIPE – UNICAMP, Brazil

Luiz Gustavo Antônio de Souza, Lecturer, Brazilian Air Force Academy, Pirassununga, Brazil

Manoel Regis L. V. Leal, Senior Researcher, Brazilian Bioethanol Science and Technology Laboratory (CTBE), Brazil

Marcelo Melo Ramalho Moreira, Senior Researcher, AGROICONE, Brazil

Marcelo Pereira da Cunha, Professor, Institute of Economics, UNICAMP, Brazil

Marcos Djun Barbosa Watanabe, Research Scientist, Brazilian Bioethanol Science and Technology Laboratory (CTBE), Brazil

Mariane Romeiro, Researcher, AGROICONE, Brazil

Mario Melgar, Senior Researcher, Centro Guatemalteco de Investigación y Capacitación de la Caña de Azúcar

Mateus Ferreira Chagas, Research Scientist, Brazilian Bioethanol Science and Technology Laboratory (CTBE), Brazil

Mohsen Mandegari, Research Scientist, Department of Process Engineering, Stellenbosch University, South Africa

Oscar Antonio Braunbeck, Professor, School of Agricultural Engineering – UNICAMP, Brazil

Otávio Cavalett, Researcher, Brazilian Bioethanol Science and Technology Laboratory (CTBE), Brazil

Patricia Osseweijer, Professor, Delft University of Technology, The Netherlands

Paulo Augusto Soares, Senior Researcher, DEDINI, Brazil

Rafael Aldighieri Moraes, Researcher, Kroton Educacional, Brazil

Ricardo Baldassin Junior, Research Scientist, Agropolo Campinas-Brasil, Brazil

Rocio A. Diaz-Chavez, Research Scientist, Imperial College London, UK

Rogério Cezar de Cerqueira Leite, Emeritus Professor, Brazilian Center for Research in Energy and Materials (CNPEM)

Rubens Augusto C. Lamparelli, Senior Researcher, NIPE – UNICAMP, Brazil

Rui Carlos da Maia, Professor of environmental engineering and disaster management at the Technical University of Mozambique and director of university extension services, Technical University of Mozambique (UDM), Mozambique

Sergio C. Trindade, Global sustainable business consultant, SE2T, USA

Terezinha de Fátima Cardoso, Researcher, Brazilian Bioethanol Science and Technology Laboratory (CTBE), Brazil

Willem Heber van Zyl, Professor, Stellenbosch University (SU), South Africa

Acknowledgements

We would like to thank the São Paulo Research Foundation (FAPESP) for the financing and the support of the Bioenergy Contribution of Latin America & Caribbean and Africa to the GSB Project – LACAf-Cane Thematic Project (Process Number 2012/00282-3) (http://bioenfapesp.org/gsb/lacaf/) from March 2013 to May 2018.

The financial support received from FAPESP allowed us to coordinate a large team composed of researchers from different parts of the world, but mainly from Latin America (Colombia and Guatemala) and Africa (Mozambique and South Africa). We have organized field trips, meetings, and workshops in many different regions where we have now the opportunity to present a great part of it in the book format.

The Editors

Foreword

In 2009, assessments of merit and need for bioenergy (fuel, electricity, and heat from plant biomass) were increasingly divergent. In response to this, a group of colleagues formed the Global Sustainable Bioenergy project, later referred to as the Global Sustainable Bioenergy (GSB) initiative. Our overall goal was to provide guidance with respect to the feasibility and desirability of sustainable, bioenergy-intensive futures. Two questions were articulated early on:

Could we? That is, is it physically possible to gracefully reconcile large-scale bioenergy production with feeding humanity, meeting demands from managed lands, and preserving wildlife habitat and environmental quality?

Must we? That is, do we have to produce bioenergy at a large scale in order to have a reasonable expectation of achieving a sustainable world?

These questions have remained a prominent focus of the GSB initiative till now, joined by an increasing interest as the project has progressed using bioenergy to advance human development.

The GSB initiative held five continental conventions in 2010 which were instrumental in sharing perspectives, developing plans for follow-up activities, and developing a network that has led to notable outcomes. Underlying the intellectual and analytical questions that project participants were asking each other was a more practical question: *What institutions would provide support to address the research questions identified in the GSB conventions?* That institution turned out to be the São Paulo Research Foundation (FAPESP), with a first project focused on sugarcane bioenergy in selected Latin American, Caribbean, and African (LACAf) countries.

This book highlights the accomplishments and expands the contributions of the LACAf project with chapters authored by many of the project leaders. A chapter is also included describing a second FAPESP-supported GSB project addressing geospatial aspects, currently in midstream. Ably edited by Luís Cortez, the LACAf project leader, the book informs paths to socially beneficial and sustainable deployment of bioenergy in many parts of the world where both human development needs and the biomass resource potential are great. I am grateful to Luis and the chapter authors for their effort, insights, and friendship.

Lee R. Lynd
November, 2017

Foreword

Reducing the emissions of greenhouse gases to retard or avoid global climate change is one of the greatest challenges of our time. The use of energy to improve the quality of life is one of the main causes of emissions, and this leads to a breadth of research on renewable (low-emission) sources of energy. Science results recognize that several combined initiatives are necessary to reduce emissions, and most scenarios point to the increased use of bioenergy as a necessary contribution to lowering emissions.

In this book, the authors study how Africa and Latin America could enhance their use of bioenergy to achieve the goal of reducing emissions and obtaining other benefits. These are developing regions where social progress will imply and require more extensive use of energy for the improvement of the quality of life. Also, in many cases, the countries in the region do not have access to oil. Fortunately, Latin America and Africa have available land and enough sunlight to harvest sugarcane, the most efficient plant so far used for the large-scale production of bioenergy.

The experience of Brazil with ethanol produced from sugarcane demonstrates the enormous possibilities of this path. Without threatening the production of food, Brazilian farmers have been able to make the production of sugarcane more effective each year, since the country started its Alcohol Program in 1975. The success of this alternative to powering automobiles was such that in some recent years Brazil used more liters of ethanol than liters of gasoline. The extensive use of sugarcane ethanol is an essential factor for Brazil's result of having 42% of its primary energy coming from renewable sources (data for 2016).

The research results discussed in the book are part of an effort by researchers from several countries, collaborating under the GSB initiative, to understand how the experience of Brazil with bioenergy could be used to assist other countries in developing their bioenergy sector. The main funding for the research was provided by the São Paulo Research Foundation (FAPESP), a research funding agency maintained by the taxpayer in the State of São Paulo, Brazil. Professor Luís Cortez and his colleagues analyze the essential points of bioenergy production, such as sustainability, productivity, and social and economic impacts.

Among the many valuable insights by the authors stands the idea that, by developing a bioenergy sector, it is possible to foster economic development, creating jobs and revenues that fuel the economy and might allow for an increase in infrastructure investments that would benefit other sectors of the economy. In this way, the reduction in emissions could be, under certain conditions, accompanied by poverty alleviation. The authors use the case of Brazil to make a compelling case on this.

The work covers an analysis of the availability of land for sugarcane production, taking into consideration the need of land for food production. Sustainability of bioenergy production is also covered, an essential topic especially if one considers large-scale production. Country-specific cases are presented for Colombia, Guatemala, Mozambique, and South Africa, and these are extremely useful in demonstrating the challenges involved in a detailed way.

I am sure that researchers and students in the bioenergy sector will benefit immensely from the ideas presented in this work.

Carlos Henrique de Brito Cruz
Science Director, FAPESP, and Professor,
Physics Institute, Unicamp

Introduction

Production of sugarcane bioenergy in Latin America and Africa

Luís A. B. Cortez, Luiz Augusto Horta Nogueira and Manoel Regis L. V. Leal

The production of modern and sustainable bioenergy can be an effective way of substituting fossil energy (e.g., petroleum and coal), while at the same time achieving other macroeconomic objectives:

- increase energy security,
- create income and jobs,
- improving local and global environmental quality.

The first question that usually arises is related to the availability of agricultural land in the world.

According to Doornbosch and Steenblik (2007), 60% of the world's potentially available land could be used for bioenergy production by 2050 (440 Mha), of which about 60% (250 Mha) will be in Latin America and 180 Mha in Africa. Since the availability of land for bioenergy will largely originate from grazing land, Figure I.1 gives an estimate by country based on FAO (Food and Agriculture Organization) data.

The total amount of bioenergy produced in these areas will partly depend on the availability of fertile land with good climatic conditions. It is clear that if food production needs to be increased to meet future needs and protected biodiversity, basically the world will also depend on Latin America and Africa for future bioenergy expansion.

The LACAf Project[1] (A Contribution to the Production of Bioenergy in Latin America and Caribbean and Africa), therefore, focused on these two continents and evaluated the production of bioenergy (e.g., bioethanol) from sugarcane. The production of bioelectricity from sugarcane has also been considered due to its important impacts on rural development and its synergy with the production of ethanol.

The Brazilian experience with bioethanol production and use

The Brazilian experience with biofuels production and use has fully demonstrated that a middle-income emergent nation can promote a competitive production and use of sustainable bioenergy, helping to alleviate unemployment (Moraes, 2010), interiorize development, and decrease air pollution in big towns and reduce greenhouse gas (GHG) emissions.

The Brazilian saga started, in the early 20th century, blending ethanol produced from sugarcane molasses into imported gasoline to reduce foreign energy dependence and at the same time help local farmers.

Later, in the 1970s, hit by two oil shocks, Brazil has implemented the world's largest renewable energy program, the Proálcool. It was not simple, it was not easy… Many mistakes were made but at the end, the Proalcool is helping Brazil improve its energy security (Figure I.2). In 1975, around

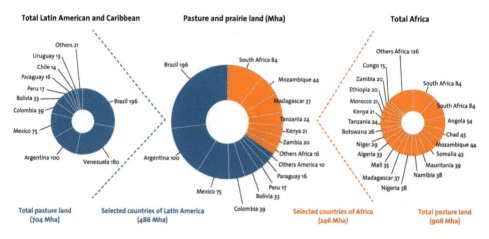

Figure I.1 Available land in selected countries in Africa and Latin America.
Source of data: FAO (2008).

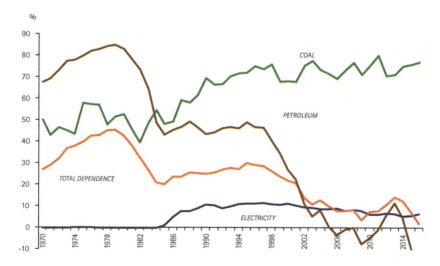

Figure I.2 Evolution of Brazilian energy dependence 1970–2015.
Source: MME (2016).

80% of all gasoline was imported, and today, the country is "technically self-sufficient" in petroleum, selling oil surplus and importing fractions of oil derivatives (Cortez et al., 2016).

Regarding food production, the impact was very positive. Brazil created a model for sugar and ethanol production in a way that today, around 50% of the sugar commercialized worldwide is from Brazil, being the first sugar exporter (SCOPE, 2013). In several other agricultural and animal production, Brazil became a reference: second in soybean exports, first in beef, first in poultry, first in coffee, first in orange juice, and now growing fast in corn production and exports (FGV, 2015). Agriculture was actually the only sector that did not go into recent crisis in Brazil. Last year only agricultural sector grew around 15%, with a record agricultural harvest.

Can Brazil become a model for Latin American and African countries? Not really. Brazil is in this way a unique country in terms of availability of land and local conditions. However, several other countries present very good conditions to produce sustainable bioenergy production and benefit from it! This book's objective is exactly to address these issues and try to understand which can be a model that suits other countries and other realities, such as encountered in Colombia and Mozambique.

Challenges in Latin America & Caribbean and Africa

The challenges encountered in these two countries (Colombia and Mozambique) are very diverse. Both are developing countries, like Brazil, but Latin America is relatively more advanced in terms of infrastructure and availability of human resources, and this makes a difference. However, Mozambique has a well-established sugarcane industry proving that what is been proposed here is not something impossible to be achieved.

Probably the biggest challenge above all is try to answer the persistent question: will bioenergy expansion threaten food production? Can both be reconciled? Then, if these questions are properly answered, we can go beyond to how much bioenergy we can produce and how this will be done.

The chapters on Part I address energy-related issues and global trends such as how bioenergy will evolve in a very competitive scenario of other renewable energies. Solar and wind technologies are spreading really fast, particularly in Europe, and US car makers are making promises about ending the Otto and Diesel engines in the next coming decades. But the solar and wind solutions are then adequate for developing countries. The development of these technologies will not require the existence of national electricity grids, simply not existent in less developed nations.

Then, the other remaining parts of the book present, in more detail, the results inferred from the LACAf Project, in which the Colombian and Mozambican cases are presented.

Part II starts with the LACAf Project analysis on why developing countries must or not adopt policies to implement modern and sustainable bioenergy in their respective energy matrix. Authors try to understand the socioeconomic impacts of bioenergy production and also the possible conflict with food security.

Then, having answered the why question, the next step is to address the issue of how much sugarcane bioenergy can be produced. Since the LACAf Project is about Latin America and Africa, we have decided to work in more detail on two countries: Colombia and Mozambique. In these countries, a detailed study was made regarding the agronomic potential for sugarcane cultivation. This part (Part III) of our book contains detailed maps for sugarcane potential for the two countries with a correspondent estimate of the total bioenergy potential considering minimum adequate land.

However, the total sugarcane bioenergy potential would not be possible if a compatible production system could not be implemented, which is discussed in Part IV. A virtual biorefinery simulation tool developed by the Brazilian Bioethanol National Science and Technology Laboratory (CTBE) was used for that end. Simulations were made and results are presented considering different technology levels for both agriculture and industry. In other chapters, different production models are discussed.

The last part of our book, Part V, refers to an assessment made by invited authors with strong experience and competence in energy field for Latin America and Africa.

Note

1 More information about the LACAf Project available at http://bioenfapesp.org/gsb/lacaf/.

References

Cerqueira Leite, R.C. de; Leal, M.R.L.V.; Cortez, L.A.B.; Griffin, M.W.; Scandiffio, M.I.G. "Can Brazil replace 5% of the gasoline world demand with ethanol?", *Energy*, 2009, 34: 665–661.

Cortez, L.A.B. (Org.); Brito Cruz, C.H.; Souza, G.M.; Cantarella, H.; van Sluys, M.A.; Maciel Filho, R. "Universidades e Empresas: 40 anos de ciência e tecnologia para o etanol brasileiro", *Blucher*, 2016: 224.

Doornbosch, R.; Steenblik, R. "Biofuels: is the cure worse than the disease?" Round Table on Sustainable Development, OECD, Paris, 11–12 Sept 2007.

FAO, Food and Agriculture Organization of the United Nations. Climdata Rainfall Database. Rome: United Nations Food and Agriculture Organization, Sustainable Development Department, Agrometeorology Group, 1997. In: BNDES & CGEE; Bioetanol de cana-de-açúcar – energia para o desenvolvimento sustentável. 1ª Edição, Rio de Janeiro, novembro – 2008.

Ferraz Dias de Moraes, M.A.; Zilberman, D. "Production of Ethanol from Sugarcane in Brazil – From State Intervention to a Free Market" Springer, 2014.

FGV. "Brazilian Agribusiness Overview" Getúlio Vargas Foundation – FGV, FGV Projetos, Number 25, 2015. ISBN 978-85=64878-28-0.

Horta Nogueira, L.A. *Perspectivas de um programa de biocombustíveis em América Central.* México, CEPAL, 2004.

Horta Nogueira, L.A. *Biocombustíveis na América Latina – situação atual e perspectivas.* Memorial da América Latina, São Paulo, 2007.

Lynd, L.R. Squaring biofuels with food. *Issues in Science and Technology*, 2009, 25(4): 8–9.

MME-Brazilian Ministry of Mines and Energy, National Energy Balance, Brasilia, DF, 2016. www.mme.gov.br/. Accessed in 2017.

Moraes, M.A.F.D. "Biofuels for social inclusion", presentation at the 3rd GSB Convention, 23–25 March 2010, São Paulo, Brazil. www.fapesp.br/eventos/2010/03/gsb/Marcia_Azanha_Moraes_11h15_240310.pdf. Accessed in 2017.

SCOPE Bioenergy & Sustainability: bridging the gaps, December 2013. http://bioenfapesp.org/scopebioenergy/index.php. Accessed in 2017.

Part I
Bioenergy overview
The context

1 The future of fuel ethanol

Rogério Cezar de Cerqueira Leite, Luís A. B. Cortez, Carlos Eduardo Driemeier, Manoel Regis L. V. Leal, and Antonio Bonomi

Disclaimer

The present analysis focuses on the future of fuel ethanol and of global energy markets. By no means is it oriented to specific countries. It was also not our intention to discuss in detail the technologies that we refer to here, but much less provide assurance as to the precise path they will follow and rather they will converge to some of the possible configurations suggested in our analysis. Finally, it is important to also note that real outcomes may end up diverging considerably from the scenarios under consideration if, for instance, oil prices take an unpredictable course.

Introduction: fuel ethanol in Brazil and the US

Brazil, as it is well known, has long been recognized as a leading country in the development and use of sugarcane bioethanol as fuel. It is difficult to say when it started, but the scientific literature presents information that registers several events concerning the adoption of ethanol as fuel since the beginning of the 20th century. The fact that Brazil did not have production and refining of petroleum within its borders until the 1950s, and that sugarcane cultivation has permeated the country's history since its discovery by the Portuguese provides a good clue as to why such innovations took place there earlier.

The success of the Brazilian experience together with the oil crisis of the 1970s created conditions that led to the implementation of the largest renewable fuel project ever since: the National Alcohol Fuel Program, known as Proálcool (The Brazilian Development Bank (BNDES) and CGEE (2008) and Cortez et al. (2016)). The investment of substantial amounts of resources in both agriculture and industry increased significantly the country's production capacity of ethanol and sugar. Brazil became the world's largest producer and exporter of cane sugar while producing huge volumes of fuel ethanol as well. In 1979, an agreement between the Federal Government, the sugar and ethanol sector, and the automobile industry sealed the pact for the years to come: to use ethanol in dedicated combustion engines. This was indeed a big step. The motivation then was not the environment per se but simply to reduce the country's dependence on oil. Two ethanol-based fuels were produced: hydrous ethanol, to be initially used directly in ethanol engines (E100) and later in flex fuel cars, and anhydrous ethanol, to be blended with gasoline (today E27) and used in adapted engines. Presently, ethanol represents about 40% of the overall liquid fuel utilized in light vehicles in Brazil. Ethanol is mainly produced using molasses and higher impurity sugars as inputs. Processing takes place at sugar mills where nearly 50% of the output ends up sold as raw sugar and the remaining 50% further is processed, through fermentation and distillation, and transformed into fuel ethanol.

At the turn of the new century, policy makers in the US began to see fuel ethanol as an alternative energy source that could help local farmers and reduce the country's dependence on oil imports (NDSU (2018) and US EIA (2018)). The 2005 Energy Policy Act imposed mandatory targets for biofuel production and discouraged the use of MTBE (methyl *tert*-butyl ether) as a gasoline additive. MTBE is an oxygenate, and its presence in gasoline is meant to raise its octane level. Due to serious environmental and health concerns related to the contamination of water by this substance, the goal of the new policy, at least in part, was to stimulate the reduction of its use. This could be achieved by using MTBE as a substitute for ethanol.

The US model is different than the one practised in Brazil where fuel ethanol is produced using sugarcane as input. In the US, the main input is corn since it is a widely cultivated crop there. With the new incentives, production of ethanol increased substantially, first by simply obtaining corn outputs from existing plantations with no need for additional land commitments. At least initially, therefore, investments were more limited to the development of innovative processes and technologies and to the construction of distilleries. Another final product of this new energy model was the so-called E85 (85% ethanol + 15% gasoline), which is used in newly produced flex fuel engines. By 2007, more than 30% of the corn harvested in the US was channeled to ethanol production.

As the production of ethanol began to take up more land and consume other resources originally used solely for food production, criticism began to grow until a heated debate ensued. One side tended to look at the problem as merely a competition for land use, equating it, perhaps rather simplistically, to a zero-sum game. It was the beginning of the food versus fuel dilemma (Rosillo-Calle and Johnson, 2010). The other side of this debate that rejects the use of land for fuel production in defense of food safety does not seem to take into account possible synergies that can result from investments in innovative technologies and the potential benefits to food production itself. It also tends to attribute biofuel production as the main cause of the food price spike of 2007/2008. From 2002 to late 2006, the price of corn rose by 16%, rice by 49%, and wheat by 35%. Such disruption affected most severely the poorest regions of the world where even small variations in commodity prices can have a strong impact on a populations' diet. An opposing view, however, suggests that other factors were much more relevant than ethanol production. The rise in oil prices, for instance, caused diesel and nitrogenous fertilizers to become more expensive, increasing production and transportation costs for cereals in general. Once the recession was over (officially in November 2001), the price of a barrel of oil would climb, year after year, from a low point of US$ 20 to a record of US$ 130 in July 2008. Other important factors were the depreciation of the US dollar; expansion of the global economy, more notably in emerging markets up to 2008; reduction of import tariffs; and adverse weather conditions in some producing countries. As crop yields suffered from droughts, some countries imposed export bans and restrictions to limit the impact on inflation. Finally, financial speculation might have also possibly played a role as investments in index funds in the agricultural futures markets that grew substantially in the US during the period.

More objectively, let us point out some potential benefits of investing in bioenergy. The US production of fuel ethanol from corn creates a particular type of residue called "dried distillers grain", or DDG, rich in ingredients that can be added to improve the quality of beef cattle's feeds. The uses of such by-products suggest great synergies between the energy and food industries, and yet are representative of just one instance of those possibilities. Similar interactions are taking place in the context of ethanol production in Brazil.

As the situation progressed in both countries in the first decade of the 21st century, production continued to boom despite international skepticism. The Brazilian ethanol and sugar industry was growing at about 10% annually, while the US production grew at an even faster pace, eventually making it the world's largest ethanol producer. By the end of that decade, however, things started to change.

As the market for subprime real estate in the US began to melt down and its ripple effects felt around the world, the global economy would suffer a significant slowdown. A critical side effect of the "Great Recession of 2008" was a reduction of investments worldwide. Financial markets began to freeze and investors ran to the safest haven known: US short-term treasury securities. It seems paradoxical since the crisis started originally in the US, but it was exactly what happened. As investors sought safe government securities, interest rates in the developed world remained at very low levels for a long time. In spite of this, investors became extremely cautious as central banks were taking bold, coordinated actions to get the crisis under control. While all of these were going on, Brazil was ramping up the exploration of its pre-salt coastal oil reserves. Given the estimates of their size and the prevailing prices of petroleum at that time, Brazil's energy future seemed safe. Investments in pre-salt would apparently guarantee great returns for the government and its foreign partners. So much so that the prevailing debate was about what to do with the enormous additional royalties that would unequivocally flow into the government coffers. The support for ethanol producers almost naturally diminished, and the sector plunged into a crisis of its own. One can argue, however, that this was a deliberate government decision, if not a significant strategic mistake. With hindsight, it seems reasonable to conclude it was wrong in not providing more support for that sector, if anything could be used as a hedge to the future trajectory of oil prices. Over the same period, the US sought energy independence of its own, although in a different direction. It provided great incentive for the domestic exploration of shale gas, a viable and abundant alternative source of energy for them. In both countries, the future for ethanol becomes much less certain.

The new investments in 2G ethanol research

The significant progress verified in the production and use of fuel ethanol was also accompanied by investments in new science and technology. The motivation was basically the limitation of available land, particularly in developed countries, and the need to produce a fuel with less greenhouse gas (GHG) emissions. Production of ethanol from fibers, using hydrolysis technology, could boost ethanol production using the same land and with smaller CO_2 generation.

Several projects were financed in the US, Europe, and Brazil. The race to more sustainable ethanol seemed to require the development of hydrolysis. Among the most important projects, we can mention the following:

- The Department of Energy (DOE) Bioenergy Research Centers: the Great Lakes Bioenergy Research Center, led by the University of Wisconsin–Madison in partnership with Michigan State University; the Center for Bioenergy Innovation, led by the DOE's Oak Ridge National Laboratory; the Joint BioEnergy Institute, led by the DOE's Lawrence Berkeley National Laboratory; and the Center for Advanced Bioenergy and Bioproducts Innovation, led by the University of Illinois at Urbana–Champaign. www.ethanolproducer.com/articles/14498/doe-provides-40-million-for-4-doe-bioenergy-research-centers
- The Joint Bioenergy Institute (JBEI). https://www.jbei.org/
- Bioenergy Research Center at University of California, Davis. http://bioenergy.ucdavis.edu/
- The National Corn-to-Ethanol Research Center (NCERC) at Southern Illinois University, Edwardsville, IL. www.ethanolresearch.com/

At the same time, in Brazil, two studies conducted by the Brazilian Center for Research in Energy and Materials (CNPEM) converged to the creation of the Brazilian National Science and Technology Bioethanol Laboratory (CTBE) in Campinas, Brazil (Figure 1.1). The first study analyzed the possibility of Brazil substituting 10% of all the gasoline consumed in the world by 2050 for sugarcane ethanol (Leite et al., 2009). The necessary volume of

Figure 1.1 View of the CTBE (in the center) at CNPEM in Campinas, Brazil, where CTBE is located.

ethanol was estimated to be nearly 205 billion liters/year, and the average land productivity for Central South Brazilian conditions was around 7,000 liters of ethanol/ha year· of sugarcane, therefore, requiring nearly 30 million hectares of land.

If second-generation 2G technology were available, part of the cane fiber could be converted to ethanol. Among the specialists, the overall yield could increase from 7,000 to 11,000 liters/ha·year, requiring around 20 million hectares of land to accomplish the same goal. Land is not a real problem in Brazil, even if the so-called natural sanctuaries such as the Amazon, the Pantanal, or the Atlantic Forest are protected. The country still has 200 million hectares dedicated to low-productive beef cattle. However, with the study, it became clear that if 2G technology was controlled, the carbon footprint could be much smaller.

Another research team was organized to understand the bottlenecks of 2G technology. With these two teams working together, the CTBE was born in 2010. It was located at CNPEM in Campinas, Brazil, where the other three national laboratories already existed. The challenge was to transform sugarcane fiber (bagasse and the leaves) into sustainable fuel bioethanol.

In 2011, BNDES launched The Industrial Innovation Plan for the Sugarcane Ethanol Sector,[1] a program aimed at encouraging industrial technical innovation in the sugar and ethanol sector. Several projects received financial support for 2G technology development: the Sugarcane Research Center (CTC), the Granbio Project, and the Raízen Project, among others. All of them racing with CTBE to be the first to produce commercial 2G sugarcane ethanol in Brazil.

All these projects, in Brazil and in the US, assumed a difficult task for themselves. Until now, no commercial and economically viable large-scale ethanol was produced from sugarcane, corn, or other biomass. Researchers involved in these projects continue to persevere with their objectives with the hope of soon proving that 2G ethanol is feasible and sustainable. For now, however, the one reality we can count on is that 1G ethanol works for both sugarcane and corn. The yields are already high and competitive, allowing producers to stand on their own even in the absence of subsidies, which is already the reality in some countries.

Global warming and perspectives for transportation markets

The International Energy Agency (IEA) document "Energy Technology Perspectives 2017: Catalyzing Energy Technology Transformations" describes scenarios until 2060 considering the effort to reduce GHG emissions. No matter what scenarios we adopt, the 2°C increase scenario (2DS), beyond 2DS (B2DS), Reference Technology Scenario (RTS), or 6°C increase scenario (6DS), there will be a need to reduce fossil energy emissions in this century. Regardless of what are the best alternative courses from here, we can at least safely assume that there is not a viable future where fossil fuels continue to be used as they have been until now. The man-made GHG emissions need to be captured and stored or the fossil fuels simply substituted.

Transportation patterns are likely to go through profound transformations in the decades to come. According to IEA (2017a, 2017b), the expected trend will be the following, as we present in Table 1.1.

Table 1.1 Energy used in different modes of transportation as an effort to maximize fuel savings and emissions – 2DS

Main transportation markets		2030 PJ	%	2040 PJ	%	2050 PJ	%	2060 PJ	%
Passenger	Air	12053	20	12554	23	12753	26	13356	28
	Road	46593	78	40321	74	34652	70	31764	66
	Rail	1046	2	1525	3	2102	4	2758	6
Freight	Road	39253	74	38707	73	37750	71	37445	70
	Rail	1831	3	1991	4	2094	4	2170	4
	Shipping	11674	22	12341	23	13385	25	14218	26

Source: IEA (2017).

Market of passenger light-duty vehicles

According to IEA (2017), regarding the market for light vehicles in the decades to come, and considering two scenarios, 2DS and RTS, the world transportation market will be dominated by a variety of vehicle types including gasoline, diesel, light vehicles powered by natural gas (CNG)/liquefied petroleum gas (LPG), gasoline hybrid, diesel hybrid, plug-in hybrid diesel, plug-in hybrid gasoline, electricity, and fuel cell electric vehicle (FCEV). Table 1.2 presents the estimated projections for LDV stock by type of vehicle for two scenarios: 2DS and RTS.

The same data are presented in Figure 1.2. Note the "plunge" of the stock of gasoline ICE and large proportion of electric cars until 2060 for both scenarios, 2DS and RTS.

Table 1.2 Global technology penetrations in LDV stock by scenario (2DS and RTS), 2020–2060

Type of vehicle	2020 2DS	2020 RTS	2030 2DS	2030 RTS	2040 2DS	2040 RTS	2050 2DS	2050 RTS	2060 2DS	2060 RTS
Gasoline ICE	628	654	647	821	552	900	424	873	257	678
Diesel ICE	157	162	163	197	141	211	105	206	55	151
CNG/LPG	26	27	29	37	42	55	43	71	26	70
Hybrids	25	26	77	83	167	190	226	334	259	553
Plug-in electric	7	3	56	26	155	68	310	130	520	264
Battery electric	10	3	58	15	135	37	264	85	428	176
Total	853	875	1,030	1,179	1,192	1,461	1,372	1,699	1,545	1,892

Source: IEA (2017).

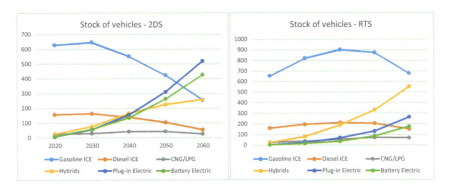

Figure 1.2 Global technology penetrations in LDV stock by scenario (2DS and RTS), 2020–2060.

Source: Based on IEA data from Energy Technology Perspectives 2017 © OECD/IEA 2017, www.iea.org/statistics, License: www.iea.org/t&c.

Table 1.3 Projections of fuel consumed (in Million Tonnes of Oil Equivalent (MTOE)) in different types of transportation in different types of energy

Types of energy	Light-duty vehicles		Urban trucks		Long-haul trucks		Shipping (sea)		Aviation	
	2010	2050	2010	2050	2010	2050	2010	2050	2010	2050
Fossil/biofuels	900	600	300	200	250	400	300	400	250	400
Hydrogen	0	100	0	75	0	0	0	0	0	0
Electricity	0	300	0	75	0	0	0	0	0	0
Total	900	1000	300	350	250	400	300	400	250	400

Source: IEA (Fulton, 2013).

Other transportation markets

We now analyze the different transportation markets, according to IEA, as a whole as well as its main trends (Table 1.3).

Even considering an optimistic scenario, the ETP 2DS, electricity and H_2 will have limited transport application if no technology breakthroughs are achieved.

The role of bioenergy in the future transportation

Using IEA (2017) data to understand the role of biofuels in the future transportation, we can reach the following conclusions:

- According to the RTS ("reference scenario"), large proportion of cars will still be powered by gasoline or diesel by 2030 (~87%) and 2060 (~44%);
- According to the 2DS ("necessary scenario"), large proportion of cars will have an electric motor (hybrid, electric, or fuel cell) by 2030 (~18%) and 2060 (~78%);
- The 2DS, a necessary scenario, will be very difficult to achieve without sustainable biofuels, considering "potential for substituting gasoline and diesel".

Considering the 2DS until the year 2075, we can expect the following development (Figure 1.3, Table 1.4).

Table 1.5 and Figure 1.4 show the projections for different liquid biofuels, which will likely be present in the energy market until 2060. Note the impressive expected participation of advanced biofuels, notably advanced biodiesel and biojet.

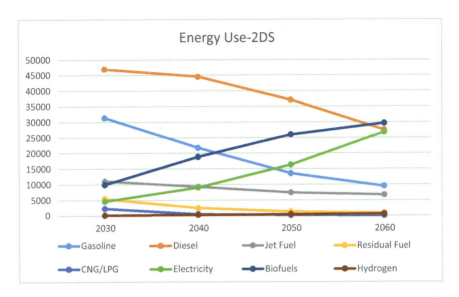

Figure 1.3 Projections of fuel consumed (in PJ) in different types of transportation in different types of energy (2DS).

Source: Based on IEA data from Energy Technology Perspectives 2017 © OECD/IEA 2017, www.iea.org/statistics, License: www.iea.org/t&c.

Table 1.4 Projections of fuel consumed (in PJ) in different types of transportation in different types of energy (2DS and RTS)

Types of energy	2030 2DS	2030 RTS	2040 2DS	2040 RTS	2050 2DS	2050 RTS	2060 2DS	2060 RTS
Gasoline	31,325	44,794	21,743	46,208	13,523	47,630	9,446	45,327
Diesel	46,974	52,050	44,539	56,611	37,110	56,237	27,460	56,425
Jet fuel	11,049	14,758	9,280	16,294	7,378	16,372	6,669	17,328
Residual fuel	5,479	9,097	2,441	10,142	1,237	11,840	814	12,754
CNG/LPG	2,285	5,995	416	7,505	93	8,947	0	9,673
Electricity	4,598	2,493	9,069	4,121	16,311	6,646	26,754	10,377
Biofuels	9,957	6,033	18,829	9,436	25,984	12,164	29,609	12,763
Hydrogen	111	14	243	36	377	66	528	119
Total	111,778	135,234	106,560	150,353	102,013	159,902	101,280	164,766

Source: IEA (2017).

Table 1.5 Energy used for different liquid biofuels – 2DS (EJ), 2020–2060

	2020	2025	2030	2035	2040	2045	2050	2055	2060
Ethanol conventional	2.61	2.76	2.56	2.34	2.56	2.60	2.60	2.90	2.68
Ethanol-advanced	0.20	0.62	1.07	1.64	2.44	3.18	3.45	3.90	3.44
Biodiesel-conventional	1.14	1.10	0.82	0.63	0.31	0.13	0.01	0.01	0.01
Biodiesel-advanced	0.27	0.98	2.22	3.90	6.50	9.13	11.54	13.97	14.53
Biogas	0.50	1.33	2.28	3.32	3.74	3.60	3.01	2.59	2.26
Biojet	–	0.28	1.00	2.10	3.27	4.60	5.37	6.03	6.69

Source: IEA (2017).

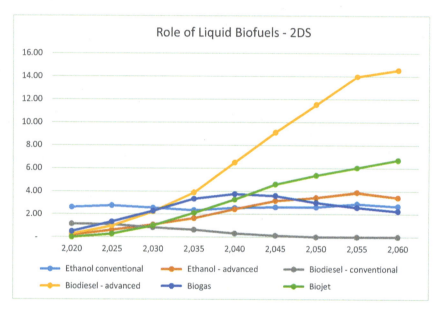

Figure 1.4 The role of liquid biofuels – 2DS, 2020–2060.
Source: Based on IEA data from Energy Technology Perspectives 2017 © OECD/IEA 2017, www.iea.org/statistics, License: www.iea.org/t&c.

Other markets for bioenergy

Adopting the most optimistic scenario (ETP 2DS), according to IEA (Fulton, 2013), biomass and waste will play the most important role in responding for almost 25% of overall energy use. According to the same source, other renewables will altogether respond for nearly 13%–15% (see Figure 1.5).

Note that besides transportation, bioenergy is also responsible for at least two major markets worldwide: providing heat for cooking (mostly using traditional bioenergy) and heat to warm homes and buildings in cold countries.

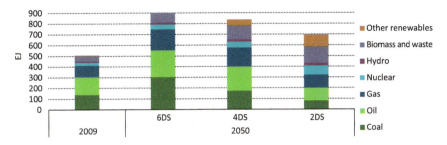

Key point: The 2DS reflects a concerted effort to reduce overall consumption and replace fossil fuels with a mix of renewable and nuclear energy sources

Figure 1.5 Participation of different energy sources in the overall energy consumption in the world, considering 2DS, 4DS, and 6DS.
Source: IEA (Fulton, 2013).

It is likely that the use of traditional biomass for cooking will tend to decrease as population wealth increases, particularly in Africa and India. However, the market for heating homes and buildings will tend to increase using the same premises, particularly in today's cold poor countries. For that purpose, countries such as Canada and the US are already producing biomass pellets and exporting them to Europe.

Technological considerations for the use of ethanol in light-duty vehicles

If we consider the promising outcomes of recent research aimed at improving fuel–engine interactions, the demand forecasts for ethanol presented previously are probably too conservative. While there have been decades of developments aimed at improving the performance of motor vehicle engines powered by fossil fuels, we are just beginning to see the same efforts and advances play out in the context of biofuels. The use of ethanol in light-duty vehicles (LDVs) is still subject to engine constraints imposed by gasoline, such as the relatively low compression ratio. Octane rating of ethanol is quite superior to that of premium gasoline. It is therefore perfectly plausible to assume that continuing developments of new engines more customized to fuel ethanol and its higher octane balance against lower energy (Martin, 2016) will potentially lead to much better performance than the current technology offers.

This relevant feature of ethanol has been demonstrated and explored by initiatives such as High Octane Fuels (HOF), sponsored by the DOE and developed by national labs in the US (Oak Ridge National Laboratory (ORNL), National Renewable Energy Laboratory (NREL), and Argone National Laboratory (ANL)) (McCormick et al., 2016; Farrell et al., 2018).

The Advanced Motor Fuels (AMF), promoted by the IEA, involving several research institutions abroad, presents similar positive results (IEA, 2017). In sum, such research initiatives indicate that there is room for efficiency gains and improvement in engines using high-level ethanol blends (>E40, typically E85) and pure ethanol, with economic and environmental benefits. Applying technology and systems more adapted to the production, distribution, and use of liquid biofuels, therefore, is almost certain to bring significant improvements.

If we take a further look into this technological frontier, we can point to the use of ethanol as a hydrogen carrier in fuel cell vehicle, which represents an important innovation and is already in advanced experimentation by some automakers, associating ethanol thermal catalytic reform to produce hydrogen using solid oxide fuel cells. Hydrogen from ethanol is simpler to produce and economically competitive compared to hydrogen from water electrolysis. In this way, all advantages of sugarcane ethanol (lowest carbon footprint, no changes needed in current logistics and fuel distribution schemes, large potential for expanding production) are associated with the most efficient vehicle technology, without requiring expensive systems of hydrogen production and compression. Indeed, the ethanol path reinforces expectations the executives of automotive industry have, 78% of which foresee that "Fuel Cell Electric Vehicles will be the real breakthrough for electric mobility" (KPMG, 2017).

In the long run, can biofuels be competitive with other energy alternatives?

The question is, be competitive on what? Cost? GHG mitigation potential?

Specialists say that it is not worth to compare biofuels costs ($/km) with fossil because the actual competition, mainly the electric alternatives mentioned before, is and will be between renewable energies and not with fossil.

Besides costs, we should also consider GHG emissions per kilometer. Cost is an important factor, but we should remember why we are making this transition to more sustainable sources of energy. So far we don't have a premium for the CO_2 mitigation potential, so we are just concerned about comparing energy costs for different alternatives, fossil and renewable.

Finally, we will need all renewables with low GHG emissions. Because of their seasonal characteristics, they can be the base for energy matrices.

Bioenergy can be an opportunity for Latin America and Africa

Considering all the projections presented, the question is, rather, whether bioenergy will still be adequate for Latin America and Africa.

First, let us say that the most important benefits of sustainable biofuels are their impacts. In contrast to what was published by nonspecialists in this field, sustainable bioenergy doesn't, if done right, compete with food production and food security. In fact, exactly the opposite occurs. Sustainable bioenergy production can improve agricultural practices, generate better jobs, and improve women conditions among other benefits. This can be verified in several developing countries, which successfully implemented biofuel projects, such as Brazil, Argentina, and Colombia in South America.

The introduction of electric cars will likely take a few decades to dominate the light vehicle market. Today, it still represents less than 1% of total sales even in developed countries. In less developed nations, it will occur but at much slower pace. It can be estimated that in a country like Brazil, with the market dominated by foreign automakers, it will take at least three to four decades until electric cars can represent more than 50% of light vehicle market, as it was discussed before. In less developed countries, this can take even more time.

Therefore, the developing countries still have a "window of opportunity" to get engaged in sustainable fuel ethanol production. Sustainable sugarcane production and use can produce not only ethanol and sugar but also fiber. Since sugarcane is composed of roughly 2/3 of fibers, it can be used to produce electricity, called bioelectricity. Electricity is a very important coproduct to promote socioeconomic development. Many underdeveloped countries in Africa don't have access to electricity. Several countries have less than 30% of population still not connected to the grid.

Also, sustainable bioenergy can be produced and integrated into other agricultural systems. For example, the integration of sugarcane ethanol production with beef production is practiced by few mills in Brazil today. Vale do Rosário Mill, located in Ribeirão Preto, has more than 20,000 heads depending on "baleme", a feed composed by bagasse, yeasts, and molasses produced by the mill (Souza, 2017). This integration of beef and ethanol production has a great potential to grow, even in a country like Brazil, where more than 200 million beef cattle occupying around 200 million hectares can meet simultaneously two objectives: improve energy and food security while protecting the environment by avoiding additional deforestation. So, the selection of a sustainable model is critical for the success of a bioenergy initiative.

The ethanol diplomacy

The preliminaries of the ethanol diplomacy were undertaken by the US with the Brazilian Government between 2005 and 2010. Both the governments signed cooperation agreements particularly involving 2G technology development. The CTBE was working with the US DOE laboratories to conduct joint research, also involving sustainability aspects.

Several European countries were also proactive in the so-called bioenergy diplomacy. The Netherlands has BE-Basic Foundation for that purpose and also to promote new business. The UK, France, and Germany were and still are very active in bioeconomy[2] initiatives. However, after the subprime crisis last decade, the overall enthusiasm for biofuels cooled down worldwide. It seems there now exists a certain skepticism around biofuels. Probably, the biofuels versus food debate had a negative impact on perception after all, jeopardizing even sustainable biofuels and their participation in the global economy.

However, there are reasons for optimism. Today, nearly all Latin American countries have regulation in imposing blending levels of ethanol with gasoline. The typical blend is 10%. Argentina is seeking to raise the blend up to 12.5%,[3] and in Colombia both ethanol and biodiesel have importation participation in the energy matrix (Ardila et al., 2016).

In Africa, several countries of the sub-Saharan region are already adopting the same pro-biofuels strategy. Zambia is a good example. Despite the optimistic scenario, much can be improved. Several studies corroborate the existing potential of biofuel production for Africa. See more in the LACAf Project website: http://bioenfapesp.org/gsb/lacaf/.

One way to boost the ethanol diplomacy is through existing sugar networks. Sugar is a commodity produced in most developing countries. Practically, in all countries of Latin America and Caribbean and Africa, sugar plays an important role in rural development while contributing to supply food to the population, an important fact that is often forgotten.

The same existing sugar network could also work for sustainable ethanol and bioelectricity production. Sugar and ethanol today are the main coproducts, together with electricity, from sugarcane mills. In this respect, practically 50% of all sugar produced from sugarcane in the world is coproduced with fuel ethanol already.

If more diplomacy is introduced in this robust sector, increasing the cooperation between the countries involved, a great market could be created for developing countries. Ethanol could be produced not only to satisfy domestic needs but also to boost export. All relevant world actors will need to join their efforts in support of this endeavor in order to decrease GHG emissions. Sustainable bioenergy can contribute to that.

Conclusions

Important questions arise about what will be the future of fuel ethanol when electric cars finally dominate the landscape. What will happen to the existing industry? They often forget that this will be a huge transition, not only involving the introduction of electric cars but also merging the two most important markets in the world: the liquid fuel market dominated by oil and the existing electricity market.

Can we sustain a way of life in a world without fossil fuels? If successful, how long will this transition take to play out? These are difficult questions to answer, and the adaptations required are probably some of the most serious humankind has ever faced. To introduce another variable in the equation, we can mention the impact of information technology. How can it affect transportation and energy uses? We simply don't know yet, because it involves a very dynamic process, with many critical variables interacting with each other in complex ways.

What we can conclude by the presented analysis is that due to their high-GHG mitigation potential, sustainable biofuels will be very critical to solve the problem. In some cases, such as maritime, aviation, and heat, there is simply no other way to accomplish GHG reduction goals. We will have to make a joint effort to achieve the 2DS by 2050 and the zero scenario by 2075. Bioenergy and biofuels will definitely be part of the solution.

For developing countries, particularly tropical ones with abundant land, like the majority in Latin America and sub-Saharan countries, this will also be an opportunity to reorganize their territory, implementing sound policies for proper use of the land, and, at the same time, be able to increase energy and food security.

The questions concerning the future of fuel ethanol and bioenergy now are as follows:

- Individually, why should the countries produce sustainable bioenergy and fuel ethanol? Do they have a good reason for that?
- Individually, how much bioenergy and fuel ethanol they should produce for themselves and to export?
- And finally, how much bioenergy and fuel ethanol should be produced to take advantage of this "window of opportunity". Can the derived impacts be positive on employment, food security, helping the transition to development?

Well, developing countries of Latin America and Africa are certainly good candidates, but they need to have good planning for that to happen.

Acknowledgments

We thank contributions from CTBE researchers for providing relevant information for this text.

Definitions and Abbreviations

ICE:	internal combustion engines
IEA:	International Energy Agency
LDV:	light-duty vehicles
Lge:	liters of gasoline equivalent
PLDV:	passenger light-duty vehicles. They can be divided into the following categories (based on Wikipedia):
Gasoline vehicles:	light vehicles powered by gasoline or gasohol (gasoline and ethanol blends)

Diesel vehicles:	light vehicles powered by diesel or diesel and biodiesel blends
CNG/LPG:	light vehicles powered by natural gas or liquefied petroleum gas
Gasoline hybrid:	typically uses two types of power, such as gasoline internal combustion engine to drive an electric generator that powers an electric motor
Diesel hybrid:	typically uses two types of power, such as diesel internal combustion engine to drive an electric generator that powers an electric motor
Plug-in hybrid diesel:	a vehicle that can be recharged by plugging it in to an external source of electricity as well as by its onboard diesel engine and generator
Plug-in hybrid gasoline:	a vehicle that can be recharged by plugging it in to an external source of electricity as well as by its onboard gasoline engine and generator
Electricity:	a vehicle that is propelled by one or more electric motors, using energy stored in rechargeable batteries
FCEV:	fuel cell electric vehicle – a vehicle that uses a fuel cell instead of a battery to power its onboard electric motor
ETP:	Energy Technology Perspectives – scenarios considered by the IEA for temperature increase in the atmosphere due to GHG:
2DS:	2°C increase scenario, the main focus of Energy Technology Perspectives – pathways to "clean energy system"
B2DS:	beyond 2DS, explores how far deployment of technologies that are already available or in the innovation pipeline could take us beyond the 2DS
RTS:	Reference Technology Scenario (Emissions under Paris pledge), considered to be approximately or equivalent to the 4DS: 4°C increase scenario, considers pledges by countries and improves energy efficiency, so-called normal scenario, incorporating announced policies
6DS:	6°C increase scenario, is largely an extension of current trends

Notes

1 A BNDES-Finep plan to support innovation in the sugar-ethanol energy and chemical sectors. www.finep.gov.br/apoio-e-financiamento-externa/programas-e-linhas/programas-inova/paiss

2 A broader concept including other sectors such as Agriculture, Food, Health, Green Chemistry, and Bioenergy.
3 www.lanacion.com.ar/1859113-subiria-el-porcentaje-de-etanol-en-la-nafta.

References

Ardila, C.J.S.; Souza, S.P.; Cortez, L.A.B., 2016. "Sustainable initiatives of Colombian palm oil-based biodiesel production". In *XXII ISAF International Symposium on Alcohol Fuels*, Cartagena, Colombia, March 2016.

BNDES and CGEE (coord.). 2008. *Sugarcane-based Bioethanol: Energy for Sustainable Development*, 304p. – Rio de Janeiro, Brazil. file:///C:/Users/Unicamp/Downloads/7bioetanol_ing%20(1).pdf.

Cortez, L. et al. (Coord.). 2016. "Universidades e empresas: 40 anos de ciência e tecnologia para o etanol brasileiro". *Editora Blucher*, 224p. São Paulo, Brazil (in Portuguese); Open Access: http://openaccess.blucher.com.br/article-list/proalcool-universidades-e-empresas-40-anos-de-ciencia-e-tecnologia-para-o-etanol-brasileiro-310/list#articles.

Dulac, J. *Global Transport Outlook to 2050: Targets and Scenarios for a Low-Carbon Transport Sector*. Based on IEA data. www.iea.org.

Farrell, J.; Holladay, J.; Wagner, R., 2018. "Co-Optimization of Fuels & Engines: Fuel Blendstocks with the Potential to Optimize Future Gasoline Engine Performance: Identification of Five Chemical Families for Detailed Evaluation". *Technical Report DOE/GO-102018-4970*. U.S. Department of Energy, Washington, DC.

Fulton, L., 2013. "The Need for Biofuels to Meet Global Sustainability Targets". *BIOEN-BIOTA-PFPMCG-SCOPE Joint Workshop on Biofuels & Sustainability* February 26th, 2013, São Paulo (FAPESP), Brazil.

IEA, 2017a. *Advanced Motor Fuels: Annual Report*, International Energy Agency, Paris.

IEA, 2017b. "Energy Technology Perspectives 2017: Catalyzing Energy Technology Transformations". www.iea.org/etp2017.

KPMG, 2017. *Global Automotive Executive Survey 2017, KPMG International*, available at https://assets.kpmg.com/content/dam/kpmg/xx/pdf/2017/01/global-automotive-executive-survey-2017.pdf.

Leite, R.C.C (Coord.). 2009. "Bioetanol Combustível: uma oportunidade para o Brasil" Centro de Gestão de Assuntos Estratégicos-CGEE, *Brasília*, 536 p.

Leite, R.C.C.; Leal, M.R.L.V; Cortez, L.A.B.; Griffin, W.M.; Scandiffio, M.I.G., 2009. "Can Brazil Replace 5% of the 2025 Gasoline World Demand with Ethanol?" *Energy* 34/5(655–661), 7 p.

Martin, J., 2016. *The Road to High Octane Fuels, Union of Concerned Scientists*, available at https://blog.ucsusa.org/jeremy-martin/the-road-to-high-octane-fuels.

McCormick, R.L.; Fioroni, G.M.; Ratcliff, M.A.; Zigler, B.T.; Farrell, J., 2016. Bioblendstocks that Enable High Efficiency Engine Designs, 2nd CRC Advanced Fuel and Engine Efficiency Workshop Livermore, CA, November 3, 2016

NDSU – North Dakota State University. "History of Ethanol Production and Policy". www.ag.ndsu.edu/energy/biofuels/energy-briefs/history-of-ethanol-production-and-policy.

Rosillo-Calle, F.; Johnson, F. X. (Editors) 2010. Food versus Fuels Dilemma: An Informed Introduction to Biofuels, Zed Books, 232p.

Souza, N. R. D. de. 2017. "Techno-Economic and Environmental Evaluation of Beef Pasture Intensification with Sugarcane Ethanol". *MSc Dissertation, School of Agricultural Engineering - UNICAMP.* Campinas, SP. Brazil.

US EIA, International Energy Outlook 2016. www.eia.gov/outlooks/ieo/pdf/transportation.pdf.

US EIA – Energy Information Administration. "Biofuels: Ethanol and Biodiesel Explained". www.eia.gov/energyexplained/index.cfm?page=biofuel_ethanol_home.

World Energy Council. Global Transport Scenarios 2050. Published 2011. www.worldenergy.org/wp-content/uploads/2012/09/wec_transport_scenarios_2050.pdf.

2 Bioenergy in the world
Present status and future

José Goldemberg

Introduction

As a fuel, bioenergy includes wood, wood wastes, straw, manure, sugarcane and other by-products from a variety of agriculture processes. Among other renewable forms of solar energy, such as wind or photovoltaics, bioenergy – energy made available from materials derived from biological sources – has the special advantage of storing the energy captured from the sun (Goldemberg and Moreira, 2017). Wind and photovoltaics are intermittent, and their use might require storage or rely on other engineering schemes. It has been used since the dawn of civilization for heating and cooking in low-efficiency primitive wood stoves. It continues to be used as "traditional biomass" in rural areas of many less developed countries as the only available option, since it can be accessed freely, or at a very low cost, and represents more than half of the total contribution of biomass. "Modern biomass" is used in the production of electricity, heat, and fuels.

The contribution of biomass to "traditional" and "modern" forms with a total of 49EJ is shown in Figure 2.1. The "traditional biomass" contribution is 28EJ, and "modern biomass" contribution is 21EJ.

The growing importance of bioenergy can be gauged by the rate of growth of its contribution to the energy supply in the period 2008–2015 (Figure 2.2). Bioenergy for transport and electricity production is growing at rates above 4% in the period 2008–2015 compared to less than 2% from fossil fuels. Such rates of growth that continue the present contribution from biomass would grow substantially from the present 49EJ and could reach 145EJ by 2050–2060 according to International Energy Agency (IEA) projections without negative effects on food production and the environment (IEA, 2017).

Worldwide estimates for bioenergy potential and consumption

In 2016, International Renewable Energy Agency (IRENA) and IEA organized a workshop of bioenergy experts from ECOFYS (WWF), German Aerospace Agency (DLR), Netherlands Environment Agency (PBL) and Oak & Ridge National Laboratory (USA) to evaluate the contribution of modern

28 *José Goldemberg*

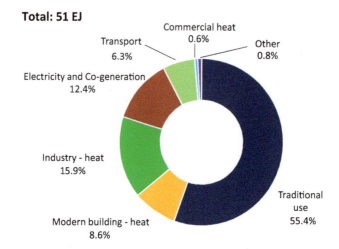

Figure 2.1 Traditional and modern biomass use in 2015 (left) and modern bioenergy growth by sector (2008–2015).
Source: Based on IEA data from Energy Technology Perspectives 2017 © OECD/IEA 2017, www.iea.org/statistics, License: www.iea.org/t&c.

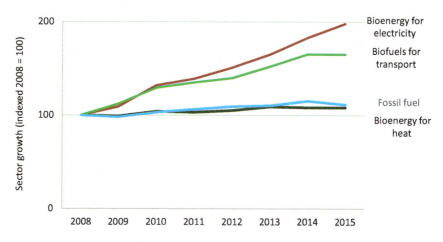

Figure 2.2 Growth rate of bioenergy to the energy supply in the period 2008–2015.
Source: Based on IEA data from Energy Technology Perspectives 2017 © OECD/IEA 2017, www.iea.org/statistics, License: www.iea.org/t&c.

biomass to the world's energy supply in 2050 (IRENA, 2016). Projections of the contribution from these organizations vary by less than a factor of two, ranging from 105EJ by WWF down to 57EJ by PBL (Table 2.1).

In Figure 2.3, one shows the average value of bioenergy contributions in four categories (electricity, industrial process heat, buildings and transport) with the span of values among the institutions (IEA, GREENPEACE, PBL, and WWF).

Table 2.1 Biomass of bioenergy use (EJ) (2050)

	Electricity	Industrial process heat	Buildings	Transport (liquid fuels)	Total
IEA	11	22	23	24	80.4
GREENPEACE	11	11	26	9	57
PBL	14	39	25	17	96
WWF	16	35	3	51	105
Total	52	107	77	101	338
Average	13	27	19	25	84
Range	11–16	11–35	3–25	9–51	57–105

Source: IRENA (2016).

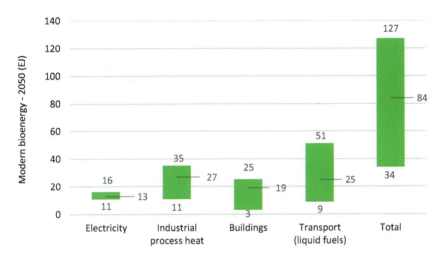

Figure 2.3 Modern bioenergy in 2050 (EJ).
Source: IRENA (2016).

According to such data, the contribution of modern bioenergy to the world's energy system will grow to a minimum of 34EJ and a maximum of 127EJ: an average value of 84EJ.

The same workshop convened by IRENA and IEA in 2016 compared the projections of the biomass potential in four categories: farm residues, forest residues, energy crops and port-consumer waste (municipal solid waste). The results are shown in Table 2.2. For 2050, IRENA projects a potential of 278EJ followed by IEA (128EJ), PBL (149EJ), and GREENPEACE 54-185EJ (Table 2.2).

The total biomass potential varies from a minimum of 140EJ to a maximum of 287EJ with an average value of 210EJ.

Table 2.2 Comparing the projections of the biomass potential in four categories: farm residues, forest residues, energy crops and port-consumer waste (municipal solid waste)

	54 Farm residues	26 Forest residues	16 Energy crops	20 MSW	210 Total
IRENA	45	50	191	–	287
IEA	56	13	144	15	228
GREENPEACE max.	88	–	97	–	185
PBL	25	16	72	26	140
Total	215	79	504	41	840
Range	25–88	13–50	97–191	15–26	140–287

Source: IRENA (2016).

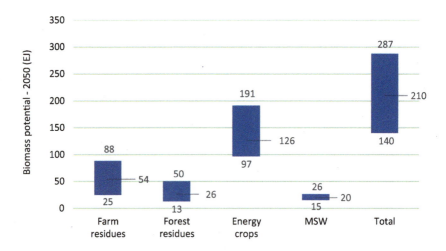

Figure 2.4 Biomass potential (2050).
Source: IRENA (2016).

Figure 2.4 shows these results in the form of graph.

The biomass potential available in all projections is higher than the consumption projections for 2050 as shown in Table 2.1.

Brazil's contribution to bioenergy production

Among the four areas of growth in bioenergy consumption in the future (electricity, industry, processes, building heating and cooking and transport liquid fuels), the most significant ones for Brazil are liquid biofuels and electricity (Goldemberg, 2006, 2017).

Ethanol from sugarcane is the most important biofuel produced in Brazil and together with the production of ethanol from corn in the United States, corresponds to 86% of this fuel in the world in 2016 (Figure 2.5).

The spectacular growth in production in both countries from 2001 to 2008 has been followed by stagnation in more recent years.

The adoption of mandates for adding ethanol to gasoline has been the key driver of the growth of consumption: more than 100 countries adopted them but the most significant ones are of course adopted by Brazil and the United States (REN21, 2016). The mixture in Brazil is set at 27% in 2016, and in the United States the mandates set is 68.6 billion liters of renewable fuels including 871 million liters of cellulosic ethanol, 7.2 billion liters of biodiesel and 13.7 billion liters of advanced biofuel (defined as biofuels which the reduction of CO_2 emission corresponds to more than 61% as compared to gasoline).

The advantage of the adoption of mandates is that they allow the introduction of an expensive fuel in the market with the expectation of decline in costs as the volume-produced increase through as known by the "learning curves".

This was the case of the production of ethanol from sugarcane in Brazil: in 1980, the cost of ethanol was US$ 0.7 per liter much higher than gasoline at US$ 0.25 per liter. The increase in production lowered these costs dramatically: for each doubling of accumulated production, the cost of production consume approximately 20%. In the year 2000, ethanol production reached a cost of US$ 0.30 per liter approximately the same of gasoline at Rotterdam prices (Goldemberg, 2006).

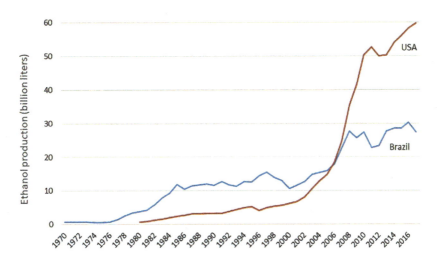

Figure 2.5 Evolution of ethanol production in Brazil and the United States (1970–2015). Source: EIA (2015).

In the period 1980–2000 subsidies and the total accumulated ethanol production 200 billion liters approximately 30 billion dollars were needed for this process of learning to occur. This is equivalent to a carbon tax of US$ 80 per ton of CO_2 which after 2004 had no subsidies. The price of gasoline was determined by the government to cover such cost.

In the United States, a "ceiling" on the amount of ethanol mixed in gasoline was adopted, which limited the amount of blending to 10% on the basis of the flawed argument that higher blends would harm the motors or increase the emissions above the levels set by EPA (Environmental Protection Agency). Presently, in the United States there is a strong debate with the auto industry on proposing to reduce blending to 9.7%, and ethanol producers aim at a 15% blend and higher, which can be technically done in optimized engines benefiting from the high-octane ethanol characteristics or by adoption of flex-fuel engines (RFA, 2018). Once the level of mixture is set, the market determines the cost of the fuel sold to consumers, and certificates can be traded among the supplies in order to comply with the mandate.

The great novelty in this area is the intention of China to increase their present blend by 10% in all its provinces in order to reduce the urban pollution. Presently, the 10% blend is adopted in only nine provinces. Ethanol is produced from corn in China, and there is a large surplus of corn production in China (Agence France-Presse, 2017).

Technological advances

One of the great challenges for the future of bioenergy is the improvement in the efficiency of biomass production, particularly sugarcane in Brazil (Goldemberg, 2017).

As is well known since 1975, the agricultural productivity of sugarcane has increased from approximately 45 tons/ha to 80 tons/ha and 6000 liters/ha. From the plot of agricultural yield versus the sugar content from 165 producing units in the southeast of Brazil shown in Figure 2.6, it can be observed that additional gains are possible. The average agricultural production of this significant group of plants is 82 tons/ha and 12.9% of fermentable sugar but productivity of 100 tons/ha and 14.5% of sugar has been already reached in a number of plants using different types of cane, irrigation, or better management.

An option tried is use of a variety of "energy cane" in the northeast with agricultural yields of 180 tons/ha and 8.5% of sugar. This increased the amount of bagasse available from energy cane (Alisson, 2017).

Table 2.3 shows comparison of the products as the cogenerated electricity one could obtain from a distillery using "energy cane".

The use of enzymatic or acid hydrolysis or other methods opens the possibility of converting fermentable sugars into cellulose and hemicellulose, which are the dominant components of bagasse and other agricultural products.

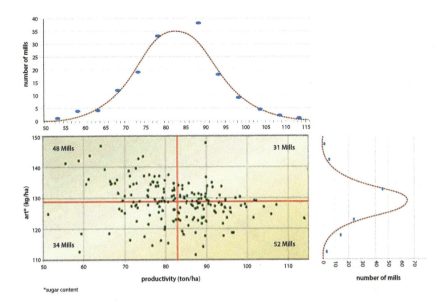

Figure 2.6 Productivity and sugar content (ATR) from 165 sugar/ethanol mills in the southeast of Brazil.
Source: Goldemberg (2017).

Table 2.3 Comparison between agricultural and industrial yields using traditional biomass and "energy cane"

Type of sugarcane	Agricultural yield (tons/ha)	Liters/ha	Electricity cogenerated (MWh)
Traditional	80	6500	5.6
Energy cane	180	6800	20

In the case of sugarcane, an increase in 60% of ethanol production could be reached, and this approach is being pursued in a number of pilot plants and a few industrial scale plants with somewhat disappointing result so far.

References

Agence France-Presse. *China quer ampliar o uso do etanol até 2020.* 13 September 2017. www.novacana.com/n/etanol/mercado/exportacao/china-ampliar-uso-etanol-2020-130917/

Alisson, E. *Levedura modificada aumenta a produção de açúcar da cana-energia.* Agência FAPESP. 14 September 2017. http://agencia.fapesp.br/levedura_modificada_aumenta_a_producao_de_acucar_da_canaenergia/26132/

EIA, U.S. Energy Information Administration, 2015.

Goldemberg, J. *The Ethanol Program in Brazil. Environmental Research Letters.* November 2006. http://iopscience.iop.org/article/10.1088/1748-9326/1/1/014008/pdf

Goldemberg, J. Atualidades e perspectiva no uso de biomassa para geração de energia. Revista Brasileira de Química. *Rev. Virtual Quim.* 2017, 9 (1), 15–28. http://rvq.sbq.org.br/imagebank/pdf/v9n1a04.pdf

Goldemberg, J. and J.R. Moreira. What's the Best Way to Use Solar Energy. *Photoenergy.* 2017, 1 (1), 1. http://medcraveonline.com/OAJP/OAJP-01-00004.pdf

IEA. Technology Roadmap-Delivering Sustainable Bioenergy. 89p. 2017. www.iea.org/publications/freepublications/publication/Technology_Roadmap_Delivering_Sustainable_Bioenergy.pdf

IRENA (International Renewable Energy Agency). *Sustainable Bioenergy Supply: Potential, Scenarios and Strategies.* Sept. 28th, 2016, Berlin, Germany. www.irena.org/menu/index.aspx?mnu=Subcat&PriMenuID=30&CatID=79&SubcatID=3779

REN21 2016. www.ren21.net/wp-content/uploads/2016/05/GSR_2016_Full_Report_lowres.pdf

RFA. Renewable Fuels Association, Blend Wall. 2018. www.ethanolrfa.org/issues/blend-wall/

3 Is there really a food versus fuel dilemma?

Frank Rosillo-Calle

The underlying reasons of why this situation has arisen are numerous, ranging from failure of the scientific community to communicate the implications to the wider audience to resistance of established views to accept the evidence-based facts and persistence of negative reporting (this seems more rewarding). Also, the narrowness of the debate, e.g. the focus in just a few feedstocks (maize, sugarcane, cereals), and the geographical dimension [although biofuels can be produced in many countries, just a few (Brazil and the USA, and to a lesser extent, the EU)], are the key players. These countries have a huge potential for increasing food production without causing any food and fuel conflict; at the same time, these countries also waste a huge amount of food.[1] It is important to emphasise that the major concerns refer to liquid biofuels. Solid biomass is widely available and comprises a large range of feedstocks, including large amount of agroforestry residues (e.g. Rosillo-Calle, 2012, 2015).

The media has further compounded this perception by emphasising the negative implications of biofuels without taking fully into account their intertwine and multidisciplinary nature. For far too long, the emphasis has been on conflicts rather than the potential mutual benefits of food and fuel production. The development of biofuels is limited by many factors, but not necessarily by these so often emphasised, such as direct land use competition or food price impacts. An important outcome emanating from this debate, however, has been the growing recognition that sustainability should apply equally to food and biofuel production.

The rapid transformation of the energy sector, in which oil and coal are no longer kings, presents biofuels with great challenges and opportunities, particularly in the transportation sector, in the drive to cut down emissions; biofuels are an important component of this energy transition. Renewable energy is increasing rapidly and becoming competitive with traditional sources, e.g. in many rural areas where biomass was the only energy alternative, can now be supplied also by solar and wind. This will be positive for bioenergy development whose focus should be on sustainability, modern applications and the use of right crops (e.g. non-food crops, agroforestry waste, cellulosic crops, sugarcane). Evidence-based, greater pragmatism and holism should be the main drivers.

Many of the bioenergy issues are covered elsewhere in this book. This chapter focuses on land use and price impacts. It is argued that the food versus fuel debate is a hollow one as it can easily be avoided. For example, currently about 4% of cropland is dedicated to biofuels (2% cereals, 32% maize, 6% sugarcane, as seen in Figure 3.1. At the same time, land use to feed one person/year was 0.45 ha in 1961 and 0.19 ha in 2015. On a global basis, this would require only 1.7 billion ha (1/3 of current land) to feed 9 billion people in 2050 (Goldewijk et al. 2017). In 2017/18, world production of cereals was 2613 Mt, of which 1065Mt was corn, used as animal feed, and 753 Mt was wheat, the two most important crops (see Figure 3.2). It is a fact that the prime use of land is for animal feed rather than for direct human consumption, benefiting primarily the wealthiest.

Based on a scientific analysis of food production and current waste (e.g. according to Food and Agriculture Organisation (FAO), one-third of global food production is lost), it could then be asserted that the *food versus fuel dilemma* is a hollow and false one. We are presiding in a world dominated by social and economic injustice, poor agricultural systems, inequality and food waste. People who go hungry, or are undernourished, are not for lack of food, it is simply because they cannot afford to purchase it. With greater social and political changes that eliminate injustices, and with a modern dynamic agricultural sector (in most countries, it is wistfully underfunded and undervalued) and populated by a young skilled labour force, the world can produce far more food. This combined with rapid changes in the energy sector (e.g. rapid increase of solar and wind), the use of agroforestry wastes (ethanol from cellulose) and sustainability of both food and fuel production, which could easily render this perceived food and fuel dilemma a thing of the past, particularly when there could be multiple other choices.

Introduction

A key aim of this chapter has been to demonstrate the hollowness of the so-called "food versus biofuel dilemma". Thus, the following pages present the general reader an overview of the global implications of biofuels from a perspective of land use and food price impacts. Many other issues related to bioenergy are covered elsewhere in this book. The chapter examines biofuels from a historical perspective, global land use, grain production and consumption and impacts on food prices, including food waste; it also considers the role of agriculture, a key-determining sector in the future development of biofuels. Finally, it tries to answer the question, "*is there really a food versus dilemma?*" and concludes that there it is not, since there could equally be other multiple and compatible alternatives that invalidate this misperceived view, based on distorted facts.

Historical perspective

The historical development of biofuels has not been exempted from pitfalls. It has had many supporters and many detractors. In fact, as far back as 1987, the author wrote a paper as an attempt to dispel some of the myths against Brazil's ethanol fuel programme (Rosillo-Calle & Hall, 1987). The food versus fuel debate is not, however, universal; it is based primarily in the EU and the USA, two key players in the industrial world. In the EU, for example, this debate is impacting negatively on policy development. Though the idea that the EU biofuels are causing an impact on the global food supply, or contributing to hunger may be a myth, it is shaping the policy. For example, European ethanol production uses a minuscule share of the EU grain harvest (only 2%, net, in 2015) – not enough to reduce grain supply to food markets or affect food prices since grain production has increased globally (see Figure 3.2) (Desplechin, 2017).

In fact, EU grain production increased between 2013 and 2015 by more than twice the amount of grain used to make ethanol in 2015. The additional grain demand for biofuels has been met through agricultural productivity increases, not by taking away supply from other uses such as food. While biofuel production has increased nearly 60% since 2008 in Europe, global food prices have decreased by 20% during the same period, according to Desplechin (2017). In 2015, ethanol production in the EU used 14 Mt of EU-grown crops and residues that were produced on less than 1% of EU agricultural land.

Yet, despite growing evidence of the low impact of biofuels, political sensitivity is pushing a move away from biofuels-based food crops, emanating from the food versus fuel debate (EURACTIV, 2017). In the USA, the debate has also been distorted primarily due to vested interests, with just 20% of the land dedicated to food production (Dale et al., 2010). In Brazil, a key player, biofuels are being actively promoted. Not only there is a land and fuel conflict, but biofuel development has been an important instrument for modernizing agriculture and increasing food production. In the rest of the world, the concern with food versus fuel is far more muted.

Global land use and biofuels

The heated debate on the use of land for biofuels production has been far overstated and simplistic, given the amount of land dedicated to such end, as clearly illustrated in Figure 3.1. Of the 1.7 billion hectares of crop area in 2016, a mere 4% was dedicated to biofuels (corn 28 Mha, soybeans/oil 21 Mha, sugarcane 5 Mha and wheat 2 Mha). The largest share corresponds to maize, mainly in the USA, with 32% (also used as feed). It is worth noting that for each ton of wheat processed into ethanol, on average, 295 kg of dried distillers' grains (DDGs) with solubles and 309 kg of maize are produced. Such by-products are an important buffer for farmers, allowing them to have greater control over prices.

According to Desplechin (2017), the EU ethanol sector does not import crops from developing countries and hence is not responsible for land grabs. In 2015, the production of crop-based biofuels (both biodiesel and ethanol) generated 28 Mt of EU-grown farm produce, bringing at least €6.6 billion in direct revenue for EU farmers. Furthermore, by-products from biodiesel

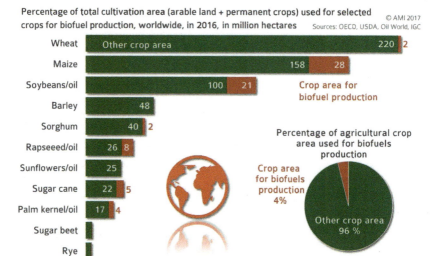

Figure 3.1 Land use in the production of biofuels in 2016 (Mha).
Source: UFOP Supply Report (2018).

production alone generate about 17 Mt of feed in the EU; to replace such feed will require 4–5 Mha of soya, which will come primarily from South America, with the consequent implications for indirect land use and declining income for the EU farmers.

In the USA, according to Dale et al. (2010), about 80% of the land is dedicated to produce animal feed, not food. According to the authors, "using less than 30% of total US cropland, pasture, and range, 400 billion litres of ethanol can be produced annually without decreasing food production or agricultural exports". This could also be achieved merely by increasing soil fertility, with significant reduction on GHGs.

Land use and biofuels are, however, complex issues that go beyond energy and food. As Kline et al. (2016) put it, "bioenergy can contribute to improved food security through production systems designed to increase adaptability and resilience of human populations at risk and to reduce context-specific vulnerabilities that could limit access to local staples and required nutrients in times of crisis".

Biofuels and grain production

Contrary to what some people may believe, food production has outpaced demand, as illustrated in Figure 3.2, which shows that world grain production increases from 1971/72 to 2017/18 harvest. It has increased from 1300

Is there really a food versus fuel dilemma? 39

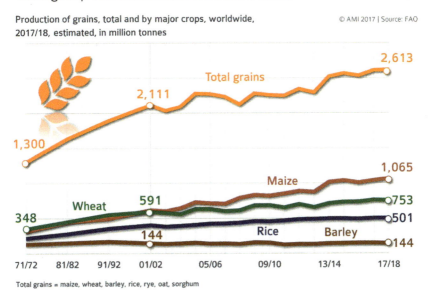

Figure 3.2 Worldwide production of total grains, and major crops, from 1971/72 to 2017/18 harvest.
Source: UFOP Supply Report (2018).

Mt to 2613 Mt, respectively. The main crop is maize, increasing from about 348 Mt to 1065 Mt, followed by wheat from approximately 348 Mt to 753 Mt over the stated period. By far the biggest human use of land is pasture, for animal feed production rather than food, as shown in Figure 3.3.

As indicated, another major implication, largely underestimated, is the huge amount of food production destined as animal feed, as illustrated in Figure 3.3. This shows different uses of grain consumption from 2010/11 to 2017/18. As can be appreciated, feed rose from 749 Mt in 2010/11 to 2105 Mt in 2017/18, representing 44% that year. Contrary, food increased from 621 Mt in 2010/11 to 695 Mt in 2017/18 (33% of total that year), followed by biofuels with 8% of total in 2017/18 or 169 Mt. Overconsumption of meat is also becoming a health issue, in addition to animal feed production, in detriment of direct food production.

There is also a huge disparity of grain per capita/person, e.g. Oceania, 1082 kg/person; the USA, 976 kg/person; EU, 724 kg/person and Africa, just 129 kg/person. The same applies per capita meat consumption (kg/person/year) with 95 kg in the USA, 70.7 kg in the EU, 63 kg in Oceania and just 4.2 kg in Africa. Globally, meat production increased from 56 Mt in 1965 to 253 Mt in 2015 (e.g. see UFOP, 2017, 2018). This huge disparity results in unhealthy diet and food waste, and undernourishment and poor diets in others.

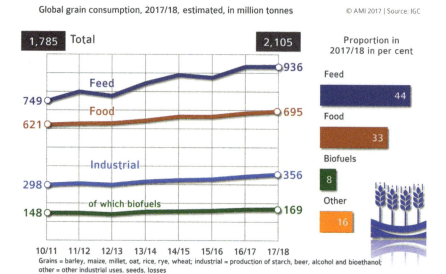

Figure 3.3 Worldwide consumption of grains and major uses, from 2010/11 to 2017/18 harvest.

Source: UFOP Supply Report (2018).

Bioenergy is quite site-specific and needs to be assessed in an integrated manner for a wide range of geographic conditions and contexts. Bioenergy goes beyond energy, embracing socio-economic development, technological and political change (e.g. see Strapasson et al., 2017). It is against this background that any serious study on food versus fuel should be carried out.

Food prices and wastes

The relationship between biofuels and food prices has been another battleground. The answer to this question is more complicated that it appears. Many studies have been carried out to prove or disprove such impacts (e.g. see FAO, 2008; Brown and Brown, 2012; Lagi et al., 2012). There is an abundant literature for the interested reader to consult, but with different outcomes due to the use of differing methodologies, use of data, scenarios, etc. It seems that, more often than not, studies have been trying to prove or disprove a particular point of view rather than entirely evidence-based. However, generally, it can be stated that a growing body of evidence indicates the impacts of biofuels on food prices are relatively weak.

The early debate on the potential impacts of biofuels on the spike in food prices during 2006–2008 was largely exaggerated, as it was often based on fragmented data, in which commodities speculation, (particularly agricultural commodities), adverse weather conditions, climate change, etc., were not fully taken on board (e.g. see Rosillo-Calle and Johnson, 2010; Malins, 2017).

For example, as can be appreciated from Figure 3.4, from 2008 to 2017, ethanol fuel production increased rapidly, while FAO Food Price Index declined sharply. This demonstrates that ethanol production is not the main driver of food prices. This chart uses the latest world ethanol fuel production data from Licht's 2017 forecast (see www.globalrfa.org/news). Despite the improvement of the data, the debate on food prices is unlikely to go away as it is not a black and white issue, but clouded with claims and counterclaims and large regional, national and global variations in crops, climate change issues, etc.

The problem with many scenarios is that they ignore, to a large extent, the potential of technology to transform agriculture and reduce the huge food waste (e.g. according to FAO, 1/3 of the world's food production is wasted (see www.fao.org/save-food). There is no question that the current agricultural systems are unsustainable, suffer from gross underinvestment and serious unequal social and economic injustice, and are hugely wasteful, which are at the core of many problems. It should not be surprising, therefore, that impacts of biofuels on food prices are politically so sensitive.

Most feedstocks used for biofuels do not come from staple food stocks. Although Malins (2017) disagrees with the assertion of the pro-biofuel lobby that biofuels produce both food and livestock feed price increase because their

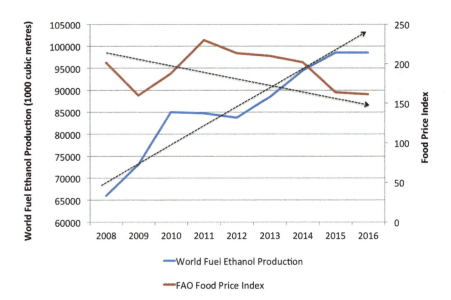

Figure 3.4 World ethanol fuel production and FAO Food Price Index, from 2008 to 2017.
Source: www.globalrfa.org/news, 31 May 2017.

contribution is not enough to compensate the food market for the material converted into biofuel.

The largest market for maize and wheat, used in biofuel production, is originated from animal feed. Sugarcane, the major source of ethanol fuel in Brazil, also provides many other benefits (e.g. heat and power, animal feed, yeasts, etc.) and has been, at the same time, the main channel for modernising agriculture and increasing food production around sugarcane-growing areas.

It would be unrealistic to say that biofuels do not cause any impact on food prices; unquestionably some impacts are inevitable, as happens with any other commodity. As Wiggins and Macdonald (2008) put it,

> Price rises hurt the poor, the urban poor more than the rural, net food buyers more than those farmers who are not sellers. But even for the poor, the effects are not necessarily that strong. A 10% rise in all food prices might overall rise poverty by 0.4% percentage points- not welcome, but hardly disastrous.

It is also important to keep in mind both the producers and consumers interests. Higher prices stimulate the farmers to produce more food, with the consequent benefits of greater availability.

Food waste

Food waste is a serious problem, particularly in the most developed countries where the price of food is relatively cheap; this has a considerable impact on food prices in general, and for this reason, it is receiving an increasing attention. In many parts of the world, 70% of the harvest is wasted, particularly vegetables (Economist, 2011). In the USA, about 40% of the food is discarded annually (Gunders, 2012). In the EU, about 88 Mt of food is wasted annually. A reduction in food waste would also ensure considerable CO_2 emission reduction and land use savings (www.euractiv.com/section/circular-economy/). This kind of food waste is scandalous.

The FAO of the UN Director General José Graziano da Silva has joined in calls for a renewal global commitment to zero tolerance for food loss and waste. The call was made at a high-level event at the 2017 72nd session of the UN General Assembly that focused on tackling food loss and waste as a pathway to achieving Sustainable Development Goal 2: Zero Hunger. He added, "Zero tolerance for food loss and waste makes economic sense e.g. for every $1 companies have invested to reduce food loss and waste, they saved $14 in operating costs".

This loss and waste occurs throughout the supply chain, from farm to fork. Beyond food, it represents a waste of labour, water, energy, land and other inputs. By reducing loss and waste along the food value chain, healthy food systems can contribute to promoting climate adaptation and mitigation, preserving natural resources and reinforcing rural livelihoods.

FAO has developed tools and methodologies for identifying losses, their causes and potential solutions along the entire food value chain, from production, storage and processing to distribution and consumption. In 2013, FAO launched a global initiative on food loss and waste called "save food" (www.fao.org/save-food). However, there is some resistance in few industrial sectors because restrictions to food waste add extra cost to producers.

The ever-increasing demand for agricultural products

The demand for agricultural products is enormous and increasing. It is not only food and fuel but many other natural products, remember the seven "Fs" – food, feed, fuel, feedstock, fibre, fertilizer and finance. Such demand cannot be met with the present stage of agricultural development in most countries. In the poorest countries, agriculture is still largely undeveloped, uses traditional methods and is dominated by small traditional landholdings with very low productivity, land ownership problems and so forth. Agriculture needs to be transformed, modernised, mechanised and populated with a skilled young labour force if it is to meet such growing demand.

This should not deny the fact that in many countries, agricultural productivity has increased significantly. For example, globally land use to feed one person/year was 0.45 ha in 1961 and 0.19 ha in 2015. On a global basis, this would require only 1.7 billion ha (1/3 of current land) to feed 9 billion people in 2050 (e.g. see Goldewijk et al., 2017). Undoubtedly, this is a major achievement but insufficient to meet new challenges, particularly those emanating from potential climate change (dealt with elsewhere in this book).

The question is, can the agricultural sector raise to this challenge? Such transformation requires a huge investment in land, people, infrastructure, etc., and a major political commitment. This will be a huge challenge particularly in the poorer countries. Is this a realistic outcome? Perhaps not.

In the specific case of the energy sector, the world is witnessing a major transformation, with many actors involved as there is not a single option but a range of potential alternatives, e.g. biomass, solar, wind, oil, natural and shale gas, coal, etc. For many decades, coal and oil were kings but this dominance is gradually being phased out; consequently, the world is facing a very mixed *energy* scenario, far more complex and difficult to predict. As for biofuels, they will be important transitional alternatives, with a major role in some countries such as Brazil and the USA.

The main challenge will be in Africa because this is where the agriculture sector is less developed with very low productivity, fragmented, problems with land ownership, lack of investment, traditional agriculture systems, poor distribution systems, etc. This is further compounded by rapid population growth, undernourishment, inequality, constant conflicts and poor health services. But, at the same time, the continent has a considerable amount of potential agricultural land waiting for cultivation (e.g. see Malabo Montpellier Panel, 2017).

Conclusions

This chapter has attempted to disprove there is not a food versus fuel dilemma. It has investigated two key factors: land use and price impact. From the previous pages, it becomes clearer that there is no dilemma in the use of land for biofuels, because (i) the proportion of land involved is very small (~4%), (ii) most agricultural land is not for direct human consumption but for animal feed and (iii) agricultural productivity is so low in most countries that direct land competition would be primarily caused by poor management practices rather than land shortages. The key fundamental problem is political, manifesting in huge social and economic inequality.

As for price impact, there is an abundant body of evidence demonstrating there is a weak link. This is not to deny there is not any impacts on food prices; such impacts are inevitable, but higher prices have also important knock-offs, both positive and negative, e.g. higher prices stimulate the farmers to produce more food; ethanol from sugarcane and maize acts as buffer for farmers to have a better price control. A greater impact derives from huge food waste.

Finally, transportation is the rapidly changing energy sector, particularly, which is also a major social, economic and technological transformation. For the first time in human history, the world is confronting a range of potential alternatives, a very mixed energy scenario. Therefore, there is no any dilemma since there are, potentially, equally multiple choices, which could be "competitive and complementary". Thus, the perceived food versus biofuel dilemma is based on incomplete, distorted, and lacks evidence-based facts.

Note

1 A good example of how this debate influenced policymaking is the EU. The EU is currently debating a reduction of the cap for first generation of biofuels from food crops from 7% in 2020 to 3.8% in 2030.

References

Brown RC & Brown TR (2012) *Why are We Producing Biofuels-Shifting to the Ultimate Source of Energy*, Bioenergy Institute, Iowa, USA.
Dale BE, Bals BD, Kim S & Eranki P (2010) Biofuels done right: Land efficient animal feeds enable large environmental and energy benefits, *Environmental Science and Technology*, 44: 8385–8389.
Desplechin E (2017) www.epure.org. *European Renewable Ethanol Association*. Ascended 30 March/17.
Economist, The (2011) Special report on feeding the world, 26th February 2011.
EURACTIV (2017) *Phasing-out biofuels: What's really at stake?* Special Report, October 2017, www.euractiv.tv/98Ru.
FAO (2008) *The State of Food and Agriculture. Biofuels: Prospects, Risks and Opportunities*, Rome, Italy.

Goldewijk KK, Dekker SC & Van Zanden JL (2017) Per-capita estimations of long-term historical land use and the consequences for global change research, *Journal of Land Use Science*, 12–15: 313–337.

Gunders D (2012) *Americans waste, throw away nearly half their food Report*, (www.nrdc.org; www.switchboard.nrdc.org/dgunders/.

Kline KL et al. (2016) Reconciling food security and bioenergy: Priorities for action, GCB *Bioenergy*, 9(3): 557–576.

Lagi M, Bar-Yam Y, Bertrand KZ & Bar-Yam Y (2012) *The Food Crises: Predictive Validation of a Quantitative Model of Food Prices including Speculators and Ethanol Conversion* (February 21012 update), New England Complex Systems Institute, Cambridge MA, USA.

Malabo Montpellier Panel (2017) *Nourished – How Africa can be a future free from hunger and malnutrition*, www.mamopanel.org.

Malins C (2017) *Thought for food-A review of the interactions between biofuel consumption and food markets*, Cerulogy, www.cerulogy.com.

Rosillo-Calle F (2012) *Food versus Fuel: toward A New Paradigm-The Need for a Holistic Approach* (Commissioned paper "Spotlight Article for ISRN Renewable Energy, Volume 2012, Article ID 954180; www.hindawi.com/isrn/re/contents.

Rosillo-Calle F (2015) Review of biomass energy: Shortcomings and concerns, *Journal of Chemical Technology and Biotechnology*, 91: 1932–1945.

Rosillo-Calle F & Hall D (1987) Brazilian alcohol: Food versus fuel? *Biomass*, 12: 97–127.

Rosillo-Calle F & Johnson F (2010), Eds + contributors. *The Food versus Fuel Debate: An Informed Introduction to Biofuels*, Zed Books, London; Paperback ISBN: 9781848133839; Hardback: ISBN: 9781848133822.

Strapasson A, Woods J, Chum H, Kalas N, Shah N & Rosillo-Calle F, et al. (2017) On the global limits of Bioenergy and land use for climate change mitigation, *GCH Biology*, doi.10.1111/gcbb/12456.

UFOP (2017) Supply Report 2016/2017, www.ufop.de (Union zur Forderung von Oel-und Proteinpflanzen e.V.), Berlin.

UFOP (2018) Supply Report 2017/2018, www.ufop.de (Union zur Forderung von Oel-und Proteinpflanzen e.V.), Berlin.

Wiggins S & Macdonald S (2008) Review of the indirect effects of biofuels: Economic benefits and food insecurity, *Nutrition*, 158. Retrieved from www.odi.org.uk.

www.euractiv.com.

www.euractiv.com/section/circular-economy/.

www.globalrfa.org/news.

4 Integrating pasture intensification and bioenergy crop expansion

Eleanor E. Campbell, John Joseph Sheehan, Deepak Jaiswal, Johnny R. Soares, Julianne de Castro Oliveira, Leonardo Amaral Monteiro, Andrew M. Allee, Rubens Augusto C. Lamparelli, Gleyce K.D. Araújo Figueiredo and Lee R. Lynd

Introduction

As global population and overall living standards increase, there is a widely recognized need to increase global food production (Godfray et al., 2010). At the same time, the climate crisis demands an increase in low-carbon bioenergy production to decarbonize the transportation sector and limit global warming (Fulton et al., 2015). Land use is an inextricable part of the problem and solution for both dilemmas.

Pastureland – which occupies 3.4 billion ha and accounts for over 60% of land use by humans (Figure 4.1A) – has received considerably less attention than cropland (1.6 billion ha) and forest (3.9 billion ha) in this context (Lambin and Meyfroidt, 2011; Ramankutty et al., 2008). Pasture is defined here to include FAO land classes for permanent meadows, pastures, and grasslands used for grazing (FAO, 2013). Vegetal (e.g. crop-based) contributions to dietary consumption vastly outweigh those of livestock products (Figure 4.1B). Further, grazing of pasture accounts for only a small share of animal products, estimated as 1% of global dietary energy consumption and 3% of global dietary protein (Woods et al., 2015). Most livestock products are generated in mixed crop-livestock systems in which grazing is supplemented by grains, forages, stovers, and other feeds (Herrero et al., 2013). Additionally, well-recognized limitations on land use change in cropland, and managed forests limit their ability to expand production (e.g. Fargione et al., 2008). These facts suggest pasture as a large land category with relatively low contributions to global food production, making it a promising area to consider productivity increases.

Increasing demands for food and bioenergy must be met either through intensification, which increases yield per unit area, or extensification, which increases land area in agricultural use (FAO and DWFI, 2015). Extensification risks deforestation and biodiversity losses in natural ecosystems,

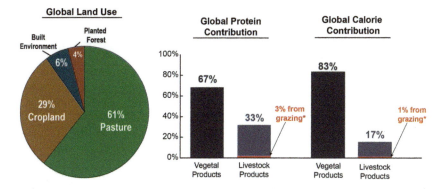

Figure 4.1 The distribution of land use by humans, as calculated from estimates by Lambin and Meyfroidt (2011), excluding natural forest and unmanaged productive lands from "Global Land Use". Also, the relative contributions of vegetal and livestock products to global protein and Calories (FAO, 2012). *Globally, grazing has been estimated to contribute 3% of dietary protein and 1% of dietary energy (Woods et al., 2015).

is ultimately limited by the quantity of arable land on the earth, and is considered the option of last resort in most of the world (DeFries and Rosenzweig, 2010). Intensification is the desired mode of increasing agricultural output, ideally while continuing to honor environmental, social, and economic objectives (Tilman et al., 2011). Studies show capacity for substantial gains in agricultural production through intensification practices that include improving labor efficiency, water and nutrient management, access to enhanced seeds, and changes in crop rotations (Mrema et al., 2008; Mueller et al., 2012).

Sustainable intensification has been proposed as a large-scale mechanism to 'make room' for expanding bioenergy production in the existing agricultural land base without impacting net food production or leading to clearing of native lands (Woods et al., 2015). In pasture systems, this raises two key questions. First, *what is the potential for sustainable intensification of livestock production on pasture?* Second, *how suitable is pastureland for sustainable bioenergy crop expansion?* At a global scale, answering these questions will allow quantification of potential for sustainable bioenergy crop expansion and identification of promising regions where this could occur. At the local level, these analyses may locate rural areas suitable for bioenergy projects. When properly configured, these projects have been shown to promote positive on-the-ground change, improving formal employment, education, and energy security (Cardoso et al., 2018; Leite et al., 2016; Moraes et al., 2015).

One metric for identifying land with capacity for sustainable intensification is the **yield gap**, or the difference between the current yield of a parcel of land and its maximally attainable yield (Mueller et al., 2012). High yield gaps indicate high intensification potential. Estimates for the global intensification potential of pastureland have not been reported in the literature. As pastureland is currently one of the biggest land uses by humans (Figure 4.1A), addressing this knowledge gap is critical to better understand the global potential for intensification of pasture-based agricultural systems, as well as their capacity to accommodate the expansion of bioenergy crops.

Globally scaled evaluation of pasture intensification is challenged by a paucity of available data. Pasture-based agriculture is widespread, complex, and regionally variable. Furthermore, definitions of pasture are not consistent and, unlike in cropping systems, there is no geographically distributed global database for tracking yields of pasture-derived animal products through time. The first global census of livestock populations (Wint and Robinson, 2007) did not distinguish between animals on pasture and animals raised in other systems (e.g. feedlots, mixed crop, and livestock). A more recent global livestock dataset is more comprehensive, but only represents the year 2000 (Herrero et al., 2013), thus missing temporal dynamics as well as more recent agronomic advances.

It is clear that pasture is a large land stock with potential to produce more food and biofuel if intensified, but also that its status and potential are not well understood. An ongoing research project in the Global Sustainable Bioenergy Initiative, referred to herein as the GSB Geospatial project, looks to address this need through a combination of global-level and landscape-level analyses alongside new techniques for evaluating the potential for bioenergy crop expansion. The GSB Geospatial project focuses on 'grazed-only' pasture, where animal occupancy and feed production overlap completely. This simplicity allows the development of an analytical framework that can be adapted across scales and eventually used to evaluate more complex pasture-based livestock systems. While 'grazed-only' pasture is not a dominant provider of total livestock products (producing only 10% and 8% of total meat and milk production, respectively (Woods et al., 2015)), it is recognized for its regional importance and key role in providing food and economic stability in many communities across the globe (Herrero et al., 2013).

In this chapter, we discuss the basis for ongoing work in global-scale pastureland yield gap and energy crop analyses, paired with landscape-scale process-based modeling and remote sensing analyses that use Brazil as a case study. Brazil is an ideal location for landscape-scale focus due to its pasture-dominated livestock system and rich long-term datasets. Furthermore, in recent decades the country has both expanded sugarcane-based ethanol production and improved livestock system performance (Figure 4.2), making it a unique case study opportunity (Adami et al., 2012; Jaiswal et al., 2017; Martha Jr. et al., 2012).

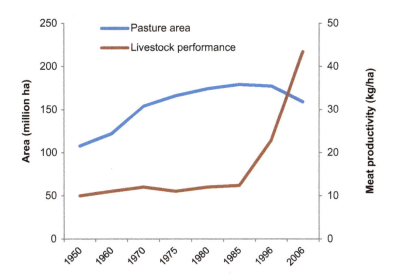
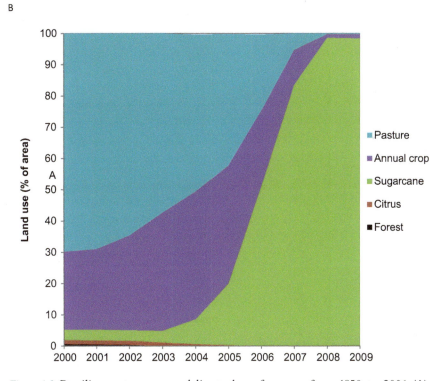

Figure 4.2 Brazilian pasture area and livestock performance from 1950 to 2006 (A). Brazilian land use change for sugarcane expansion from 2000 to 2009 (B). Adapted from Adami et al. (2012); Martha Jr. et al. (2012).

Pastureland intensification for bioenergy: addressing key uncertainties

Global potential for pasture intensification

To calculate the yield gap of an agricultural system, one needs to know current *versus* attainable product yields. At a global scale, this requires two data layers: one that identifies pasture land area around the world (Figure 4.3A), and one that describes the productivity of each unit of area. Given these data, a "climate-binning" method can be used to compare agricultural productivity between land areas of similar climate characteristics. This analytical approach has been previously applied to row crops (Mueller et al., 2012). In this method, spatially explicit datasets for climate variables (e.g. total annual precipitation and growing degree-days) are used to define climate 'bins' (Figure 4.3B).

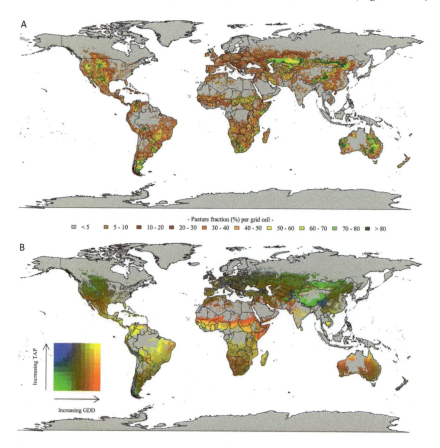

Figure 4.3 Global spatial distribution of pasturelands from Ramankutty et al. (2008) (A); and global spatial distribution of climate space defined in two dimensions: total annual precipitation and growing degree-days, as reported by World-Clim (B).

Global pasture areas are then assigned to bins according to similar climate characteristics. This technique aims to control for climate's strong association with pasture growth, allowing the remaining variation in production to be attributed to factors other than climate (e.g. management practices). The distribution of livestock production within a climate bin can thus be interpreted as the spectrum of achievable performance within a given climate 'space'.

Animal productivity, as measured by animal product yields per unit area and time, is a function of both plant productivity and the efficiency of plant matter conversion to harvested animal product (e.g. meat, milk, wool). At a global scale, direct reporting of livestock production data is extremely limited, and so yields must be estimated. Since yield gap calculations are very sensitive to yield data, it is useful to compare results obtained with various estimates of this important parameter. Herrero et al. (2013) – widely credited for the best available global datasets on ruminant production – provide one estimate of global livestock yields from pasturelands. An alternative approach is to model the aboveground net primary production (ANPP) of pastureland, and then translate plant growth into livestock meat and milk estimates. Among multiple available models, the National Center for Ecological Analysis and Synthesis (NCEAS) model (Equation 4.1) for non-tree plants predicts grass ANPP quite well globally and is only a function of precipitation (Del Grosso et al., 2008).

$$ANPP = 4000(1 - exp(-0.0000477 * TAP)) \tag{4.1}$$

In the above equation, ANPP is expressed in $g\ C\ m^{-2}\ year^{-1}$, and TAP is the total annual precipitation in millimeters.

ANPP can be converted to Calories of meat and milk using literature values for forage utilization, animal maintenance, and feed conversion rates. Pastureland ANPP and associated livestock production estimated from these simple modeling approaches can be considered conservative, as they do not consider management practices that improve biomass production in pasture systems (e.g. replanting grass species, nutrient management). Calculating potential livestock production through both ANPP modeling and third-party livestock production datasets (e.g. Herrero et al., 2013) enables the estimation of pastureland intensification potential via multiple pathways. The maps resulting from these studies highlight regions with large potential for sustainable intensification, as well as those in which production is above natural carrying capacity and attention is needed to avoid pasture degradation.

Energy crop production potential on land made available by pasture intensification

The total global production of biofuel in 2015 was 80 million tons oil equivalent with major contributors being the USA (42%) and Brazil (24%) (BP, 2017). In the coming years, advances in cellulosic ethanol technology (see Chapter 9) are expected to enable the commercial production of feedstocks to be diversified from sugarcane, coarse grains, and vegetable oils to include perennial grasses (Long et al., 2015).

Biofuels will continue to play an important role in the future energy matrix to meet the goals of the Paris Climate Accord (Jaiswal et al., 2017), and feedstock demand will likely be met via a variety of energy crops. An ideal energy crop should have low nutrient requirements, be high yielding on poor soil, be suitable for conversion via biological or thermochemical processes, and provide greenhouse gas (GHG) benefits and other ecosystem services (Davis et al., 2014). Some new candidates for ideal energy crops include miscanthus, energy cane, switchgrass, willow, and agave (Davis et al., 2014). Ongoing research efforts aim to identify regions and energy crop types with the greatest potential for development.

A major challenge associated with the expansion of energy crops is the limited number of crop field trials. The paucity of basic research makes modeling and prediction of yield difficult on land areas dissimilar to those where field trials were conducted. The inability to predict crop yields may result in failure of biofuel projects, as observed in the case of *Jatropha curcas* (Edrisi et al., 2015). Recent attempts have been made to fill this gap by accumulating bioenergy crop information for plant traits and yields (LeBauer et al., 2018), which can better address the need for observed data, guide basic research, and help in the development of predictive models. Using process-based modeling approaches instead of statistical models to predict yields and identify new areas for field trials or commercial projects is another pathway to overcome current limitations of experimental data (Jaiswal et al., 2017; Wang et al., 2015). Geospatial analyses of global yield maps based on model predictions for ideal energy crop candidates, in conjunction with land use mapping (e.g. Figure 4.4A), can be used to identify suitable areas for biofuels that do not compete with food production.

The agave plant is of particular interest in the global context (Davis et al., 2011, 2014) because of its sturdiness and ability to grow in arid and hyperarid regions, as well as in the vast availability of low-performing pastureland in these regions (Figure 4.4A). Development of an Agave model that can be used for global-scale simulations in arid and hyperarid regions is critical to its advancement as a bioenergy feedstock (e.g. Figure 4.4B). It also considerably widens the scope of global pastureland areas where integration of pasture intensification and bioenergy crop expansion may be considered (Figure 4.4A).

Sustainability of pasture intensification with expanded bioenergy production

Remote sensing

The evaluation of global potential for pasture intensification requires aggregation of global-scale datasets that do not always reliably represent specific

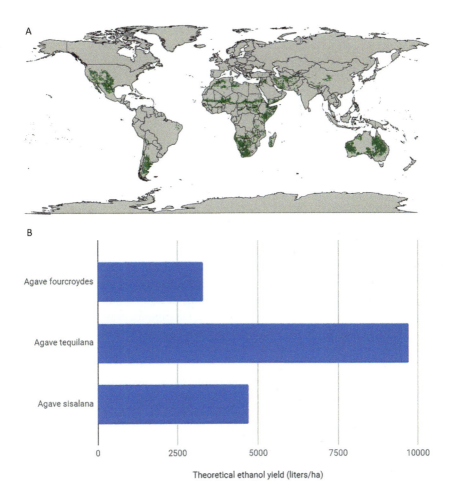

Figure 4.4 Global spatial distribution of livestock production systems under arid and semiarid climates in a spatial resolution of 5 × 5 arcmin, totaling 745 million hectares (A), and associated theoretical ethanol yields from varying species of Agave (B).

Source: Wint and Robinson (2007).

countries or regions (FAO, 2013; Herrero et al., 2013). Brazil has a well-developed pasture-dominated livestock system and, as one of the largest beef exporters worldwide (USDA, 2016), is a global influencer in livestock and pasture systems (Strassburg et al., 2014). It is an ideal location to pair global-scale analyses of pasture intensification and energy cropping with landscape-scale 'ground-truthing' and sustainability analyses that more accurately represent regional dynamics.

Remote sensing techniques allow efficient mapping, and quantification and assessment of pastureland cover at national and regional levels (Ali et al.,

2016). Ongoing work in Brazil (e.g. Parente et al., 2017) uses remote sensing approaches to compare pastureland maps at different spatial and temporal resolutions, to evaluate land use change, and to identify the impact of database-specific referential techniques on pasture definition and classification. At regional and local levels, pasture systems can be classified according to management style (e.g. extensive, integrated, or intensified systems) as well as the quality of the pasture (e.g. degraded *versus* undegraded). Both of these classifications are important when evaluating potential for pasture intensification.

The wide spectral range of different remote sensing technologies allows them to be adapted to a variety of uses. The MODIS (Moderate Resolution Imaging Spectroradiometer) sensor aboard the Terra and Aqua satellites collects data across 36 spectral bands every 1–2 days that can be used to identify land characteristics (National Aeronautics and Space Administration, 2018). For example, ANPP can be estimated using the MODIS sensor by integrating biotype information, daily meteorological data, fraction of absorbed photosynthetically active radiation (FAPAR) and leaf area index (LAI). Comparing ANPP results from different sources (i.e. MODIS *versus* Equation 4.1) can improve the understanding of climate's influence on pasture areas through time (Zhang et al., 2017). Changes in grassland ANPP can signal land degradation or influence by climate change (Paruelo et al., 1997). Using MODIS and other such resources, remote sensing can inform inputs to crop and agroecosystem models (Liu et al., 2014). It is also useful for validating these models because of its capability to (1) monitor large areas with high temporal resolution, (2) identify land use changes through time, and (3) provide data for key agronomic variables (e.g. LAI, ANPP, and normalized difference vegetation index (NDVI)).

Process-based agroecosystem modeling

Combining pastureland intensification with increased bioenergy production requires land use change as well as changes in agricultural management practices. This raises the possibility of negative impacts on soil carbon and GHG emissions (Mello et al., 2014), important metrics for agroecosystem sustainability. These concerns are highly relevant in Brazil, which is undergoing pastureland intensification alongside the expansion of sugarcane production to meet growing global demands for sugar and biofuels (Figure 4.2).

The impacts of intensification, particularly on soil carbon and GHG emissions, will be affected by complex interactions of multiple factors including climate, soil type, and land use history. Determining pathways of sustainable expansion of bioenergy production with pasture intensification requires approaches that are tailored to regional agriculture based on associated environmental, chemical, and physical soil characteristics. While direct measurement of these variables is preferred, they are often logistically difficult to gather. Therefore, process-based agroecosystem models (e.g. Daycent;

Del Grosso et al., 2002) play an important role in assessing the sustainability of future scenarios. These types of models use climate, soil, and land management information to simulate changes in plant growth processes as well as decomposition and soil organic matter dynamics, among others, in order to predict changes in soil carbon and GHG emissions. In the past, agroecosystem modeling approaches have been applied successfully for this purpose in Brazil (Galdos et al., 2009; Oliveira et al., 2017).

Quantifying soil organic matter and soil GHG emissions associated with current land use and management practices is the first step in understanding the sustainability implications of changes related to pasture intensification paired with increased bioenergy crop production. This information is needed to explore and suggest viable, regionally relevant pathways to use pasture intensification as a mechanism for sustainable crop-based bioenergy expansion.

Moving beyond the GSB Geospatial project

At the time of publishing this chapter, global enthusiasm in bioenergy has waned since peaking circa 2005–2008 (Herrera, 2006; Service, 2007). Many factors contributed to this downturn in support: falling oil prices, underperforming technologies, food versus fuel concerns, and environmental pessimism on the possible negative impacts of biofuels. Indeed, promises that were driven by unprecedented funding opportunities and pro-biofuels policies often failed the test of reality (Lynd et al., 2017). High expectations were prematurely placed on cellulosic ethanol, too, leading to high levels of investment and increasingly optimistic claims by pioneers (Lynd et al., 2017). But companies failed to produce at costs competitive with rapidly decreasing oil prices, and thus nearly all cellulosic biofuel start-ups have either failed or are surviving at a value far below their initial public offering price (Reboredo et al., 2016).

Despite these troubles, bioenergy and biofuels remain crucial to a sustainable future (Fulton et al., 2015) and a component of strategies to address environmental, social, and economic goals (e.g. sustainable energy production, rural development) without repeating mistakes of recent history. Second-generation (2G) biofuel technologies continue to improve, though more investment is necessary and warranted (see Chapter 9). Ideally, pasture intensification with bioenergy could emerge as part of a virtuous cycle: driving biofuel production by lowering feedstock costs, while being driven by the growing feedstock demand of new 2G producers.

As the potential for pasture intensification and bioenergy crop expansion is better understood, a critical next step is to include social and economic dimensions in order to better understand its transformative potential and the possibilities for effective, sustainable change. Localized, regionally specific analyses will be key to these efforts, as implementation and application of these findings will ultimately happen on the ground in collaboration with local stakeholders, and under local and regional constraints. Bioenergy practitioners must simultaneously have in view the big picture (*what is the benefit to the world?*) and the local context (*will this strategy work to benefit these specific people given their needs and constraints?*). Just as the global pasture analyses need to be "ground-truthed" with local, farm-level observations, we must ground-truth the application of the work in both physical and social dimensions. This means adoption of a nuanced approach with local stakeholders and regional expertise at the heart of decision-making, to advance positive change in land management and land use.

This chapter introduces the potential role of pasture intensification in sustainably meeting rising demands for food and bioenergy feedstock. It also presents the basis for the research approach of the GSB Geospatial project, and a review of pathways to address key uncertainties. There are good reasons to expect tremendous global potential for pasture intensification (Figure 4.1). However, past experience with bioenergy is evidence that complex systems with land, food, bioprocess, and societal components are prone to unexpected challenges. We expect future research to generate sizeable estimates of global pasture intensification potentials, scientifically driven suggestions for sustainable land management, and global maps of attainable energy crop yields. These insights, if paired with the development of better bioprocessing technology and a ground-truthed approach to implementation, may suggest a viable way forward in meeting the food and energy demands of the coming decades.

Acknowledgments

This research was supported by FAPESP grant numbers 2014/26767-9, 2016/20307-1, 2016/08741-8, 2016/08742-4, 2017/06037-4, and 2017/08970-0.

References

Adami, M., Rudorff, B.F.T., Freitas, R.M., Aguiar, D.A., Sugawara, L.M., Mello, M.P., 2012. Remote sensing time series to evaluate direct land use change of recent expanded sugarcane crop in Brazil. *Sustainability* 4, 574–585. doi:10.3390/su4040574.

Ali, I., Cawkwell, F., Dwyer, E., Barrett, B., Green, S., 2016. Satellite remote sensing of grasslands: from observation to management. *J. Plant Ecol.* 9, 649–671. doi:10.1093/jpe/rtw005.

BP, 2017. Statistical Review of World Energy | Energy economics | BP [WWW Document]. bp.com. URL www.bp.com/en/global/corporate/energy-economics/statistical-review-of-world-energy.html (accessed 10.26.17).

Cardoso, T.F., Watanabe, M.D.B., Souza, A., Chagas, M.F., Cavalett, O., Morais, E.R., Nogueira, L.A.H., Leal, M.R.L.V., Braunbeck, O.A., Cortez, L.A.B., Bonomi, A., 2018. Economic, environmental, and social impacts of different sugarcane production systems. *Biofuels Bioprod. Biorefining* 12, 68–82. doi:10.1002/bbb.1829.

Davis, S.C., Dohleman, F.G., Long, S.P., 2011. The global potential for Agave as a biofuel feedstock. *GCB Bioenergy* 3, 68–78. doi:10.1111/j.1757-1707.2010.01077.x.

Davis, S.C., LeBauer, D.S., Long, S.P., 2014. Light to liquid fuel: theoretical and realized energy conversion efficiency of plants using Crassulacean Acid Metabolism (CAM) in arid conditions. *J. Exp. Bot.* 65, 3471–3478. doi:10.1093/jxb/eru163.

DeFries, R., Rosenzweig, C., 2010. Toward a whole-landscape approach for sustainable land use in the tropics. *Proc. Natl. Acad. Sci.* 107, 19627–19632. doi:10.1073/pnas.1011163107.

Del Grosso, S., Ojima, D., Parton, W., Mosier, A., Peterson, G., Schimel, D., 2002. Simulated effects of dryland cropping intensification on soil organic matter and greenhouse gas exchanges using the DAYCENT ecosystem model. *Environ. Pollut.* 116, Supplement 1, S75–S83. doi:10.1016/S0269-7491(01)00260-3.

Del Grosso, S., Parton, W., Stohlgren, T., Zheng, D., Bachelet, D., Prince, S., Hibbard, K., Olson, R., 2008. Global potential net primary production predicted from vegetation class, precipitation, and temperature. *Ecology* 89, 2117–2126. doi:10.1890/07-0850.1.

Edrisi, S.A., Dubey, R.K., Tripathi, V., Bakshi, M., Srivastava, P., Jamil, S., Singh, H.B., Singh, N., Abhilash, P.C., 2015. Jatropha curcas L.: a crucified plant waiting for resurgence. *Renew. Sustain. Energy Rev.* 41, 855–862. doi:10.1016/j.rser.2014.08.082.

FAO, 2013. *Fao statistical yearbook 2013.* Food & Agriculture Organization of the United Nations, Rome.

FAO and DWFI, 2015. *Yield gap analysis of field crops: methods and case studies.* FAO Water Reports No. 41, Rome, Italy.

Fargione, J., Hill, J., Tilman, D., Polasky, S., Hawthorne, P., 2008. Land clearing and the biofuel carbon debt. *Science* 319, 1235–1238. doi:10.1126/science.1152747.

Food and Agriculture Organization of the United Nations. (2012). FAOSTAT. Rome, Italy: FAO

Fulton, L.M., Lynd, L.R., Körner, A., Greene, N., Tonachel, L.R., 2015. The need for biofuels as part of a low carbon energy future. *Biofuels Bioprod. Biorefining* 9, 476–483.

Galdos, M.V., Cerri, C.E., Cerri, C.C., Paustian, K., van Antwerpen, R., 2009. Simulation of sugarcane residue decomposition and aboveground growth. *Plant Soil* 326, 243–259. doi:10.1007/s11104-009-0004-3.

Godfray, H.C.J., Beddington, J.R., Crute, I.R., Haddad, L., Lawrence, D., Muir, J., Pretty, J., Robinson, S., Thomas, S.M., Toulmin, C., 2010. Food security: the challenge of feeding 9 billion people. *Science* 327, 812–818. doi:10.1126/science.1185383.

Herrera, S., 2006. Bonkers about biofuels. *Nat. Biotechnol.* 24, 706–755. doi:10.1038/nbt0706-755.

Herrero, M., Havlik, P., Valin, H., Notenbaert, A., Rufino, M.C., Thornton, P.K., Blummel, M., Weiss, F., Grace, D., Obersteiner, M., 2013. Biomass use, production, feed efficiencies, and greenhouse gas emissions from global livestock systems. *Proc. Natl. Acad. Sci.* 110, 20888–20893. doi:10.1073/pnas.1308149110.

Jaiswal, D., De Souza, A.P., Larsen, S., LeBauer, D.S., Miguez, F.E., Sparovek, G., Bollero, G., Buckeridge, M.S., Long, S.P., 2017. Brazilian sugarcane ethanol as an expandable green alternative to crude oil use. *Nat. Clim. Change* advance online publication. doi:10.1038/nclimate3410.

Lambin, E.F., Meyfroidt, P., 2011. Global land use change, economic globalization, and the looming land scarcity. *Proc. Natl. Acad. Sci.* 108, 3465–3472. doi:10.1073/pnas.1100480108.

LeBauer, D., Kooper, R., Mulrooney, P., Rohde, S., Wang, D., Long, S.P., Dietze, M.C., 2018. BETYdb: a yield, trait, and ecosystem service database applied to second-generation bioenergy feedstock production. *GCB Bioenergy* 10, 61–71. doi:10.1111/gcbb.12420.

Leite, J.G.D.B., Leal, M.R.L.V., Nogueira, L.A.H., Cortez, L.A.B., Dale, B.E., da Maia, R.C., Adjorlolo, C., 2016. Sugarcane: a way out of energy poverty. *Biofuels Bioprod. Biorefining* 10, 393–408. doi:10.1002/bbb.1648.

Liu, Z., Wang, L., Wang, S., 2014. Comparison of different GPP models in China using MODIS image and ChinaFLUX data. *Remote Sens.* 6, 10215–10231. doi:10.3390/rs61010215.

Long, S.P., Karp, A., Buckeridge, M.S., Davis, S.C., Jaiswal, D., Moore, P.H., Moose, S.P., Murphy, D.J., Onwona-Agyeman, S., Vonshak, A., 2015. Feedstocks for biofuels and bioenergy, In: Souza, G.M., Victoria, R.L., Joly, C.A., Verdade, L.M. (Eds.), *Bioenergy & Sustainability: Bridging the Gaps*. SCOPE, Paris, pp. 302–347.

Lynd, L.R., Liang, X., Biddy, M.J., Allee, A., Cai, H., Foust, T., Himmel, M.E., Laser, M.S., Wang, M., Wyman, C.E., 2017. Cellulosic ethanol: status and innovation. *Curr. Opin. Biotechnol.* 45, 202–211. doi:10.1016/j.copbio.2017.03.008.

Martha Jr., G.B., Alves, E., Contini, E., 2012. Land-saving approaches and beef production growth in Brazil. *Agric. Syst.* 110, 173–177. doi:10.1016/j.agsy.2012.03.001.

Mello, F.F.C., Cerri, C.E.P., Davies, C.A., Holbrook, N.M., Paustian, K., Maia, S.M.F., Galdos, M.V., Bernoux, M., Cerri, C.C., 2014. Payback time for soil carbon and sugar-cane ethanol. *Nat. Clim. Change* 4, 605–609. doi:10.1038/nclimate2239.

Moraes, M.A.F.D., Oliveira, F.C.R., Diaz-Chavez, R.A., 2015. Socio-economic impacts of Brazilian sugarcane industry. *Environ. Dev.* 16, 31–43. doi:10.1016/j.envdev.2015.06.010.

Mrema, G.C., Baker, D.C., Kahan, D., 2008. *Agricultural Mechanization in Sub-Saharan Africa: Time for a New Look*. Food and Agriculture Organization on the United Nations, Rome.

Mueller, N.D., Gerber, J.S., Johnston, M., Ray, D.K., Ramankutty, N., Foley, J.A., 2012. Closing yield gaps through nutrient and water management. *Nature* 490, 254–257. doi:10.1038/nature11420.

National Aeronautics and Space Administration, 2018. MODIS land Mission [WWW Document]. URL https://modis-land.gsfc.nasa.gov/index.html (accessed 2.12.18).

Oliveira, D.M.S., Williams, S., Cerri, C.E.P., Paustian, K., 2017. Predicting soil C changes over sugarcane expansion in Brazil using the DayCent model. *GCB Bioenergy* n/a-n/a. doi:10.1111/gcbb.12427.

Parente, L., Ferreira, L., Faria, A., Nogueira, S., Araújo, F., Teixeira, L., Hagen, S., 2017. Monitoring the brazilian pasturelands: a new mapping approach based on the landsat 8 spectral and temporal domains. *Int. J. Appl. Earth Obs. Geoinformation* 62, 135–143. doi:10.1016/j.jag.2017.06.003.

Paruelo, J.M., Epstein, H.E., Lauenroth, W.K., Burke, I.C., 1997. ANPP Estimates from NDVI for the Central Grassland Region of the United States. *Ecology* 78, 953–958. doi:10.2307/2266073.

Ramankutty, N., Evan, A.T., Monfreda, C., Foley, J.A., 2008. Farming the planet: 1. Geographic distribution of global agricultural lands in the year 2000. *Glob. Biogeochem. Cycles* 22, GB1003. doi:10.1029/2007GB002952.

Reboredo, F.H., Lidon, F., Pessoa, F., Ramalho, J.C., 2016. The fall of oil prices and the effects on biofuels. *Trends Biotechnol.* 34, 3–6. doi:10.1016/j.tibtech.2015.10.002.

Service, R.F., 2007. Biofuel researchers prepare to reap a new harvest. *Science* 315, 1488–1491. doi:10.1126/science.315.5818.1488.

Strassburg, B.B.N., Latawiec, A.E., Barioni, L.G., Nobre, C.A., da Silva, V.P., Valentim, J.F., Vianna, M., Assad, E.D., 2014. When enough should be enough: improving the use of current agricultural lands could meet production demands and spare natural habitats in Brazil. *Glob. Environ. Change* 28, 84–97. doi:10.1016/j.gloenvcha.2014.06.001.

Tilman, D., Balzer, C., Hill, J., Befort, B.L., 2011. Global food demand and the sustainable intensification of agriculture. *Proc. Natl. Acad. Sci.* 108, 20260–20264. doi:10.1073/pnas.1116437108

USDA, 2016. USDA ERS – USDA Agricultural Projections to 2025 [WWW Document]. URL www.ers.usda.gov/publications/pub-details/?pubid=37818 (accessed 10.26.17).

Wang, D., Jaiswal, D., Lebauer, D.S., Wertin, T.M., Bollero, G.A., Leakey, A.D.B., Long, S.P., 2015. A physiological and biophysical model of coppice willow (Salix spp.) production yields for the contiguous USA in current and future climate scenarios. *Plant Cell Environ.* 38, 1850–1865. doi:10.1111/pce.12556.

Wint, W., Robinson, T., 2007. *Gridded Livestock of the World*. Food and Agriculture Organization of the United Nations.

Woods, J.; Lynd, L. R.; Laser, M.; Batistella, M.; Victoria, D. de C.; Kline, K.; Faaij, A. 2015 Land and Bioenergy. *In: Souza, G. M.; Victoria, R. L.; Joly, C. A.; Verdade, L. M. (Ed.). Bioenergy & Sustainability: bridging the gaps.* Paris: SCOPE, pp. 258–301.

Zhang, Y., Song, C., Band, L.E., Sun, G., Li, J., 2017. Reanalysis of global terrestrial vegetation trends from MODIS products: browning or greening? *Remote Sens. Environ.* 191, 145–155. doi:10.1016/j.rse.2016.12.018.

5 The role of the sugarcane bioeconomy in supporting energy access and rural development

Francis Johnson and Rocio A. Diaz-Chavez

Introduction

A key difference between the bioeconomy and the fossil economy lies in the spatial aspects of resource supply and demand. Since solar radiation and biomass are more widely distributed compared to fossil fuels, the source of raw materials is much less concentrated. Rural areas stand to benefit with the shift from fossil economy to bioeconomy because there are greater opportunities to obtain value-added products in these areas in comparison, for example, to an oil refinery where products are concentrated in locations normally near urban centers. At the same time, rural areas will have increased local availability of bioenergy compared to the common case in which energy is transported long distances to rural areas (for both liquid fuels and electricity). The combination of these effects has sometimes been termed the "double dividend" that bioeconomy offers to rural areas over the long term (Johnson and Altman, 2014). Such effects cross different levels of development and are thus applicable in low-, medium- and high-income economies alike. The benefits for low-income areas or countries are potentially the greatest of all since it is precisely these areas or countries that have poor access to modern energy services and/or spend a much higher share of their income on energy.

The case of an agro-industry like sugarcane that is concentrated for climatic reasons in developing countries is therefore of particular interest in this context, for several key reasons. First, there are many different products (and processes) derived from sugarcane as a feedstock or resource, so that it offers flexibility and diversity that improves the competitiveness of farmers and rural areas and is seen as a key feature of future bioeconomies (Johnson et al., 2017; McKay et al., 2016). Second, in a developing country context, agro-industries can help to solve the chicken-and-egg problem that hinders development, namely that stable industrial demand is needed for a thriving energy infrastructure but yet investors are reluctant to support industrial development where energy supplies are unreliable. Agro-industries come with their own energy through agricultural residues that are readily available and cost-effective. Third, agro-industrial and agro-energy investments create economic linkages across many different small businesses, suppliers and end users. Such linkages become motors for local and regional growth and development.

In this chapter, we provide an overview on the role of the sugarcane bioeconomy in supporting energy access and rural development, which are in turn connected to the broader aims embodied in the Sustainable Development Goals (SDGs). Poverty reduction, food security and other fundamental development aims must be pursued alongside goals or targets aimed at improved energy access, particularly in rural areas of developing countries where poverty is often deepest (Herrmann and Grote, 2016). Again here the bioeconomy can play a unique role since it relies strongly on raw materials sourced from rural areas but also on the expertise and experience of farmers who can essentially form the backbone of future bioeconomies. Each section below considers some different aspects and contributions of the sugarcane bioeconomy: energy access, livelihoods, poverty reduction and rural development. The developments in Latin America and Africa are briefly reviewed since sugarcane has played a rather important role in these regions in various aspects. Some overall conclusions on the role of the sugarcane bioeconomy are then provided.

Implications for energy access

The contribution to energy access associated with sugarcane can be direct or indirect. The indirect contributions arise primarily in two ways. The first of these is related to economic growth and development in general, which leads to trade and investment in infrastructure, including especially electric power plants (Batidzirai and Johnson, 2012). The most detailed studies have been in Brazil and suggest a significant contribution to economic growth (Deuss, 2012). Furthermore, the benefits at local level are spread more widely compared to many other economic sectors: the employment rate goes up, wages increase and productivity in other agricultural sectors goes up due to spillover effects (Assunção et al., 2016).

The second is related to the increase in income of agricultural workers and sugarcane outgrowers, who might otherwise have low buying power. In general, agricultural workers are not as well paid as workers in the manufacturing and service sectors that often expand as countries undergo structural changes with economic development. However, the agricultural side of the sugarcane sector normally experiences higher wages compared to other agricultural crops or agribusinesses. Sugarcane outgrowers in the African context and to some extent wage earners as well (at sugarcane estates) have been found to have significantly higher incomes compared to control groups in the same general areas (Herrmann, 2017; Mudombi et al., 2016). The additional income allows the workers to expand their reliance on electricity and modern fuels, since these are highly valued as households climb out of poverty.

The direct contribution to energy access would occur when energy production at the sugarcane factory or ethanol distillery or other associated facilities becomes available in the neighboring community, thereby increasing energy access in the area. The improved access could occur through electricity production from bagasse or could also be from the use of ethanol or pellets (made from bagasse) as a cooking fuel to replace fuelwood or charcoal. In some cases, it might arise from associated facilities such as a biogas plant that uses vinasse (organic waste stream) from the ethanol distillery as input. In theory, energy provision could thereby be expanded considerably around a sugar factory and ethanol distillery (Leite et al., 2016). However, in reality, there is normally little or no local use of ethanol or bagasse-based electricity or pellets because the logistical infrastructure is designed on a regional basis, i.e. to provide electricity to the grid or to use ethanol in transport. In the African context, workers and small farmers (sugarcane outgrowers) are normally much better off compared to control groups in the area but at the same time they may not have electricity access because there is no grid or any equivalent local distribution infrastructure (Mudombi et al., 2016).

Poverty reduction, health and rural development

The aforementioned indirect contributions of the sugarcane bioeconomy to energy access goals can go hand in hand with more direct contributions to poverty reduction and rural development. Agro-industries such as sugarcane are often among the few sectors where significant economic value-added stays in rural areas even as urbanization accelerates with economic development. By stimulating economic growth and innovation, the effects can reverberate across the local or regional economy. Furthermore, because an agro-industry such as sugarcane is connected to the broader bioeconomy and is far more sustainable in environmental terms than fossil-driven growth, the resulting structural shifts (i.e. expanding the role of agriculture rather than reducing its significance) has profound implications for development pathways. The ethanol industry in Brazil was perhaps the first to provide strong empirical evidence that improving sustainability for one stream of resources (ethanol as a renewable fuel) could occur in synergy with improvements in the sustainability in production of sugar and other resources or products (Goldemberg et al., 2008). Such knock-on effects in which sustainability becomes more deeply embedded in regional and national economies highlight the importance of investing in extremely productive crops such as sugarcane.

A key issue in the sugarcane sector in low- and medium-income countries (where the sector is overwhelmingly concentrated for climatic reasons) is the issue of mechanization. With or without mechanization, the sugarcane sector can contribute to rural development, while the differences arise more in terms of equity, i.e. the distribution of benefits and who benefits the most. Sugarcane outgrowers tend to rely on manual harvesting, either directly or by hiring wage laborers, and wages are normally significantly higher than other options in rural areas. A transition to mechanical harvesting entails various trade-offs that need to be analyzed, also because the sugarcane productivity losses may occur with mechanization, arising from the billeting and cleaning in the harvester and soil compaction and ratoon damage (Norris et al., 2015).

The practice of burning sugarcane before the harvest has negative impacts on the environment and human health that negatively affect overall development. Brazil has almost phased out the burning of sugarcane but it remains common in southern Africa. Norris et al. (2015) estimated that 39% of cane in the 14 largest sugarcane-producing countries is mechanically harvested. Some regions (e.g. East Africa and Southeast Asia) harvest green cane manually, which is even more labor-intensive than harvesting burnt cane and presents other health risks due to snakes and worker injuries from the dense crop areas. Some associated health problems are related to lung function and the respiratory system (VIB, 2017). The VIB report (2016) also referred to links between sugarcane production and malaria. It appears that the use of the insecticide Malathion in sugarcane fields has been associated with the increasing resistance of mosquitos particularly with those acting as malaria vectors (VIB, 2017). Migrant workers have also been associated to some public health issues, such as in the case of Zambia where most cane cutters constitute are migrants and have contributed significantly to high HIV infection rates.

Socioeconomic impacts and livelihoods

The historical legacy of sugarcane makes it one of the most controversial crops when looking at human history over centuries-long time frames. It is a crop that was associated with the most degrading conditions yet at the same time leading to significant changes in local socioeconomic conditions. There are a wide range of examples in the literature that link the production of sugarcane with both positive and negative impacts. These impacts depend critically on local conditions, including climatic, environmental, socioeconomic and political.

Sugarcane is the highest yielding crop in tonnage worldwide (1.9 billion tons) occupying only 2% of the world's cropland (VIB, 2017). The VIB report (2017) indicates that Africa contributes only 5% of the current global sugarcane production and 83% of this is in sub-Saharan Africa. Sub-Saharan Africa has adequate climatic conditions to produce and expand the production of sugarcane, and consequently sugarcane has often been viewed as promoting both sustainable development and economic competitiveness throughout the region (Johnson and Seebaluck, 2012).

The associated socioeconomic impacts are more related to the production of the feedstock itself rather than the industrial processes where labor needs are lower and modernized efficient processing minimizes losses, which in turn generally minimizes negative impacts (Diaz-Chavez, 2014). The indirect socioeconomic impacts associated with the bioenergy from using ethanol as a biofuel or the cogeneration of electricity in the sugarcane mills are less studied but there are some examples particularly in Brazil where the ethanol program has been running for more than 20 years and where census data allows to compare changes related to the sector (Azanha et al., 2014; Diaz-Chavez et al., 2014).

Because of the physical characteristics of sugarcane (once cut, it degrades easily) and the associated logistics and costs, normally plantations are close to the sugar mills. This was partly historic and partly because of logistics for modern sugarcane mills globally. Therefore, most of the environmental and socioeconomic impacts (both positive and negative) are localized near the sugar mills. The overall benefits of ethanol use as a biofuel are more related to the GHG savings it may have depending on the type of production, as discussed in other chapters in this book.

In terms of livelihoods, the most dominant aspect is the one related to the labor intensity and associated impacts such as working conditions, land property, job creation and wages. There are different forms to produce ethanol but a common one is to use the excess of molasses to produce ethanol and keep the production of sugar. The International Labour Organisation (ILO, 2017) reported that the sugarcane sector generates significant levels of rural employment, supporting some 100 million livelihoods, which is due to the high dependence globally on manual harvesting. The industry has a range of work settings from formal employment to informal and seasonal labor force, across the key producing countries. The sector also presents a variety of production forms with some integration between cultivation and milling and other systems working independently with outgrowers.

IRENA (2017) reported a bioethanol employment increment among all leading producers in 2017, and for biofuels overall noted worldwide employment of over 1.7 million, an increase of 2% from the previous year. These jobs refer mainly to the first stages of the value chain on the agricultural side (planting and harvesting different types of feedstock). There are fewer jobs, but with higher wages, in the construction of fuel processing facilities and in the operation and management on the industrial side. In order to improve the operations and labor conditions, mechanization processes have been implemented in some countries such as United States and Brazil, and therefore the labor intensity has been reduced (IRENA, 2017). Brazil is still the country with the largest liquid biofuel labor force estimated at 783,000 jobs.

Other than Brazil, other significant biofuel production countries with a significant share of ethanol from sugarcane include Colombia and Philippines. In the case of Colombia, IRENA estimated a total of 85,000 jobs in the country (without making a difference between biodiesel and ethanol) but the National Federation of Biofuels estimated more than 191,200 biofuel jobs in 2015 (IRENA, 2017). The Philippines' National Bioethanol Board estimated around 20,000 jobs (Biofuels International, 2015 in IRENA 2017).

In sub-Saharan Africa, there is no robust data on the number of jobs created particularly on the sugarcane industry. The best data reported is from the South Africa Sugar Association (SASA), which estimates that 79,000 direct jobs and 350,000 indirect jobs are associated with production and processing in South Africa (SASA, 2014 in Hess et al., 2016). Following this data, Hess et al. (2016) estimated based on jobs generated by tons of sugarcane produced that there may be as many as 1.8 million jobs associated with the industry in SSA.

Sustainability indicators and certification

Sustainability certification can and does address social and economic aspects as well as technical and environmental aspects. Certification has been pursued through various schemes, of which the two most relevant for sugarcane are the Roundtable on Sustainable Biomaterials (RSB) and Bonsucro. The RSB operates according to 12 principles guiding sustainable use of biomass, biofuels and biomaterials across different uses and sectors (RSB, 2017). Bonsucro was founded in 2008 and focuses only on sugarcane, drawing its membership and partners from across the entire supply chain and now certifying over 25% of sugarcane areas across the world (Bonsucro, 2017). These schemes are recognized by the European Commission as well as by many other national, regional and international organizations. Although somewhat less important in connection to poverty reduction and low-income countries, prioritizing the advanced industrial development of products from sugarcane is also viewed as a long-term sustainability issue (O'Hara and Mundree, 2016).

There are references in the literature regarding working conditions which vary largely from one country to another, and since the bioethanol surge around 2008, several sustainability standards include issues related to working conditions (Diaz-Chavez, 2014; Hess et al., 2017). ILO (2017) has reported on the working conditions of sugarcane production focusing on child labor and forced labor. In the case of children, they are driven mainly by poverty conditions, and they are employed on the application of agricultural pesticides. Forced labor is identified in the main five producing countries including Brazil, and it is related to some form of power over disadvantage people or minorities. Inequality in gender is also observed in the sector. The working conditions are still a main issue in countries where the harvest is dominated by manual cut and extended working hours as well as the conditions of migrant workers (ILO, 2017). The ILO has reported some forms to improve the working conditions of the sugarcane sector (production), particularly through the engagement of all relevant stakeholders and the enforcement of the agreements in signatory countries of the ILO conventions (ILO, 2017).

Experiences in Africa and Latin America

Many countries in Africa and Latin America are sugarcane producers. A review of the production indicates that worldwide sugar is produced in 120 countries, and global production now exceeds 165 million tons a year (Hess et al., 2016). Approximately 80% is produced from sugarcane, which is largely grown in tropical countries. The remaining 20% is produced from sugar beet, which is grown mostly in the temperate zones of the northern hemisphere (Hess et al., 2016).

Figure 5.1 shows the percentage of production of sugarcane by region from the last 20 years in average. It can be observed that most of the production is in the tropical areas of the world in Asia and America. The production in Africa has not been too large, and even with the biofuels surge, there is little evidence that this has influenced the expansion of the crop, in spite of the significant potential in sub-Saharan Africa due to suitable climate, available land and low labor costs.

The world production in terms of both tons and area has been growing with a clear spur after 2004 when ethanol production also increased (Figure 5.2).

But the production of sugarcane from 10 countries is presented in Figure 5.3. However, this production does not represent sugar or ethanol production, which vary considerably not only due to differences in yields and conditions but also since a few countries and regions such as Brazil and Malawi have put special focus on ethanol but many have not.

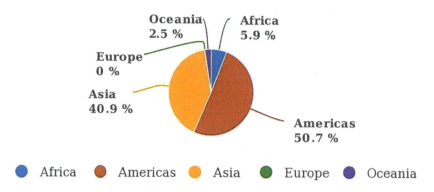

Figure 5.1 Sugarcane production by region (1994–2016).
Source: FAOSTAT (2018).

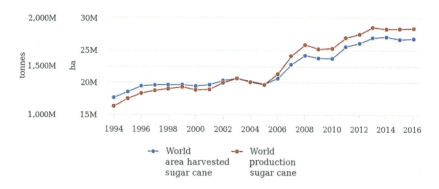

Figure 5.2 Global area and production of sugarcane over time (1994–2016).
Source: FAOSTAT (2018).

The role of the sugarcane bioeconomy 67

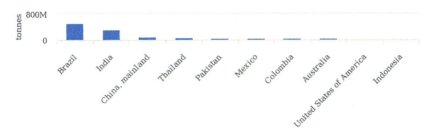

Figure 5.3 Top 10 sugarcane producers (1994–2016).
Source: FAOSTAT (2019).

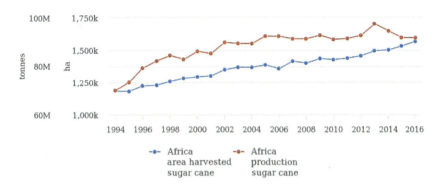

Figure 5.4 Production and area in Africa over time (1994–2016).
Source: FAOSTAT (2018).

In Latin America, sugarcane production is highly dominated by Brazil (over 500 M tons of production) followed by Mexico (50 M tons) and Colombia (35 M tons) in average produced in the period of 1996–2016.

The three main producers in Latin America have worked on policies to determine the production areas according to agro-ecological condition. Nevertheless, these areas refer to the expansion because historically the main producer areas have existed since colonial times.

In Africa, the total production in tons and the area has also increased in the past 20 years as can be observed in Figure 5.4. Nevertheless, this is not comparable with the total world production which is heavily dominated by the top 10 producer countries.

In sub-Saharan Africa, the production was dominated by South Africa (more than 18 M tons), Kenya (over 6 M tons) and Sudan (5 M tons) in 2014.

While the production of sugarcane and the relationship with the land use and food security have dominated the literature in the past 10 years in sub-Saharan Africa, in Latin America has been more focused in issues of working conditions and land expansion particularly in Brazil.

In sub-Saharan Africa, there is little evidence of the expansion of sugarcane due to ethanol production (VIB, 2016). Food and Agriculture Organization (FAO) also produced a work on agro-ecological zones for SSA to determine rain-fed regions and areas for sugarcane irrigation. Despite this effort, it is undoubtable that the effects of climate change on the sector are important mainly to the issue of water. Changes in the water cycle and water availability are the major problems for the sugarcane production in this region (VIB, 2017).

From the point of view of bioenergy with ethanol and other renewable energy carriers from sugarcane, these are still little exploited in sub-Saharan Africa. The reasons for this are varied but include financial barriers, lack of technical expertise, land availability and government policies. Large-scale biofuel production and electricity cogeneration still need a large impulse in the region. The two main examples are Malawi and Mauritius, where sugarcane has proven to be a viable way to achieve energy security, but apart from these two countries, there is limited experience with large-scale production and use of biofuels in sub-Saharan Africa. This may need to be the way forward if a better input from biomass is indeed a main goal for African countries, especially those which are landlocked and have to import fuels.

Conclusions

The sugarcane bioeconomy that has emerged around the world is due mainly to the high productivity of the sugarcane crop in combination with the global interest in a transition to renewable energy and bio-based materials. However, relatively few countries have made major investments or long-term commitments, with Brazil being the primary example in Latin America and Malawi the main case in the African context. Research has shown significant evidence of contributions from sugarcane to poverty reduction, livelihoods and rural development. There is less evidence that sugarcane has contributed directly to energy access improvements, although such improvements can be indirectly attributed to sugarcane investments due to the resulting increase in incomes, infrastructure and development options as well as spillover effects to other agricultural sectors associated with innovation and the economic linkages that are common in an agro-industry such as sugarcane. There remains tremendous potential in Africa and Latin America to expand and employ sugarcane not only as an efficient source of renewable energy and bio-based materials but also a driver for sustainable rural development and poverty reduction.

References

Assunção, J., Pietracci, B., Souza, P. (2016). *Fueling Development: Sugarcane Expansion Impacts in Brazil.* INPUT-Iniciativa para o Uso da Terra, 55 (Climate Policy Initiative).
Bonsucro (2017). www.bonsucro.com/.

Deuss, A. (2012). The economic growth impacts of sugarcane expansion in Brazil: an inter-regional analysis. *Journal of Agricultural Economics*, 63(3), 528–551.

Diaz-Chavez, R. (2014). Chapter 2. Indicators for socio-economic sustainability assessment. In: Janssen, R. and Rutz, D. (Eds.). *Socio-Economic Impacts of Bioenergy Production*. Springer.

Diaz-Chavez, R., Colangeli, M., Morese, M., Fallot, A., Azanha, M., Sibanda, L., Mapako, M. (2015). Social considerations. Chapter 21. In: Souza, G. M., Victoria, R., Joly, C., and Verdade, L. (Eds.). *Bioenergy & Sustainability: Bridging the Gaps* (Vol. 72, p. 779). Paris: SCOPE. pp. 514–539.

FAOSTAT (2018). *Crops: Sugar Cane*. www.fao.org/faostat/en/#data/QC/visualize Date: Feb 2018.

Goldemberg, J., Coelho, S.T., Guardabassi, P. (2008). The sustainability of ethanol production from sugarcane. *Energy Policy*, 36(6), 2086–2097.

Herrmann, R.T. (2017). Large-scale agricultural investments and smallholder welfare: a comparison of wage labor and outgrower channels in Tanzania. *World Development*, 90, 294–310.

Herrmann, R., Grote, U. (2015). Large-scale agro-industrial investments and rural poverty: evidence from sugarcane in Malawi. *Journal of African Economies*, 24(5), 645–676.

Hess, T.M., Sumberg, J., Biggs, T., Georgescu, M., Haro-Monteagudo, D., Jewitt, G., Ozdogan, M., Marshall, M., Thenkabail, P., Daccache, A., Marin, F., Knox, J.W. (2016). Sweet deal? Sugarcane, water and agricultural transformation in Sub-Saharan Africa. *Global Environmental Change*, 39, 181–194.

ILO (2017). *Child Labour in the Primary Production of Sugarcane*. International Labour Organisation. Available: file://icnas4.cc.ic.ac.uk/radc/downloads/CLP_Sugarcane%20Report_20171019_WEB.pdf.

IRENA (2017). *Renewable Energy and Jobs Annual Review 2017*. International Renewable Energy Agency. Available: www.irena.org/DocumentDownloads/Publications/IRENA_RE_Jobs_Annual_Review_2017.pdf. Date: February 2018.

Johnson, T.G., Altman, I. (2014). Rural development opportunities in the bioeconomy. *Biomass and Bioenergy*, 63, 341–344.

Johnson, F.X., Leal, M.R.L.V., Nyambane, A. (2017). Sugarcane as a renewable resource for sustainable futures, In: Rott, P., Ed. *Achieving Sustainable Cultivation of Sugarcane*, Vol. 1 (Chapter 15), Burleigh-Dodds, Cambridge, UK.

Johnson, F.X., Seebaluck, V. (2012). *Bioenergy for Sustainable Development and International Competitiveness: The Role of Sugar Cane in Africa*. Routledge/Earthscan, London, UK.

Leite, J.G.D.B., Leal, M.R.L.V., Nogueira, L.A.H., Cortez, L.A.B., Dale, B.E., da Maia, R.C., et al. (2016). Sugarcane: A way out of energy poverty. *Biofuels, Bioproducts and Biorefining*, 10, 393–408. doi:10.1002/bbb.1648.

Moraes, M.A.F.D., Oliveira, F.C.R., Diaz-Chavez, R.A. (2016). Socio-economic impacts of Brazilian sugarcane industry. *Environmental Development*, 16, 31–43.

McKay, B., Sauer, S., Richardson, B., Herre, R. (2016). The political economy of sugarcane flexing: Initial insights from Brazil, Southern Africa and Cambodia. *The Journal of Peasant Studies* 43(1), 195–223.

Mudombi, S., von Maltitz, G. P., Gasparatos, A., Romeu-Dalmau, C., Johnson, F. X., Jumbe, C., et al., 2016. Multi-dimensional poverty effects around operational biofuel projects in Malawi, Mozambique and Swaziland. *Biomass and Bioenergy*, doi:10.1016/j.biombioe.2016.09.003.

O'Hara, I.M., Mundree, S.G. (2016). (Eds.). *Sugarcane-Based Biofuels and Bioproducts*. Wiley Blackwell: New Jersey, USA.

RSB (2017). *Roundtable on Sustainable Biomaterials*. Available online: http://rsb.org/about/what-we-do/the-rsb-principles/.

VIB (2017). *Sugar Cane in Africa*. International Plant Biotechnology Outreach, The Netherlands. www.vib.be/en/about-vib/plant-biotechnews/Documents/vib_fact_Sugercane_EN_2017_1006_LR_single.pdf.

6 Is the sugarcane capital goods sector prepared to respond to large-scale bioenergy supply in developing countries?

José Luiz Olivério and Paulo Augusto Soares

Introduction

We should have an outlook of future ethanol consumption as a fuel, even knowing that market projections are questionable and require periodic updates so as to effectively represent reality and include technological advances that certainly will occur over the years. Here, we consider as a general rule that world demand for ethanol in the next decade will increase as a gasoline additive for light vehicles, because this alternative does not require significant investments in existing gasoline distribution networks. However, ethanol can also be used as a fuel, named E100 or hydrous ethanol, in flex or dedicated vehicles, as already used in large scale in Brazil.

The following highlights can be considered valid for the next decade (ExxonMobil, 2017):

- Vehicles fueled with conventional gasoline will still be the most popular because of their mobility, low cost, and increased efficiency;
- The increased efficiency of gasoline vehicles will be achieved by a combined effect of enhanced onboard technology and improved fuels quality (Anti-Knock Index average 95);
- Fuel consumption in light vehicles is expected to decline in North America/European Union by the end of the next decade, but will increase in developing regions in South America/Africa/Asia, resulting in a small increase of gasoline consumption by 2035.

Anhydrous ethanol as gasoline additive has been widely used in the U.S. and Brazil, with over a decade of consolidated experience, and is well accepted by the automotive industry. In Brazil, the use of additive ethanol dates back to the 1930s, and current blends are 27% (E27) with regular gasoline and E25 with premium gasoline; in the U.S., mixing up to 15% is permitted, but in recent years the average has been around E10 (EIA, 2016).

The choice of the gasoline additive should be based on some key aspects. It must (i) be environmentally friendly, (ii) reduce total GHGE – greenhouse gas emissions per kilometer traveled, (iii) enhance the gasoline octane rating, and (iv) be price competitive.

Figure 6.1 indicates the octane rating gains with each of the main kinds of additives, which compete today with ethanol.

In terms of costs, ethanol is cost-effective, similar to petroleum base gasoline without additive (U.S. Grains Council, 2017).

The need to reduce GHGE and environmental and human health impacts urged governments to review the use of additives to be blended with petroleum gasoline, resulting in ethanol as a preferential choice due to its capacity to boost octane rating, reduce emissions, improve the combustion efficiency, and cause low environmental and health impacts, combined with a relatively low cost and the possibility to create new jobs in the country. These were the reasons that led the U.S. to adopt ethanol as the main gasoline blend option. The Optima Program (DOE, 2015) is evaluating the use of mid-level blends (E25 to E40) as standard fuel.

Gasoline specifications set the oxygen content as 2%–3.6% in several countries to enhance the efficiency of Otto cycle engines. Ethanol requires almost half of the methyl *tert*-butyl ether (MTBE) volume to achieve a specified value.

It is relatively safe to assume that the use of ethanol as a universal additive, as E10, is feasible for almost all countries, considering environmental protection, emissions reduction, low toxicity, and low costs, and this blend level will meet the minimum oxygen content required for good quality oxygenated gasoline. The adoption of E10 will require diversification of ethanol

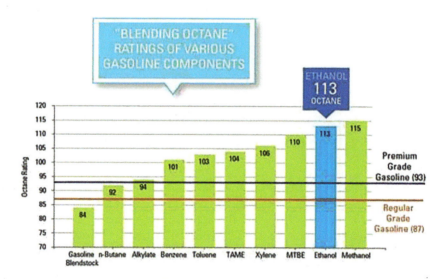

Figure 6.1 "Blending octane" ratings of various gasoline additives/components. Sources: EESI (2016) and NREL (2000).

production worldwide, including feedstock and production technologies, but for simplification reasons and its superior environmental qualities, we will consider that sugarcane will be the main feedstock. Furthermore, E10 meets the U.S. and EU market demands, and is also appropriate in the developing countries.

Gasoline consumption in 2016 was 1,344 billion liters and will have a slight increase by 2035, with a global peak demand in the following decade (2030/40). Peak consumption will be close to the present one.

Ethanol consumption as an additive and/or fuel has grown significantly in recent years. In 2016, consumption was 97 billion liters.

Assuming a total 5% growth of fuel consumption in the world (gasoline + ethanol) in Otto cycle engines in the coming decade (0.5% y/y), and that by the end of this decade the ethanol content will be 10%, we will have an approximate total fuel consumption of 1,513 billion liters/year, i.e., a total ethanol production increase of 54.3 billion liters/year, which is expected to occur between 2030 and 2040.

Scenario of required new "Future Mills"

The possibility of using various feedstocks, the development of new agro-industrial technologies, and the introduction of plants for ethanol production from cellulosic materials make it difficult to estimate each type of plant, which is the reason why we adopted a typical Brazilian sugarcane mill with a milling capacity of 2.4 million tons of sugarcane/crop (TCC) dedicated to produce ethanol at a rate of 200 million liters of ethanol/crop, which we call "equivalent-mill" (EM).

The required number of new EMs to be built to supply an increase of 54.3 billion liters/year will be 275 EMs in one decade or 30 EMs/year on average, as shown in Figure 6.2 – fast/proposed Scenario.

The proposed scenario of 30 new EMs/year on average, from the viewpoint of the capital goods industries (also termed "equipment industry"), can be considered conservative if we take into account the results already achieved in the U.S. and Brazil in past decades.

Based on recent years, we calculated the historical number of EMs that was necessary to supply the increased global ethanol consumption; peak occurred in 2008, when the equipment industry had capacity to implement 77 EMs in the U.S. and Brazil and 84 EMs in the world.

Based on these historical consumption increase, we can assume that the global equipment industry will be able to deliver up to 60 EMs/year. For conservative reasons and considering the complexity of implementing the agricultural activity in various countries, we admit that this number could be reduced to 30 future EMs/year, which meets the proposed scenario.

Thus, the use of 10% ethanol blend in gasoline in a period of 10 years, starting from current 6.7%, is totally viable, considering the entire range (E5–E100) of fuel ethanol mixtures.

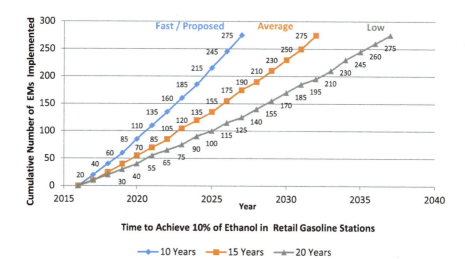

Figure 6.2 Scenarios to achieve 10% ethanol.
Source: by author – Paulo A. Soares.

The expansion of the Brazilian sucroenergy industry

Brazilian sugarcane sector expansion in the past four decades was the world's most representative, being a reference for our conclusions. Today, Brazil leads the production of cane, sugar, and ethanol (see evolution in Figure 6.3).

- Sugarcane: from 68 (1975/1976) to 672 million TCC (2015/2016);
- Sugar: 6 (1975/1976) to 38 million tons (2012/2013);
- Ethanol: 550 million (1975/1976) to 30 billion liters (2015/2016).

First-great-leap: sugarcane production increased to meet a growing demand for ethanol (implementation of Proálcool[1]). The great increase in sugarcane processing and ethanol production was the key factor for the development of new technologies in these sectors. The sugar production-related technology remained outdated due to nonexistent growth demand.

Second-great-leap: sugarcane production increased due to increase of sugar production for exports, and Brazil could equal or even surpass the world state of the art.

Third-great-leap: resulted from increased ethanol production to meet domestic demand and sugar production for exports. In 2002, when PROINFA[2] was launched, institutional mechanisms were created for the marketing of surplus bioelectricity produced from bagasse/cane straw (cane vegetable residues). These accelerated growths created new incentives for technological developments.

Is sugarcane sector respond to large-scale bioenergy? 75

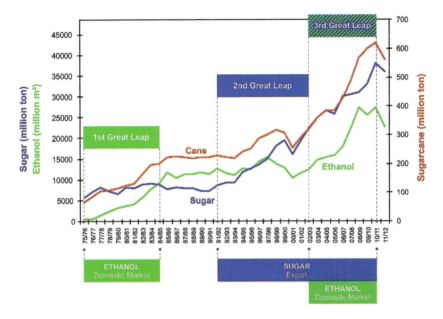

Figure 6.3 Brazilian sugarcane/sugar/ethanol production and the three great leaps.
Source: Olivério and Boscariol (2013), adapted from DATAGRO (2017).

It is also to be noted the influence of sustainability on technological evolution and mills performance.

Design development: from "Sugar-Mill" to the "Sucroenergy-Mill"

During the "first-great-leap", an "Ethanol-Process" was integrated to the plant, forming the "Sugar-Ethanol-Mill". Later on, an exclusive "Ethanol-Producing-Mill" was developed and implemented for the first time in the world.

During the "third-great-leap", novel technologies were incorporated into the mills design, resulting in the "Sugar-Ethanol-Bioelectricity-Mill" or the "Sucroenergy-Mill" (more details in Olivério and Boscariol, 2013; Olivério et al; 2010a; Olivério and Ribeiro, 2006).

Profile of the Brazilian sugarcane mills

Before the "third-great-leap", 324 "Old-Mills" were operating in Brazil, and in 10 years, 117 "New-Mills" were built (Table 6.1).

Table 6.2 shows the profiles of "Total-Mills" (old and new) and "New-Mills", the latter being more focused on ethanol and designed to produce bioelectricity (80%), though partially implemented (35%).

Capacity varies in "Total-Mills" from 300,000 to 10 million TCC.

Table 6.1 "New-Mills" installed at "third-great-leap"

Crop	"New-Mills"
2005/06	10
2006/07	19
2007/08	25
2008/09	30
2009/10	22
2010/11	8
2011/12	3
Total	117

Source: CNI (2012).

Table 6.2 "Total-Mills"/"New-Mills" classified by products (2012)

Products	"Total-Mills"	"New-Mills"
Sugar/ethanol (%)	60	25
Ethanol (%)	35	75
Sugar (%)	5	–
Total (%)	100 (441 mills)	100 (117 mills)
Bioelectricity	15% operational	35% operational 80% designed

Source: CNI (2012)/Dedini.

Technological evolution of the mill's industrial sector

Recent investments resulted in the important technological developments of equipment/processes/plants/complete mills (Olivério and Boscariol, 2013) and productivity/efficiency gains (Table 6.3):

Sugar/ethanol – Yield/efficiency improvements, maximizing sugar extraction, minimizing losses, and optimizing transformation of juice into sugar/ethanol enable optimum production/TC. Equipment manufacturers, especially Dedini, developed key technologies to use when a new mill project aims to maximization (Olivério et al., 2010a).

Bioelectricity – Similarly, two kinds of key technologies enable maximum bioelectricity/TC

- Minimum energy consumption (electricity/steam) in sugarcane mill, resulting in maximum surplus bagasse/straw;
- Maximum available energy utilization of sugarcane/mill (bagasse/straw, biogas from stillage), with maximum energy efficiency.

Table 6.3 Sugarcane mill evolution performance as a function of the available technology and equipment

Parameters	Dedini products	1975 – Beginning Proalcohol	2015 – Today state of the art
1. Production/equipment capacities increase			
Crushing capacity (TCD) – 6×78″	Vertical shredder/milling tandem	5,500	15,000
Fermentation time (h)	Batch/continuous fermentation	24	6–8
Beer ethanol content (°GL)	Ecoferm	6.5	Up to 16
2. Efficiency/yield increase			
Extraction (% sugar) – six milling units	Milling tandem/modular diffuser	93	97/98
Fermentation (%)	Ecoferm	80	92
Distillation (%)	Destiltech	98	99.5
3. Optimizing energy consumption/efficiency			
Total steam consumption (kg steam/TC)	Dedini technology	600	320
Steam consumption – anhydrous (kg steam/liter)	Split feed + molecular sieve	4.5	2.0
Boiler: efficiency (% LHV) Capac. (ton/h)/pressure/(bar)/temperature (°C)	AZ/AT/single drum boilers types	66	89
		60/21/300	400/120/540
Biomethane from stillage (Nm^3/liter bioethanol)	Methax	–	0.1
4. Global parameters			
Total yield (liter of hydrous bioethanol/TC)	DEDINI technology	66	87
Surplus bagasse (%) – bioethanol mill	DEDINI technology	Up to 8	Up to 78
Surplus bioelectricity – bioethanol mill, 12,000 TCD (fuel: bagasse) (MW)	DEDINI technology	–	50.7
Surplus bioelectricity – bioethanol mill, 12,000 TCD (bagasse + 50%/*100% straw*) – (MW)	DEDINI Technology	–	84/112
Stillage production (liter stillage/liter bioethanol)	Ecoferm/**DCV**	13	5.0/0.8
Intake water consumption (liter water/liter bioethanol)	Water-Mill	187	(–)3.7
Crop duration (months)	RGD/Flex-Mill: Cane+Sorghum+Corn	6	8/11

TC = tons cane; TCD = tons cane/day; LHV = based on bagasse low heat value; Capac. = steam production; Ecoferm = ethanol fermentation system up to 16 °GL; Destiltech = Flegma recirculation ethanol distillation plant; RGD = 24-hour system of engineering/technical assistance/retrofit/optimization/spare parts; Methax = Stillage Biodigestion Plant producing biogas/biomethane; DCV = Evaporative Stillage Concentration Plant.
Source: Olivério and Boscariol (2013).

How the capital goods industry meets the need for growth of the sucroenergy sector

The expansion already occurred was fully accomplished by the equipment industry with almost 100% in-house development and supply. During Proalcohol, more than 300 ethanol mills were built in 10 years (Olivério, 2007). In the most recent growth, 117 "New-Mills" were installed.

For the future, many aspects need to be examined. The first is whether the equipment industry is ready to meet an even more rapid stage of growth. The second is whether it will also meet the evolution with new bioproducts entering the market with significant demand. The third is whether it will be able to lead the technological evolution that has been seen globally with clear impacts on the sucroenergy sector.

For a safe and reliable analysis, we should review the respective "capabilities" and "competitiveness" of this industry to face the new challenges. We understand that in this case "capability" means "to meet market needs", i.e., to satisfy technological/engineering needs, having industrial/manufacturing and financial capacity as well as guarantees. Likewise, "competitiveness" means the capacity "to meet the client's needs", i.e., to offer quality, delivery times, and prices competitively (see Olivério, 2007).

Future capability

Industrial/manufacturing capability: in Proalcohol, supply capacity reached 60 equivalent ethanol mills/year of 600,000 TCC, and total processing capacity of 36 million TCC. In the "third-great-leap", capacity has risen to 30 EMs of 2.4 million TCC, total of 72 million TCC (Table 6.1), only considering the Brazilian equipment industry. Other countries such as India, China, and Pakistan also have a solid industry; and Australia, South Africa, and the U.S. have some mill's sectors.

Additionally, "Future-Mills" supply capacity may increase, as long as investors can trust the market growth.

Considering the above and based on our own assessment, in 3–4 years, the world can have capacity to build 60 EMs/year with a capacity of 2.4 million TCC each, equivalent to 144 million TCC, which represents a supply increase of 12 billion ethanol liters/year, sufficient to meet the global biofuels market expansion.

Financial capability and guarantees derive from diverse agents:

- "Future-Mills" investors,
- Equipment/plants suppliers,
- Financing agents,
- Buyers/traders of bioproducts/sugar/ethanol/bioelectricity.

To meet a faster expansion scenario, more favorable conditions should be offered to ensure more rapidity, agility, and safety, such as

a   Agile lines of credit for analysis/approval/financing of "Future-Mills", defining more suitable guarantees for new "turnkey" projects, e.g., the "project finance" concept;
b   Financing and guaranteeing solutions through securitization of long-term contracts for the supply of bioproducts/certificates of emissions reduction (carbon credit), such as "PPA-Power-Purchase-Agreement", already used in Brazil as guarantee in PROINFA auctions; "EPA-Ethanol-Purchase-Agreement"; "SPA-Sugar-Purchase-Agreement"; and "CCPA-Carbon-Credits-Purchase-Agreement".

Technological/engineering capability can be determined by the capacity and expertise to execute each and all of the following stages/solutions, which have been successfully accomplished by the equipment industry:

- Conception/development,
- Definitions/specifications,
- Execution,
- Commissioning/guarantee.

In agreement with each one of these four stages, this industry has mastered all stages of technology/engineering/supply/operation of the plant:

- Process development/technology/process engineering,
- Engineering: basic/equipment/interconnections/automation/control,
- Fabrication/erection/assembly,
- Start-up/performance/technical assistance.

Future competitiveness

Regarding **quality**, the equipment industry already meets – and is able to continue to meet – all international quality systems/codes/norms applied to mills: ASME/DIN/JIS/ISO, and laws/regulations.

Regarding **delivery times**, typical lead time for turnkey plants in Brazil is 20–30 months, which is satisfactory considering that the agricultural sector needs 36 months for 60% design capacity and 60 months for full capacity.

Regarding **price**, there is a strong competition between the equipment/plants suppliers in domestic markets, such as Brazil and India, which also compete internationally.

On the other hand, sugar/ethanol/bioelectricity are in balance with the very competitive commodities market. If the equipment industries were not competitive, the sugarcane sector would not have achieved the expansion it did in recent years.

Drivers of evolution trends for products, capacities, and technologies

To determine if the equipment industry can supply "Future-Mills", the design profile evolution of the 117 "New-Mills" was examined (Olivério and Boscariol, 2013) to identify the trends for this development. Conclusion is that they will be designed according to five drivers of evolution trends for products, capacities, and technologies:

1. Increased capacity/productivity of the mills/equipment,
2. Increased efficiencies/yields,
3. Increased sustainability,
4. Synergy/integration,
5. Products with higher added value from both sugarcane and mill.

The conclusion was that the equipment industry already met these evolution drivers. Special attention will be given to the "increased sustainability" driver, which is a major focus of this book.

Driver 3 – increased sustainability

Sustainability is today mandatory in all human activities; in the sucroenergy sector, demand for compliance is even greater: bioenergy is presented as "green", clean, renewable, improving air quality, and mitigating GHG.

Some sustainable solutions have already been employed: process effluents/residues (stillage/cake/ash/soot) are recycled and replace chemical fertilizers; water consumption decreased from 15.0 (1975) to 1.8 m^3/TC (2005).

To design sustainable projects, Dedini developed a systemic approach and conceived the DSM – Dedini Sustainable Mill – introduced at the Ethanol Summit 2009 (Olivério, 2009) – as a macro-machine designed to optimize the accomplishment of the three sustainability pillars: economic, social, and environmental (Table 6.4).

DSM's developed technologies enable maximum production of six bioproducts: biosugar, bioethanol, bioelectricity, biodiesel, biofertilizer, and biowater in an integrated design, minimizing emissions while maximizing ethanol contribution to GHGE mitigation. The DSM can be implemented gradually, e.g., the Barralcool Mill, Brazil, that produces four BIOs since 2006, with a pioneering biodiesel plant designed and supplied by Dedini and integrated to the mill (Figure 6.4).

DSM meets two concepts: optimization and zero (see Figure 6.5).

DSM compared with a traditional mill offers important benefits (Olivério and Boscariol, 2013):

a   Better ratios of "clean energy output" to "fossil energy input" from seven to ten;
b   Maximizes bioethanol/TC; so, more ethanol replaces gasoline, preventing more GHGE;

Table 6.4 The three pillars of sustainable development define the DSM characteristics.

DSM Characteristics

Economic	Environmental	Social
• DSM is competitive in a free market, without subsidies • Maximizes productivity/yields/efficiencies, minimizing end products costs • Maximizes bioproducts/TC	• DSM: no wastes (minimizes consumption), no harm or pollution (air, water, energy, materials, raw materials, biodiversity); minimum/zero emissions, effluents, residues, odors • Complies with standards/regulations, reducing/eliminating environmental impacts; contributes to agricultural sustainability • ISO 14001 friendly	• DSM: equipment, processes, materials, installations are located/moved/operated meeting the best practices/standards/regulations, providing comfort, hygiene, safety, and good health conditions • Minimum physical effort (ergonomy), correct man-machine interaction • Automation through intelligent software, MES level, integrated to ERP System • SA-8000 friendly

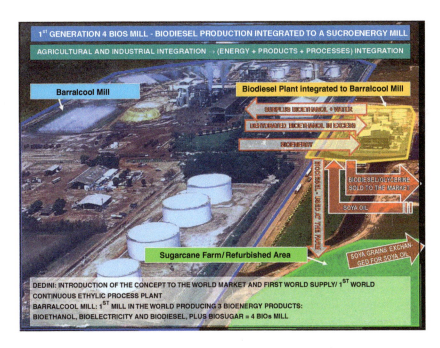

Figure 6.4 DSM can be implemented in stages and Barralcool Mill is a first-generation four BIOs Mill.

Source: Cortez et al. (2016)

Figure 6.5 DSM: the six BIOproducts Mill is designed to meet the optimization/zero concepts reaching maximum GHG mitigation effect.
Source: Cortez et al. (2016)

c   Maximizes bioelectricity/TC, preventing more fossil fuels utilization and diminishing emissions;
d   Integrated biodiesel production (Figure 6.4) with agricultural integration (soybean cropping in rotation with sugarcane); industrial integration (integrated biodiesel plant uses soybean oil and bioethanol as feedstocks – replacing fossil methanol – and bagasse energy utilization); ethyl biodiesel replaces diesel at the mill and abroad, in all cases avoiding emissions (Olivério et al., 2007);
e   The mill becomes water self-sufficient, recycling only the water contained in the sugarcane and also produces water to be exported: the biowater. A typical mill requires 23 liters water/liter ethanol, while DSM exports 3.7 liters (Olivério et al., 2010b);
f   BIOFOM – Organomineral biofertilizer – production using all residues as feedstocks, replacing 70% of the chemical fertilizers and mitigating emissions (Olivério et al., 2010c);
g   The DSM incorporates the most advanced concepts of occupational hygiene/safety.

The DSM bioethanol replacing gasoline has a mitigating effect as significant as 112% (132% using 50% of straw as energy), while the traditional ethanol avoids 89% (Olivério et al., 2010a). The DSM ethanol mitigation potential is higher than 219%!

The DSM results from various technologies integration, resulting in 10 patent applications and 8 filed by Dedini.

The DSM (Cortez, 2016) was awarded two national prizes and international recognition when Dedini was invited to introduce it in the Plenary Session of the XXVII ISSCT International Congress, Mexico (Olivério et al., 2010a). Water self-sufficiency/biowater production was highlighted in the specialized literature, being also cited by the then Minister of Mines and Energy in the opening speech of the Biofuels International Conference, Brazil, 2008 (Olivério, 2014).

In conclusion, ethanol from DSM is a real "bioethanol premium", a superior fuel than current "advanced biofuel", with more sustainability and allowing more gains in carbon credits (Cortez, 2016).

Possible impacts of sucroenergy programs in developing countries

In the 1980s, Proalcohol aroused great interest worldwide, refueled in the recent sucroenergy expansion.

Numerous international delegations visited Brazil and Dedini to learn about such development. When the visitors were from developing countries able to produce sugarcane, part of the Dedini program included a draft presentation of the possible impact that the implementation of a DSM would have on the respective country. Such approach engaged the visitors' interest because it highlighted, though briefly and roughly, the benefits that the country would have in developing the sugarcane sector and in building mills.

But the recommendation was that the country itself should develop its own model and calculate the corresponding impacts, and that Dedini, and certainly Brazil, would be ready to collaborate in these programs.

It was always emphasized that someone else's experience is a guide, not a solution.

As example, Figure 6.6 was part of the presentation to the Zimbabwean official visit in April 14, 2013: an "EPZ – Ethanol Project in Zimbabwe" was proposed by installing new DSMs.

A possible EPZ was considered, implementing five DSM (six BIOs Mill) in 10 years – 12,000 TCD and 36,000 ha each – using 4.5% of arable land and 0.5% of total area.

This approach caused a great interest, mainly employment generation (28,100/56,200 direct/indirect jobs), increased GDP (12.2% of 2011 GDP), better income distribution, substitution of imports with local products, improved trade balance (31% of 2011 deficit reduction), creation of strong currency via exports, and strategic benefits to other African countries. Only as a reference, if the CO_2 reduction emissions were calculated using DSM Brazilian parameters, the mitigation using the EPZ ethanol would be 15.4 million tons of CO_2 in 10 years.

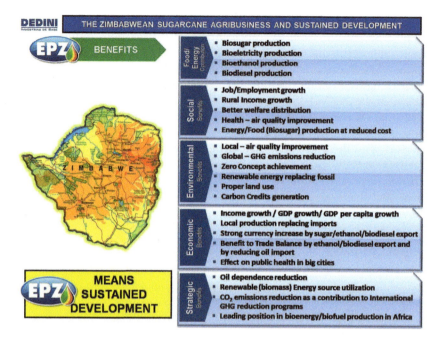

Figure 6.6 Impacts in building an EPZ.
Source: Dedini.

Conclusion

An E10 worldwide program was considered, and this biofuel was shown as the most adequate additive to oxygenate gasoline, considering environmental protection, emissions reduction, better air quality, and low costs.

This chapter begins with a question and, to answer it, we demonstrated that over the years the capital goods industry has effectively met the expansion of the sucroenergy agribusiness, and is prepared to fulfill new world demands, having already developed/supplied solutions that will be part of "Future-Mills" design, meeting the five drivers of evolution trends for products, capacities, and technologies. Special emphasis was given to the "increased sustainability" driver.

It was emphasized the importance of developing specific projects for building programs of bioproducts mills for each sugarcane-producing country, and were included as simple, unpretentious example, some data used during the visit of the Zimbabwe delegation.

Considering all these aspects, the conclusion is that "the equipment industry has the necessary competences, capabilities and competitiveness to fully meet the future market and clients' demands": updated and innovative technologies; a suitable supply capacity; and competitive quality, delivery time, and prices.

Finally, this is a challenge that the equipment industry is ready to take up and have the capacity to deal with it successfully.

Notes

1 Proalcohol – 1975: Brazilian alcohol/ethanol program introduced ethanol as a fuel to the Energy Matrix, either blended or replacing (100% ethanol) gasoline.
2 PROINFA – 2002: Brazilian program for alternative sources for electricity production, including biomass.

References

CNI (2012); *Bioethanol – The Renewable Future, National Confederation of Industry/Sugar Energy national Forum*, Brasilia, 76 p (Rio+20 Sectorial Fascicle).

Cortez, L.A.B. (ed.) (2016); *Proálcool 40 anos – Universidades e empresas: 40 anos de ciência e tecnologia para o etanol brasileiro*. Blucher, 2016.

DATAGRO (2017); Informativo - Year 2016- Ner. 12P, 2017, SP.

DOE, (2015); *U.S. Department of Energy (April 2015), Optima: Co-Optimization of Fuels and Engines*, SAND 2015–2142 M, www.energy.gov/sites/prod/files/2015/05/f22/optima_sand2015-2142m.PDF, accessed on 03/10/2017.

EESI (2016); *Fact Sheet (March 2016), A Brief History of Octane in Gasoline: From Lead to Ethanol*, www.eesi.org/papers, accessed on 02/10/2017.

EIA (2016); *Today in Energy* (May 4, 2016), www.eia.gov/todayinenergy/detail.php?id=26092#, accessed on 03/10/2017.

ExxonMobil (2017); *Outlook for Energy; A View to 2040*, http://corporate.exxonmobil.com/en/energy/energy-outlook/charts-2017, accessed on 02/10/2017.

NREL (2000); *National Renewable Energy Laboratory* – NREL/SR-580–28193 (May 2000), Review of Market for Octane Enhancers, www.nrel.gov/docs/fy00osti/28193.pdf, accessed on 03/10/2017.

Olivério, J.L. (2007); *Os mercados interno e internacional na ótica da indústria nacional de equipamentos, Comissão Especial de Bioenergia da Secretaria de Desenvolvimento do Governo do Estado de São Paulo*, São Paulo, 30/October/2007.

Olivério, J.L. (2009); *DSM – Dedini Sustainable Mill*. Conference in Ethanol Summit. São Paulo: Unica, 2009.

Olivério, J.L. (2014); *Maximizando a sustentabilidade com novas tecnologias, ou USD – Usina Sustentável Dedini*. Workshop Internacional Sobre a Cadeia Sucroenergética, Piracicaba, ESALQ/USP, 24 Jul. 2014.

Olivério, J.L., Barreira, S.T. and Rangel, S.C.P. (2007); *Integrated Biodiesel Production in Barralcool sugar and alcohol mill*. Proceedings of XXVI ISSCT – International Society of Sugar Cane Technologists Congress, Durban, South Africa, July 30 – August 3, 2007.

Olivério, J.L., Boscariol, F.C. (2013); *Expansion of the Sugarcane industry and the New Greenfield Projects in Brazil from the view of the equipment industry*. Proceedings of XXVIII ISSCT – International Society of Sugar Cane Technologists Congress, São Paulo, Brazil, June 24–27, 2013.

Olivério, J.L., Boscariol, F.C., César, A.R.P., Gurgel, M.N.A., Mantelatto, P.E and Yamakawa, C.K. (2010b); *Water production plant*, Proceedings of XXVII ISSCT – International Society of Sugarcane Technologists Congress, Vera Cruz, Mexico, March, 7–11, 2010.

Olivério, J.L., Boscariol, F.C., Mantelatto, P.E. et al. (2010c); *Integrated production of organomineral biofertilizer(BIOFOM*) using byproducts from the sugar and ethanol agro-industry, associated to the cogeneration of energy*, Proceedings of XXVII ISSCT – International Society of Sugarcane Technologists Congress, Vera Cruz, Mexico, March, 7–11, 2010.

Olivério, J.L., Carmo, V.B. and Gurgel, M.A. (2010a); *The DSM – Dedini Sustainable Mill: a new concept in designing complete Sugarcane Mills*. Proceedings of XXVII ISSCT – International Society of Sugarcane Technologists Congress, Vera Cruz, Mexico, March, 7–11, 2010.

Olivério, J.L., Ribeiro, J.E. (2006); Cogeneration in Brazilian sugar and bioethanol mills: past, present and challenges. *International Sugar Journal*, 108, 1291, July 2006.

U.S. Grains Council (2017); *Ethanol Market and Pricing Data* (September 26, 2017), www.grains.org, accessed on 03/10/2017.

7 Energy medium- and long-term perspectives

Are we moving toward an "all-electric model"?

Cylon Gonçalves da Silva

Disclaimer

The title of this article was suggested by the Editors, who requested a short piece on electric vehicles. The author tried to comply with the best of his abilities. However, it must be said that he believes ethanol (first and second, maybe nth generation) and biomass to be absolutely strategic energy resources for Brazil, that need to be further researched and deployed. In an extremely uncertain world, energy independence must be an essential and permanent goal for our country. The author believes, furthermore, that Brazil, unfortunately, does very little – practically none to speak of – research on artificial photosynthesis and batteries. The former has the potential to substantially replace the biological route to fuels and chemicals. Brazil must be in this game, if nothing else because it could wipe out its natural advantages. Researching batteries is important, because the future of large transportation sectors is electric, whether we like it or not. For this to make sense environmentally, electricity will have to come from renewable sources. Which brings us back to biofuels and biomass and artificial photosynthesis (of which photovoltaics, the conversion of the Sun energy into electricity is an essential component). I suppose now we can begin.

Preliminary considerations

This short article is written in the spirit of a Boy Scout (be prepared) and with the cynicism of advancing age (evade the answer to the question). I shall use a trick. Instead of asking what the future will be, I will describe a present that could have been. It makes my task slightly easier. I will avoid having to guess all sorts of uncertain growth rates, regulatory changes, incentives outlooks, and technological advances.

In these preliminaries, the first step is a quick review of the relevant information about the present global energy landscape. As I intend to focus on battery electric vehicles, I will try to pay careful attention to its numbers. I count on the reader to detect and let me know mistakes large and small. The second step, in violation of my rule (no talk about the future), will be an equally brief look into the future using the only reliable information we have about it: demographics. At least, it will help set the scale for further discussions.

And then I will go completely electric.

There will be few conclusions. I will leave the hard work to the reader... and the future.

What is the present situation?

Information about the world energy supply and consumption is abundant and well known. Whether it is always reliable is a matter; however, I will not go into at this moment.

According to the International Energy Agency (IEA, 2016a), the annual global total primary energy supply in 2015 was about 560 EJ (10^{18} Joules), with fossil energies – coal, oil, and gas – representing 80% of the total. The total amount of electricity generated was stated to be about 24,000 TWh (10^{12} Watt-hour) or 86 EJ[1] (physical units conversion).

Also, according to IEA, the total production of electricity from renewables, including hydro, reached about 6,000 TWh in 2016, that is 25% of the total generation.[2] Note that we are talking about energy, not installed capacity (power).

For the transportation sector (all modes included), oil represents the largest share of the final consumption, followed by gas, then coal, respectively: 102 EJ, 1.1 EJ, and 0.26 EJ. Of the total electricity final consumption, only 1.5% goes into the transportation sector. This, obviously, does not mean battery electric vehicles, but vehicles that run off the grid (buses, trains, mostly). We must take into account the fuels that produce this electricity: coal, gas, and oil. They contribute to the transportation sector, via electricity, respectively: 0.437 EJ, 0.231 EJ, and 0.46 EJ (again using physical conversion units – grossly estimating thermodynamic efficiencies, we could multiply these numbers by 3, but this would not affect too much the final accounting for the transportation sector, where oil dominates). Except for coal, the indirect contribution (electricity) of fossil fuels to the transportation sector can be considered negligible. For the numbers used, see IEA (2016a).

The total number of passenger cars and light commercial vehicles in 2015 is given, respectively, in round numbers, as 950 million and 335 million.[3] The vast majority runs on internal combustion engines (ICEs) powered by fossil fuels.

Electric cars, presently, come in several flavors: pure electric running exclusively on battery power, hybrid plug-in (fairly sizeable battery, rechargeable from the grid, plus an ICE), hybrid (small battery recharged by power regeneration and the ICE), and fuel cell vehicles. According to an IEA Report (IEA, 2016b), 4.7 million hybrid electric vehicles and 0.7 million hybrid plug-in and pure electric vehicles were in circulation in the world (the IEA statistics does not distinguish between these two types). In other words, a very small fraction of hybrid and electric cars compared to the estimated 950 million were in circulation.

Lithium

Lithium batteries are the predominant power source for electric vehicles, although there is a lot of research on alternatives.[4]

In the imaginary present that I will describe below, I could have assumed that, over a century of technological evolution of batteries, alternative battery technologies would have been developed, cheaper, less dangerous, and less dependent on a single major resource (Li metal). However, much of the science needed to improve batteries has been developed only since World War II and its techniques and results applied to batteries more intensively since the 1970s. So, our imaginary present, in terms of battery technologies, might not be qualitatively very different from our real present. In this scenario, Li metal becomes something like oil, and we must ask ourselves about production, reserves, resources, country distribution, exhaustion of reserves, etc. Unlike oil, in a crunch, Li metal can be recycled, at least partially, and in our imaginary present, the recycling techniques have been perfected. We start this discussion from our real present.

We recall that resources are all the presumed quantities of Li available. Reserves, according to the definition of the United States Geological Survey, are "those parts of resources that have a reasonable potential for becoming economically available within planning horizons beyond those that assume proven technology and current economics." There is a fair amount of guesswork involved in building tables like the one below, not only because of geological and prospection uncertainties, but also because governments and companies have little interest in presenting their real numbers for their strategic resources. So, they should be taken partly as numbers we can work with and partly as wishful thinking. We should also mention that there is something like 10,000 more Li in the oceans than on land (Loganathan et al., 2017). The problem is the cost of accessing these resources.

The United States Geological Survey (USGS) has annual estimates for the global production, reserves, and resources for Li metal.[5] The total production of Li metal in 2016 was 32,500 metric tons, excluding the production of the US, which was not reported. The total estimate reserves were 14,000,000 metric tons, and the estimated resources were 48,300,000 metric tons, not including the resources in Portugal, which have only recently begun to be prospected on a large scale (and already embroiled in a legal battle[6]).

We have to remember that, today, only about 40% of the world Li production is used in batteries. The other main user is the glass industry. Batteries and glass are responsible for about 2/3 of Li consumption.

Is there a future?

Let me make a rough estimate of how much energy will be needed in the future. Countries with a high Human Development Index (HDI)[7] use an average share of total primary energy of 190 GJ per capita per year and an average of 10 MWh of electricity per capita per year.[8] The world population is expected to reach about 10 billion in 2050.[9]

If the whole world population could have access to the standards of living of high-HDI countries, with the same levels of primary energy and electricity use, the world would need, in 2050, 1,900 EJ per year – about 3.4 times the current level of primary energy supply – and 100,000 TWh of electricity – about 4 times the current level of electricity production. Gains in energy efficiency have not been taken into consideration, although they should be an essential component of energy policies.

Before I discuss numbers any further, let me consider the bigger picture. In 2015, there were an estimated 1.3 billion cars and commercial vehicles in circulation in the world. Their impact on the environment was huge, not only for the fuel they consumed, their CO_2 emissions, and the pollution they produced, but also for the space they occupied in parking and garages, streets, roads, and highways. At the modest average speed of 20 km/h, with a percentage of utilization (utilization factor) of 50%, the average car would do 88,000 km in a year. The standard for cars is more like 20,000 km in a year,[10] which makes their utilization factor (the percentage of hours that the car is being used over the total number of hours in a year) a little over 11%. For 90% of the time, cars serve the only purpose of occupying precious real estate and are a financial burden to their owners. What a nonsense! One would imagine that a rational species would have found a better way to use them a long time ago. Because, in fact, humans don't need cars. They need transportation that is convenient, cheap, and comfortable. Conflating the private car with transportation is the product of a specific society at a specific historical period. There is nothing sacred about it. Hopefully, by 2050, we can have more transportation with far less cars. I would say, in fact, that we must have, if we wish, to increase our chances of survival as a civilization.

By increasing the utilization factor of cars by a modest factor of 2, we would need half as many cars to provide the same personal transportation as today, under a different usage model. What we really need urgently is to innovate the economics, the sociology, and the politics rather than the technology of personal transportation.

Uber and similar solutions point somewhat in the right direction but are awful social solutions: they run on cheap labor and contribute to increase wealth inequality in the world. We will probably have to bring in autonomous cars before long.

However, from the energy point of view, 1 billion cars running 20,000 km a year and 500 million running 40,000 km a year will consume the same amount of fuel and produce the same amount of pollution. The matter of fuels and prime movers still needs to be reconsidered between now and 2050, preferably sooner rather than later.

Gone electric in 1900

In the early 1900s, the option of electric cars was still on the table. The discovery of cheap oil, however, killed this option and the ICE began its irresistible rise to power (pun unintended). It does not mean that electric cars were a bad option in comparison to early ICE cars. They were, however, a much more expensive one. And so, they died or, at least, went into suspended animation. To be revived after the oil shock of 1973. Oil giveth, and (lack of) Oil taketh away.

But, let us imagine that, by an accident in the evolution of planet Earth, the Carboniferous had never existed and that the electric car established itself. Of course, I am ignoring the very real possibility that without the Carboniferous, there would have been no Industrial Revolution. Small detail.

Although the present below is imaginary, I have tried to use real numbers and technologies for 2017. These real numbers allow me to estimate the cost of filling up a "tank" to run about 600 km ($11 at the upper limit[11]) and the amount of electricity needed to run the all-electric fleet (12,500 TWh), which turns out to be, approximately, twice the amount of renewable electricity actually generated in the real world today.

The reader will argue that all this is utopic. In a certain sense, he is right. I am presuming a rational behavior of which the human species is demonstrably incapable. On the other hand, there is nothing utopic about the numbers that I use. The imaginary present that I describe is a possible, albeit slightly doubtful, real future. The game is not over yet.

What the (imaginary) present looks like [at this point, the reader has to understand I am moving to a virtual reality]

Thanks to a rational use of cars to maximize transportation needs and minimize exclusive private use or ownership, we have 500 million electric cars in circulation today, with an average range of 600 km. This requires 140 kWh (1 kWh = 10^3 Wh) of battery storage per car – roughly 20 kg of Li per car. The good news is that one full tank can cost as low as $5, about one cent per kilometer, from renewable electricity sources.

The total battery storage capacity in circulation is then 70 TWh (1 TWh = 10^{12} Wh) or 10 million tons of Li. Electric cars are very reliable and require little maintenance because they have 100 times less moving parts than an ICE car (the electric motor, for instance, has one moving part, and the electric car has no gearbox). Even so, roughly 5% of the world fleet needs to be produced each year to accommodate growth and replacement. In 2017, 25 million new cars will be produced, requiring correspondingly 3.5 TWh storage capacity in batteries to be built, consuming 500,000 tons of Li. With recycling technologies becoming more and more efficient, most of this Li is being reclaimed from old batteries. We are still well within the comfort zone of Li resources in the world.

A very large battery factory (announced by Tesla in the real world) can produce 50 GWh per year of storage capacity. One hundred such factories are in operation today, which means that there is an oversupply in the market, which helps to keep the prices moderate. The total investment in these factories (again based on cost estimates announced in the parallel real world) has been a modest 500 billion dollars.

Each year, the combined fleet of 500 million cars, running 100,000 km per year, covers 50 trillion km (10^{12} km). They consume between 10,000 TWh and 12,500 TWh because of varying efficiencies in the recharging processes. The global footprint of solar plants producing this annual amount of electricity is about 350,000 km^2. (A National Renewable Energy Laboratory (NREL)-based estimation is that 3.5 acres (1.42 ha) are needed in the US to produce 1 GWh per year of solar electricity – see Ong et al. (2013). To be on the safe side, I have multiplied this estimate by a factor of 2.)

The global electric car fleet runs on electricity produced by renewables, thanks to the 12 trillion dollars invested globally over the last quarter of a century to deploy mostly wind and solar power plants. To recall, the average cost of generating capacity over this period was US$2 per W_{peak}, and the system has achieved an overall utilization factor of 25% (these are cautious estimates). The foresight of our ancestors, who, at the beginning of the last century, opted for electric cars has come to fruition more than a century later, in a society that has chosen to give priority to transportation needs rather than to highly underutilized private cars.

The investments in manufacturing plants, battery plants, and renewable electricity wind farms and solar installations are measured in tens of trillions of dollars. Compensating these investments has been a reduction in investments in road and highways infrastructures (less cars); less destruction and better use of expensive urban spaces for the use and parking of cars; less noise and air pollution, and better environments; drastically reduced CO_2 emissions; more R&D-generating innovations, employment, and economic growth. Overall, it seems that our ancestors' rejection of the ICE for cars, the choice of electricity, and the emphasis on transportation services instead of universal private ownership of cars was not such a bad idea after all.

Conclusions

Back to the (real) present [attention: we are leaving the virtual reality of the previous paragraphs]

In the real world, in 2016, renewables produced 6,000 TWh of electricity. This shows that the numbers given above are within reach of modern technology with reasonable investments. *I emphasize, however, that in the construction of the imaginary present entered not only technology, but a new vision of the society of what cars are really for.* Increasing the utilization factor by a factor of 2 can half the number of cars in our cities. By increasing the utilization factor by a factor of 3, 2 out of 3 cars will disappear. Look out of the window for a moment, at the cars clogging our streets and the waste of space for parking lots and garages and imagine how much better the quality of our lives would be if, by magic, 2/3 of these cars disappeared suddenly.

I did not mention autonomous cars in the (imaginary) present. But, clearly, they are there, providing a good share of the transportation needs of society. They will be there in 2050, when the discussion about alternative fuels included a discussion about alternative ways of life. Surely, in the imaginary present less cars are being manufactured, but new industries and services provide for economic growth with a higher quality of life.

Post-conclusions

I could have continued this article with a discussion on liquid fuels and ICEs (biofuels I need not discuss for they are considered by first-rate experts in this volume) versus electricity and battery technologies/electric motors (present and future). In particular, I could have gone into safety,[12] charging times, infrastructure, costs of vehicles, and the propulsion for other modes of transportation: heavy trucks and machinery, ships, and planes. Or mentioned fuel cells (although they are clearly losing the competition with batteries for the personal car – technology, fuel, and distribution infrastructure play no small role in this competition). Or analyzed costs and impacts, including environmental and land use, considered the redefinition of the electric grid and electricity markets, caused by renewable electricity technologies deployed at a very high penetration. Or, yet, looked at electricity produced from biofuels for the cars in the future. One should note that bioethanol offers the possibility of negative emissions for the transportation factor. However, I feel that the main message would be lost. And that is

> We need innovations to zero the carbon emissions from the transportation sector as quickly as possible, but not only innovations based on hard science and technology; we need societal innovations to stop the mad destruction of our cities and environment with ever more cars, more congestion, more and bigger roads, fueled by the unsustainable model of universal private ownership of cars.

What humans really need is transportation. Cars can provide an important fraction of this need. I have tried to show that modest increases in the utilization factor of these machines can go a long way to help us create a better future. And that electric cars, quite certainly also autonomous cars, will occupy a significant niche, even if not the biggest one, in the transportation services of the future.

I feel that, as long as we limit ourselves to a technical discussion of the issue of transportation, we are drowning the baby in the bathwater.

Why buying an electric car is not yet a rational economic decision

The Ford Fiesta SFE FWD is rated at a fuel economy of 35 mpg (14.9 km/l). The Hyundai Ioniq Electric is rated at 136 mpg equivalent in electricity (57.8 km/l or 6 km/kWh). These are the numbers that a normal consumer would look at.[13]

For traveling a 200,000 km (124,301 miles) journey, the Hyundai would need 33.3 MWh (914 gallons or 3,460 l equivalent of gasoline). The Fiesta would need 3,551 gallons (13,440 l). At 11¢/kWh, the Hyundai owner will pay an electricity bill of $3,666. The owner of the Fiesta, at $2.7/gallon, would pay a fuel bill of $9,588.[14] For prices of electricity and gasoline in the US, see, respectively, endnotes [15] and [16]. Different markets will have different numbers, but the reader easily can adapt the calculation.

On balance, the Hyundai owner would "save" $5,922. Let us suppose that this saving would be invested in buying battery storage – sort of equalizing the higher cost of batteries for the electric vehicle with the higher cost of fuel for the ICE car – everything else being the same (which of course, it is not). Notice that I have not included any special subsidies for electricity and fuel, except those already present in the markets. Table 7.1 shows how much $5,992 buy in battery storage capacity, for different costs of the stored kWh.

What Table 7.1 shows us is the sort of calculation a reasonably educated consumer might make, if she tried to figure out if the EV she was buying was cheap or expensive compared with the ICE alternative. At the present day reported costs of battery storage (around $250 per kWh[17]), we can see that her savings would buy a 24 kWh battery (all the other costs being the same) or a range of only 140 km. For a more comfortable range of 470 km, she would need an additional 55 kWh in battery storage (from 24 to 79 kWh) that would cost her an extra $13,750. Everything else being the same, our potential customer would cross the road and buy herself a new fossil fuel car.

Table 7.1 "Battery buying power" from the savings on fuel expenses for the EV driver

Battery cost ($/kWh)	Storage kWh/$5.922	Range (km)
500	12	71
400	15	89
300	20	118
250	24	142
200	30	178
100	59	355
75	79	474
50	118	711

The customer will come back when the cost of battery energy storage falls below $100/kWh (sometime next decade?). At that time in the future, electric cars will become a rational economic option. However, much sooner than that early adopters who do not care so much about price as about owning the latest gadget will have flocked in the hundreds of thousands to electric cars for their other advantages (quiet, clean, cheap to maintain, and naturally, government subsidies). Governments eager to clean the urban environment and reduce CO_2 emissions will give subsidies, not to forget to support a new car industry at home and a new generation of political donors. Of course, when the real crowds come in to buy, there will be no more subsidies – even governments cannot afford to spend that much money.

Notes

1. This number does not reflect the total share of electricity production use of primary energy.
2. https://c1cleantechnicacom-wpengine.netdna-ssl.com/files/2013/06/MTrenew 2013_f1.jpg.
3. Statista.com.
4. The literature on batteries is too extensive to be reviewed here. Just to cite one recent review article on Li-ion batteries: "The development and future of Li-ion batteries", G. E. Blomgren, *Journal of The Electrochemical Society*, 164 (1) A5019–A5025 (2017). There is also a fairly recent volume of *Chemical Reviews*, 114, (2014), with several important papers on alternative battery chemistries. The interested reader may use these clues and Google Scholar to expand his knowledge on the subject.
5. See, for instance, the universal reference https://en.wikipedia.org/wiki/Lithium.
6. www.publico.pt/2017/09/24/economia/noticia/corrida-ao-litio-em-portugal-ja-se-transformou-num-caso-de-policia-1786478.
7. http://hdr.undp.org/en/content/human-development-index-hdi.
8. Numbers calculated by the author using World Bank data: http://databank.worldbank.org/data/home.aspx.
9. www.un.org/development/desa/en/news/population/world-population-prospects-2017.html.
10. See, for instance, www.carinsurance.com/Articles/average-miles-driven-per-year-by-state.aspx, where the average number of miles for a car in the US is quoted as 13,476 (21,600 km).
11. This translates to 6¢/kWh. An NREL report calculates the levelized cost of PV electricity in the US, in 2017, to be between 4¢ and 6¢/kWh for utility scale production (*"U.S. Solar Photovoltaic System Cost Benchmark: Q1 2017 "*, Editors Fu et al., NREL 2017. An earlier report, *"2015 Cost of Wind Energy Review"*, Editors C. Mone et al., NREL, from 2013, estimates the LCOE of wind energy at 2¢/kWh.
12. For a recent development that could potentially make Li ion batteries much safer, see "4.0 V Aqueous Li-Ion Batteries", Congyin Yang et al., Joule 1, 122–132, September 6, 2017.
13. For these numbers, see: www.fueleconomy.gov/feg/best-worst.shtml.
14. Instead of calculating net present values, I assume that discount rate and inflation balance each other. A rigorous calculation would be just as much of a guessing game.
15. www.eia.gov/electricity/monthly/epm_table_grapher.php?t=epmt_5_6_a.
16. www.eia.gov/dnav/pet/pet_pri_gnd_dcus_nus_a.htm.
17. See, for instance, the IEA Report "Global EV Outlook 2017", www.iea.org.

References

IEA, Key World Energy Statistics, 2016a.

IEA, "Hybrid and Electric Vehicles", Ed. Gereon Meyer, 2016b, www.ieahev.org.

Loganathan, P.; G. Naidu and S. Vigneswaran, "Mining valuable minerals from seawater: a critical review", *Environ. Sci. Water Res. Technol.*, 2017, 3, 37–53.

Ong, S.; C. Campbell; P. Denholm; R. Margolis and G. Heath, "Use Requirements for Solar Power Plants in the United States", NREL Technical Report NREL-TP-6A20–56290, June 2013.

8 Comparative assessment of non-ethanol biofuel production from sugarcane lignocelluloses in Africa, including synfuels, butanol and jet fuels

Johann F Görgens and Mohsen Mandegari

Although the integration of biofuel-producing biorefineries into sugar mills, with associated improvements in the energy supply and utilisation efficiencies, provides substantial improvements in biomass (lignocellulose) supply, these economic benefits are not sufficient to overcome low market prices for biofuels. Methanol, gasoline and diesel fuel products obtained by gasification-synthesis processes have not been economically viable at fossil-fuel-equivalent market prices, while fermentative butanol production from lignocelluloses offers weaker economic performances. From an economic point of view, preferred options for non-ethanol biofuel production from lignocelluloses should focus on products that presently fetch a premium price in the market, e.g. jet fuels, and/or enjoy substantial financial support from governments, to ensure a viable investment case. Furthermore, utilisation of existing gas-to-liquid (GTL) and coal-to-liquid (CTL) facilities for partial replacement of fossil fuels with lignocelluloses offers attractive opportunities for biofuel production to be explored further. In all cases, the design of biorefineries should give preference to energy self-sufficient facilities that only make use of bioenergy, with no fossil-energy supplementation, to secure the intended environmental benefits of the biofuel products.

Introduction

The production of biofuels from sugarcane usually defaults to bioethanol, considering the current production volumes and potential to expand this industry on a global scale. Furthermore, commercial strategies are underway to expand the production of ethanol from sugarcane by conversion of lignocelluloses. However, several alternative biofuels may be produced from sugarcane lignocelluloses that may offer improved market access, especially for synfuels, *n*-butanol and jet fuels. These biofuels have properties closer or identical to conventional fossil-derived fuels, and often have "drop-in" status, thus avoiding the fuel supply chain complexities and fuel blending limits of bioethanol. The present chapter considers the technological, economic and environmental aspects of these alternative biofuels, especially for their production from sugarcane lignocelluloses in Africa. The impact of annexing such biofuel production to existing sugar mills, with integration of facilities for steam and electricity production, is also considered.

Integration of lignocellulose conversion to biofuel into existing sugar mills

Conventional and ageing sugar mills, especially in parts of the developing world, are both energy-intensive and often very inefficient in the manner that available lignocelluloses (mostly bagasse) are converted into steam and electricity (Dogbe et al., 2018). Sugar mills are mostly energy self-sufficient, although the low conversion efficiency typically results in little bagasse available for further conversion (Dogbe et al., 2018). The primary unit operations in the sugar mill responsible for energy inefficiencies, in order of priority, are the cogeneration unit (steam and electricity), following by evaporation and crystallisation (Dogbe et al., 2018).

Liberation of sugarcane lignocelluloses from existing sugarcane agriculture and milling, for conversion into alternative biofuels, can be achieved through two combined strategies. First, the cogeneration system in the existing sugar mill can be replaced by a new, high-pressure and efficient combined heat and power (CHP) plant, which will serve both the process energy (steam, electricity) requirements of the existing sugar mill and the new biofuel production facility, often resulting in surplus electricity production for sale (Farzad et al., 2017; Mandegari et al., 2017a). Second, the implementation of so-called green cane harvesting (GCH), as an alternative to sugarcane burning before harvesting, can liberate approximately 50% of the agricultural residues (non-stems) from sugarcane, for transport to the sugar mill and conversion to energy products (Farzad et al., 2017).

Implementation of the aforementioned strategies provides opportunity to develop energy self-sufficient biorefineries, using only bioenergy inputs for the production of biofuels from sugarcane lignocelluloses, as depicted in Figure 8.1. The integration of such biorefineries into existing sugar mills provides efficiency and economic advantages to biofuel production through both lignocellulose supply and energy cogeneration (Mandegari et al., 2017a, 2017b).

Comparative economic and environmental assessment of alternative biofuel production options requires the application of suitable simulation and modelling tools, to accurately represent the performances of process technologies. Process design, flow sheeting and simulation with software tools such as AspenPlus™ have proven invaluable in generating comparative technical descriptions of process technologies, through mass and energy balances derived from published data on process performance (Mandegari et al., 2016). Development of AspenPlus™ simulations of alternative processes

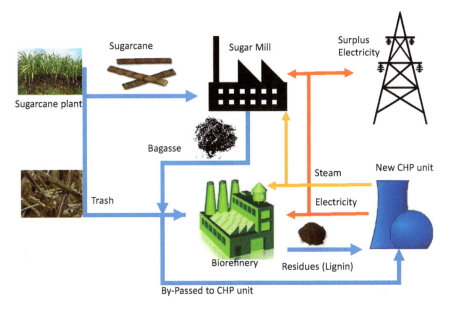

Figure 8.1 Annexed biorefinery for biofuel production, integrated into an existing sugar mill.
Source: by the author.

also provides the technical information required for subsequent economic assessments (typically investment viability assessed through capital and operating costs, as well as revenues) and environmental impact assessments (Farzad et al., 2017; Mandegari et al., 2016). For example, these methods allow for the impacts of alternative methods of energy supply to biorefineries, such as the use of bioenergy only compared to the co-combustion of coal with biomass materials, to be assessed in terms of its relative economic and environmental performances (Mandegari et al., 2017a, 2017b). The techno-economic and environmental assessments on alternative biofuel production from sugarcane lignocelluloses have been performed using these simulations tools, as further described below.

Butanol production from sugarcane lignocelluloses

n-Butanol has superior fuel properties to ethanol, due to both a high calorific value and fewer restrictions on its blending with fossil-derived fuels for road transport (Gottumukkala et al., 2017). However, the fermentative production of butanol remains economically more expensive than ethanol, due to limitations of the acetone-butanol-ethanol (ABE) process typically applied for this purpose. These limitations include substrate inhibition, end-product inhibition, by-product (acetone, ethanol) formation and low butanol titres

(Gottumukkala et al., 2017). These have significant impacts on the energetic and economic cost of butanol production and recovery (Aneke and Görgens, 2015). Technological developments have therefore focussed on both production strain improvement, to reduce by-product formation and increase the butanol titre, and process development, such as the application of gas stripping for *in situ* butanol removal from the fermentation vessel (Gottumukkala et al., 2017). Of relevance to the present discussion, are opportunities to produce butanol from lignocelluloses, through pretreatment-hydrolysis-fermentation methods typically applied for cellulosic ethanol production (Gottumukkala et al., 2017; Naleli, 2016).

Despite recent advances in ABE fermentation, the majority of techno-economic assessments confirm that butanol production from lignocelluloses is economically less attractive than ethanol (Farzad et al., 2017; Mandegari et al., 2017a, 2017b), while also being significantly more expensive than butanol production from sugars or molasses (Naleli, 2016; Van der Merwe et al., 2013). Comparison of alternative, stand-alone flow schemes for butanol production from lignocelluloses confirmed the importance of *in situ* butanol removal, for example, through gas stripping, to avoid the energy penalties associated with the purification of low concentrations of butanol from a fermentation product, which may result in an economically viable process (Naleli, 2016). A particular advantage to the use of lignocellulose for butanol production is the opportunity for an energy self-sufficient process, which requires no external inputs of fossil fuels, with associated environmental benefits (Naleli, 2016). The fermentative production of butanol from lignocellulose, through integration with an existing sugar mills, has not overcome these economic limitations (Farzad et al., 2017).

Synfuels production, including methanol and Fischer-Tropsch

The gasification of lignocellulosic biomass to produce synthesis gas (also known as syngas) is a critical step in the production of synfuels such as methanol and Fischer-Tropsch (FT) products. A wide range of gasification technologies have been under development, as reviewed elsewhere, with a focus on reducing tars formation and the requirements for gas cleanup (Farzad et al., 2016). A gasification-synthesis process converting pure biomass into liquid fuels is known as biomass to liquids (BTL). From an industrial point of view, the partial substitution with biomass of coal and natural gas streams fed to existing CTL of GTL facilities offers several technical and economic advantages (Farzad et al., 2016). Such co-feeding of biomass to existing industrial facilities, referred to as BGTL or BCTL, will benefit from substantial investments in capital equipment as well as the existing industrial gasification and synthesis technologies, provided that biomass co-feeding does not disrupt the existing CTL and GTL plants in terms of syngas yields and quality (Farzad et al., 2016).

The large differences in the chemical compositions and thermal degradation kinetics of biomass in comparison to coal (Aboyade et al., 2011, 2012, 2013a, 2013b) may require the pretreatment of lignocellulose before co-feeding to CTL facilities, typically through thermal treatments such as torrefaction or pyrolysis (Carrier et al., 2011; Nsaful et al., 2017). Such upgrading will also increase the energy density of lignocellulose, potentially offering more opportunities for the transport of sugarcane residues to existing CTL or GTL facilities.

Alternatively, the co-processing of product gas from a dedicated lignocellulose gasification unit, with the existing processing of natural gas in the reforming step of GTL facilities, offers several economic and environmental advantages for the resulting BGTL facility, through the exploitation of process synergies (Baliban et al., 2012; Gardezi et al., 2013). Whereas existing CTL and GTL facilities primarily produce synfuels in the form of products from FT synthesis, with various industrial catalysis options available, new facilities for BTL synfuels production may also produce bio-methanol (Amigun et al., 2010).

The economic and environmental impacts of stand-alone processes for lignocellulose conversion to synfuels, with no integration to existing industrial facilities, have been considered for gasification-FT (G-FT) processes. The G-FT process was shown to have a higher energy conversion efficiency than a bioethanol process, partly due to utilisation of the lignin components in lignocellulose for fuel production, and provided that sufficient internal heat integration was applied to these processes (Leibbrandt et al., 2011; Petersen et al., 2015b). Furthermore, the overall process conversion efficiencies could be improved by optimisation of the syngas composition, resulting in an increase in the co-production of electricity (Leibbrandt et al., 2013). The increased conversion efficiencies of the G-FT process, compared to cellulosic ethanol production, have been shown to increase the environmental benefits derived from lignocellulose conversion (Petersen et al., 2015b).

The economics of FT- and methanol-synfuels production from lignocelluloses have been considered for biorefineries integrated into existing sugar mills (Farzad et al., 2017; Mandegari et al., 2017b; Petersen et al., 2015a). These facilities were designed to be energy self-sufficient, thus utilising only bioenergy with no supplementation of energy supply with fossil fuels. However, neither FT- nor methanol-synfuels production from lignocelluloses, even when integrated with sugar mills to maximise synergistic benefits, has shown to be economically viable, without substantial improvements in market prices and/or government incentives. The production economics of both methanol and FT syncrude are hampered by the market perception that these should be priced according to fossil-fuel equivalents. According to this approach, FT syncrude is priced similarly to crude oil, with the associated large fluctuations in market prices.

Comparisons between the production of synfuels, when integrated with sugar mills, and that of butanol, both from lignocelluloses and based on Petersen et al. (2015a), Petersen et al. (2015b) and Farzad et al. (2017), are provided in Tables 8.1 and 8.2. The lower heating/cooling of the methanol and FT syncrude processes resulting in a higher overall conversion efficiency, although with a lower surplus of electricity (Table 8.1). The biochemical process for butanol production had substantially higher process energy demands, while also having a larger surplus of electricity co-production. Overall, the production of methanol provided a more attractive economic case that FT syncrude or butanol (Table 8.2), with all of these options for lignocellulose conversion, hampered by poor market prices and price parity to fossil-derived products (Farzad et al., 2017).

There is a lack of process simulations and economics for the co-gasification or processing of lignocelluloses with coal or natural gas in BGTL and BCTL facilities. These simulations and economics are highly dependent on the underlying performances of the existing industrial facilities. However, it is

Table 8.1 Comparing the conversion efficiencies and energy demands for the conversion of sugarcane lignocellulose to synfuels or n-butanol in a biorefinery integrated into an existing sugar mill

	Unit	Butanol	Methanol	FT syncrude
Feedstock				
Bypass to boiler	t/h	26.00	22.75	19.50
	%	40	35	30
To biorefinery	t/h	39.00	42.25	45.50
Products				
Ethanol	t/h	0.35	–	–
Butanol	t/h	4.61	–	–
Methanol	t/h	–	12.76	–
Syncrude	t/h	–	–	5.80*
Acetone	t/h	1.50	–	–
Surplus electricity	MW	4.30	0.50	1.80
Total production[a]	t/h	6.46	12.76	5.80
	t/t[c]	0.16	0.30	0.13
Energy demand[b]				
Cooling	MW	73.10	26.20	33.90
Heating	MW	61.70	2.60	2.40
Power	MW	13.70	14.20	13.70

* Density of syncrude = 634.8 kg/m^3.
[a] Total production of chemicals.
[b] Heat and power demand of sugar mill is excluded.
[c] Production yield: tonne of product(s) per tonne of biomass fed to biorefinery "exclusive of feedstock bypassed to CHP".

Table 8.2 Comparing the economics of the conversion of sugarcane lignocellulose to synfuels or *n*-butanol in a biorefinery integrated into an existing sugar mill

	TCI	Fixed operating cost	Variable operating cost	Total sales	IRR
	(million $)	(million $/year)			%
Butanol	269	5.8	17.2	40.2	4.8
Methanol	233	8.5	7.9	56.5	16.7
FT syncrude	234	8.1	8.2	12.6	11.5

TCI: total capital investment, IRR: internal rate of return.

expected that such industrial options will substantially reduce the production costs of synfuels, provided that sufficient proximity between biomass supply and conversion facilities can be achieved.

Aviation biofuel production

Aviation accounts for a major portion of the global emissions of greenhouse gases (GHGs) and has therefore become an international priority for the implementation of biofuels (Diederichs et al., 2016; Mandegari et al., 2017b). Existing biofuel production and use in commercial aviation makes use of hydro-processed esters and fatty acids (HEFA) produced from vegetable oil. Alternative methods for aviation biofuel production from lignocellulose include FT (portion of total product yield is kerosene), cellulosic ethanol (converted via the Alcohol-to-Jet route, A2J) and direct liquefaction (pyrolysis, produces a crude bio-oil that requires upgrading and refining) (Diederichs et al., 2016). Whereas synthetic jet fuel from the FT process and HEFA is both fully certified for international use in aviation, the A2J and direct liquefaction products remain to be certified.

A comparison of the economics of bio-jet production from lignocelluloses, vegetable oil and sugarcane syrups was reported by Diederichs et al. (2016), based on simulations of stand-alone conversion processes that are not integrated into sugar mills. Technical and economic performances for the GFT, cellulosic ethanol (L-ETH), a hybrid gasification-fermentation process for lignocellulose conversion to ethanol (SYN-FER), sugarcane juice to ethanol (S-ETH) and HEFA processes are presented in Tables 3 and 4. Ethanol intermediates were converted to jet fuel by means of the A2J route.

Economic assessments were based on similar quantities of the total bio-jet fuel production, and demonstrated substantially lower process energy demands for the HEFA process, due to simpler processing methods compared to alternatives considered (Table 8.3). The simpler processing required for HEFA production from vegetable oils is balanced against the higher cost of the feedstock, although overall the lowest cost of aviation biofuel was still obtained with this process (Table 8.4). The preferred process for aviation

Table 8.3 Comparing the conversion efficiencies and energy demands for the production of aviation biofuels in a stand-alone conversion facility, not integrated into an existing sugar mill

Parameter		*Process for aviation biofuels production*				
		L-ETH	SYN-FER	GFT	HEFA	S-ETH
Feed	Lignocellulose (dry MT/hr)	77.88	77.88	77.88	–	–
	Vegetable oil (MT/hr)	–	–	–	14.87	–
	Sugar cane (dry MT/hr)	–	–	–	–	63.27
	Trash (dry MT/hr)	–	–	–	–	31.16
Energy	Cooling demand (MW)	145.58	202.32	109.09	12.43	156.48
	Heating demand (MW)	88.75	81.99	39.33	7.62	72.92
	Electric power demand (MW)	23.79	25.94	42.11	2.6	15.46
Product	Jet fuel (MT/hr)	7.75	7.94	7.93	7.14	7.76
	Naphtha (MT/hr)	–	–	2.39	1.71	–

GFT: gasification-FT, L-ETH: cellulosic ethanol, SYN-FER: hybrid gasification-fermentation process for lignocellulose conversion to ethanol, S-ETH: sugarcane juice to ethanol and HEFA: hydro-processed esters and fatty acids.
Source: Diederichs et al. (2016).

Table 8.4 Comparing the required MJSPs for the production of aviation biofuels in a stand-alone conversion facility, not integrated into an existing sugar mill

Parameter		*Process for aviation biofuels production*				
		L-ETH	SYN-FER	GFT	HEFA	S-ETH
Variable OPEX (million $/year)	Raw materials and disposal	120.23	69.75	62.51	112.60	85.36
	By-product credits	24.77	13.03	38.16	25.44	38.28
Fixed OPEX	(million $/year)	24.78	22.09	27.85	10.52	18.92
FCI	(million US$)	482.6	409.7	565.5	161.4	324
TCI	(million US$)	532.7	452.5	623.9	179.4	358.3
MJSP	($ per kg jet fuel)	3.43	2.49	2.44	2.22	2.54

OPEX: operational costs, FCI: fixed capital investment, TCI: total capital investment, MJSP: minimum jet selling price, as required for a sufficient return on investment, GFT: gasification-FT, L-ETH: cellulosic ethanol, SYN-FER: hybrid gasification-fermentation process for lignocellulose conversion to ethanol, S-ETH: sugarcane juice to ethanol and HEFA: hydro-processed esters and fatty acids.
Source: Diederichs et al. (2016).

biofuel production from lignocellulose was gasification-FT, due to higher conversion efficiencies. These economics for a stand-alone, biomass-only process are expected to improve substantially when implemented in BCTL or BGTL facilities. Furthermore, HEFA is presently traded internationally at prices two to three times larger than fossil-fuel equivalents (Diederichs et al., 2016), indicating that the GFT processes may be near-commercial readiness, especially for BGTL and BCTL options.

Environmental impacts of biofuel production from lignocelluloses

The production of biofuels from lignocelluloses, especially in situations where by-products such as harvesting residues and bagasse from existing agricultural activities are used, is expected to yield substantial environmental benefits. However, the magnitude of these benefits may vary significantly according to primary drivers such as overall conversion efficiencies and the possible supplementation of bioenergy with fossil-fuel energy sources, to provide steam and electricity to meet the conversion process demands.

The environmental benefits from biofuel production from lignocelluloses often follow conversion efficiencies, where processes with lower efficiency also show greater environmental impacts (Petersen et al., 2015b). Careful consideration should be given the use of externally sourced chemicals in the conversion processes, which are often derived from fossil fuels and may result in substantial environmental burdens, in particular for biochemical conversion processes (Farzad et al., 2017; Petersen et al., 2015b). However, previous studies indicated that a 10% increase in the biomass output and its availability (per unit land area) for biorefinery processing incurred more CO_2 savings than a 10% increase in the yield of the bioconversion methods. This is consistent with our research finding from the contribution analyses above, which suggested that improvements in agrochemical utilisation efficiency in agricultural practice (e.g. lower fertiliser inputs per unit biomass harvested) would lead to overall environmental savings particularly on abiotic depletion and ozone depletion (Farzad et al., 2017; Khoo et al., 2016).

Whereas all sugar mills have the potential to be energy self-sufficient by using bioenergy only, the operational disruptions and inefficiencies in sugarcane conversion often result in the supplementation of bioenergy supply with coal (Dogbe et al., 2018). A similar practice may also be applied in biorefineries producing biofuels from lignocelluloses, in particular when considering the substantial economic benefits that can be realised (Mandegari et al., 2017a). However, these economic benefits come at the expense of reduced environmental benefits, where any increase in the supplementation of bioenergy with coal directly increases environmental impacts in acidification, eutrophication, global warming potential and ecotoxicity. Opportunities for biofuel production in a self-sufficient manner, using only bioenergy feeds, should therefore be sought, with maximisation of conversion efficiencies to realise economic potential.

Conclusions

Whereas the integration of biofuel-producing biorefineries into sugar mills, with associated improvements in the energy supply and utilisation efficiencies, provides substantial improvements in biomass (lignocellulose) supply, these economic benefits are not sufficient to overcome low market prices for biofuels. Methanol, gasoline and diesel fuel products obtained by gasification-synthesis processes have not been economically viable at fossil-fuel-equivalent market prices, while fermentative butanol production from lignocelluloses offers weaker economic performances. Similar to bioethanol, these drop-in fuels will require substantial government support through financial incentives to advance their market uptake.

However, there are a few except where, from an economic point of view, the preferred options for non-ethanol biofuel production may overcome these limitations. Biofuel production from lignocelluloses should prioritise those products that presently fetch premium prices in the market, e.g. jet fuels where international cooperation and policymaking has already established a viable biofuels market. Furthermore, utilisation of existing GTL and CTL facilities for partial replacement of fossil fuels with lignocelluloses offers attractive opportunities for biofuel production, to be explored further. In all cases, the design of biorefineries should give preference to energy self-sufficient facilities that only make use of bioenergy, with no fossil-energy supplementation, to secure the intended environmental benefits of the biofuel products. Furthermore, the role of improvements in agrochemical utilisation efficiency in agricultural practice is undeniable for environmental benefits.

References

Aboyade AO, M Carrier, EL Meyer, JH Knoetze, JF Görgens (2012). Model fitting kinetic analysis and characterisation of the devolatilization of coal blends with corn and sugarcane residues. *Thermochimica Acta* 530: 95–106.

Aboyade AO, JF Görgens, M Carrier, EL Meyer, JH Knoetze (2013a). Thermogravimetric study of the pyrolysis characteristics and kinetics of coal blends with corn and sugarcane residues. *Fuel Processing Technology* 106: 310–320.

Aboyade AO, M Carrier, EL Meyer, JH Knoetze, JF Görgens (2013b). Slow and pressurized co-pyrolysis of coal and agricultural residues. *Energy Conversion and Management* 65: 198–207.

Aboyade AO, TJ Hugo, M Carrier, EL Meyer, R Stahl, JH Knoetze, JF Görgens (2011). Non-isothermal kinetic analysis of corn cobs and sugar cane bagasse pyrolysis. *Thermochimica Acta* 517: 81–89.

Amigun B, JF Görgens, JH Knoetze (2010). Biomethanol production from gasification of non-woody plants in South Africa: Optimum scale and economic performance. *Energy Policy* 38: 312–322.

Aneke M, JF Görgens (2015). Evaluation of the separation energy penalty associated with low butanol concentration in the fermentation broth using entropy analysis. *Fuel* 150: 583–591.

Baliban RC, JA Elia, V Weekman, CA Floudas (2012). Process synthesis of hybrid coal, biomass, and natural gas to liquids via Fischer–Tropsch synthesis, ZSM-5 catalytic conversion, methanol synthesis, methanol-to-gasoline, and methanol-to-olefins/distillate technologies. *Computers & Chemical Engineering* 47: 29–56.

Diederichs GW, M Alimandegari, S Farzad, JF Görgens (2016). Techno-economic comparison of biojet fuel production from lignocellulose, vegetable oil and sugar cane juice. *Bioresource Technology* 216: 331–339.

Dogbe ES, MA Mandegari, JF Görgens (2018). Exergetic diagnosis and performance analysis of a typical sugar mill based on Aspen Plus® simulation of the process. *Energy* 145: 614–625.

Farzad S, MA Mandegari, JF Görgens (2016). A critical review on biomass gasification, co-gasification, and their environmental assessments. *Biofuel Research Journal* 12: 483–495.

Farzad S, M Alimandegari, M Guo, KF Haigh, N Shah, J Görgens (2017). Multi-product biorefineries from lignocelluloses: A pathway to revitalisation of the Sugar Industry? *Biotechnology for Biofuels* 10:87. doi:10.1186/s13068-017-0761-9.

Gardezi SA, B Joseph, F Prado, A Barbosa (2013). Thermochemical biomass to liquid (BTL) process: Bench-scale experimental results and projected process economics of a commercial scale process. *Biomass and Bioenergy* 59: 168–186.

Carrier M, T Hugo, JF Görgens, JH Knoetze (2011). Comparison of slow and vacuum pyrolysis of sugar cane bagasse. *Journal of Analytical and Applied Pyrolysis* 90 (1): 18–26.

Gottumukkala LD, K Haigh, JF Görgens (2017). Trends and advances in conversion of lignocellulosic biomass to biobutanol: Microbes, bioprocesses and industrial viability. *Renewable and Sustainable Energy Reviews* 76: 963–973.

Khoo HH, WL Ee, V Isoni (2016). Bio-chemicals from lignocellulose feedstock: sustainability, LCA and the green conundrum. *Green Chemicals*, 18: 1912–1922.

Leibbrandt NH, JH Knoetze, JF Görgens (2011). Comparing biological and thermochemical processing of sugarcane bagasse: An energy balance perspective. *Biomass and Bioenergy*, 35 (5): 2117–2124.

Leibbrandt NH, AO Aboyade, JH Knoetze, JF Görgens (2013). Process efficiency of biofuel production via gasification and Fischer-Tropsch synthesis. *Fuel* 109: 484–492.

Mandegari MA, S Farzad, JF Görgens (2017a). Economic and environmental assessment of cellulosic ethanol production scenarios annexed to a typical sugar mill. *Bioresource Technology* 224: 314–326.

Mandegari MA, S Farzad, JF Görgens (2017b). Recent trends on techno-economic assessment (TEA) of sugarcane biorefineries. *Biofuel Research Journal* 15: 704–712.

Mandegari MA, S Farzad, JF Görgens (2016). Process design, flowsheeting, and simulation of bioethanol production from lignocelluloses. In: *Biofuels: Production and Future Perspectives*, Eds. Singh, A. Pandey and E. Gnansounou, pp. 262–288, New York: CRC press, Tylor & Francis Group.

Naleli K (2016). *Process modelling in production of biobutanol from lignocellulosic biomass via ABE fermentation*. MEng thesis, Stellenbosch University. Available at http://scholar.sun.ac.za/handle/10019.1/98620

Nsaful F, F-X Collard, JF Gorgens (2017). Influence of lignocellulose thermal pre-treatment on the composition of condensable products obtained from char devolatilization by means of thermogravimetric analysis – thermal desorption/

gas chromatography – mass spectrometry. *Journal of Analytical and Applied Pyrolysis* 127: 99–108.

Petersen AM, S Farzad, JF Görgens (2015a). Techno-economic assessment of integrating methanol or Fischer-Tropsch synthesis in a South African sugar mill. *Bioresource Technology* 183: 141–152.

Petersen AM, JH Knoetze, JF Görgens (2015b). Comparison of second-generation processes for the onversion of sugarcane bagasse to liquid biofuels in terms of energy efficiency, pinch point analysis and life cycle analysis. *Energy Conversion and Management* 91: 292–301.

Van der Merwe AB, H Cheng, J Görgens, JH Knoetze (2013). Comparison of energy efficiency and economics of process designs for biobutanol production from sugarcane molasses. *Fuel* 105: 451–458.

9   2G 2.0

Lee R. Lynd

Introduction

The LACAf project described in this book has been a prominent part of the Global Sustainable Bioenergy (GSB) initiative, which was launched in 2009. At that time, global biofuel production from easily converted 'first-generation' (1G) feedstocks (corn and other starch-rich grains, sugar-rich crops such as sugarcane and sugar beets, or oil seeds) had been expanding rapidly during the prior decade, and a next wave of further expanded biofuel production was anticipated due to the emergence of cost-effective technology for the conversion of 'second-generation' (2G), lignocellulosic, feedstocks. The prospect of rapid expansion gave rise to concerns about potential consequences of land being used for the production of biofuel feedstocks – including displacement of food production, habitat loss, and further marginalization of the rural poor. The impetus for the GSB initiative was primarily to address these concerns.

As described in prior chapters, the LACAf project has made substantial progress in documenting the potential for bioenergy to positively impact economic and human development in the countries selected for study. The ongoing geospatial project (Chapter 4) is well positioned to address the issues related to pasture intensification, thereby sparing land that could be used for bioenergy production. A substantial and diverse body of work outside the GSB initiative, for example, the comprehensive SCOPE study (Souza et al., 2015), has also explored and articulated paths by which bioenergy can offer social and environmental benefits. Yet widely disparate assessments of the feasibility and desirability of large-scale bioenergy persist, indicating a need for further effort and experience-based validation in this vein.

Acknowledging the ongoing importance of social and environmental opportunities and challenges related to bioenergy and biofuels, this chapter addresses technology for the production of second-generation biofuels. Such '2G' biofuels were at the center of high expectations for bioenergy a decade ago, but have fallen far short of expectations and today receive a small fraction of the investment they once did (Lynd, 2017). Important contextual factors for biofuels have also changed. In particular, new technologies for accessing unconventional oil and gas resources have led to resurgent production and lower-than-anticipated prices, thereby making it more difficult for biofuels to be cost-competitive. As well, gains in battery and vehicle technologies have led to substantial deployment of electric vehicles, to which much of the world is now looking for decarbonization of the light-duty transport sector.

In light of these developments, it is natural and indeed desirable to reassess commitments and strategies pertaining to 2G biofuels. Two questions frame such a reassessment: *Is renewed effort to develop and deploy 2G biofuel technology warranted?* and *How should such an effort be configured?* In this chapter, I present the view that a renewed effort is indeed warranted, and offer perspectives on how such a '2G 2.0' effort can build on lessons from '2G 1.0' in order to accelerate technological progress and biofuel deployment.

Is renewed effort to develop and deploy 2G biofuels warranted?

An affirmative answer to this question requires a positive assessment of (a) the need and potential benefits of 2G biofuels, and (b) 2G biofuel production technology and prospects for economic viability.

Need and potential benefits of 2G biofuels

Production of bioenergy feedstock requires land, which is a limited resource with many important uses. Accordingly, most low-carbon energy scenarios assume that providing energy services from biomass should only be considered when it is difficult to do so from other sources. Despite this 'other renewables first' approach, biomass averages a quarter of primary energy supply in five prominent low-carbon energy scenarios for 2050 compiled by Dale et al. (2014). The IEA Sustainable Bioenergy Roadmap (Brown and Le Feuvre, 2017) states,

> …modern bioenergy is an essential component of the future low carbon global energy system if global climate change commitments are to be met. This is especially the case since bioenergy can play an important role in helping to decarbonise sectors for which other options are scarce, such as in aviation, shipping or long haul road transport. However, the current rate of bioenergy deployment is well below the levels required within IEA long term climate models. Acceleration is urgently needed to ramp up the contribution of bioenergy across all sectors notably in the transport sector where consumption is required to triple by 2030.

In the IEA two-degree scenario (2DS) (Brown and Le Feuvre, 2017), modern bioenergy provides nearly 17% of final energy demand and almost 20% of cumulative carbon savings in 2060.

Fulton et al. (2015) project that about half of the transport demand in the IEA 2DS scenario extended to 2075 is difficult to meet with electricity or hydrogen, and that aviation, ocean shipping, and long-haul trucking will still require liquid fuels. Failure to decarbonize this fraction of transport demand would result in twice the emissions compatible with a two degree mean global temperature increase even if the rest of the economy were 100% decarbonized. Bioenergy is widely regarded as one of the most attractive options for taking carbon out of the air (A better life..., 2016). Such 'negative' carbon emissions are increasingly seen as essential for climate change mitigation (Smith et al., 2016). Since the cost of capturing and concentrating CO_2 is greater than the cost of subsurface storage (Ruben et al., 2015), the nearly pure CO_2 streams available from biofuel production are among the most advantageous applications for carbon capture and storage.

About 50 EJ of global biofuel production will be needed for the half of anticipated global transport demand, expected by Fulton et al. (2015) to be difficult to decarbonize without biofuels. This is less than low-end estimates for global biomass resource availability according to authoritative compilations of literature studies by Slade et al. (2014) and the IPCC (Smith et al., 2014). An added perspective is provided by plotting land requirements as a function of biomass energy supply for various crops and representative yields (Figure 9.1). As shown therein, the production of 50 EJ would require over 500 million hectares planted in unmanaged grass or forest but less than 50 million hectares planted in energy cane. According to Brazil's Agroecological Zoning (2010), 50 million hectares is about half the estimated Brazilian land currently used for agriculture, livestock, and pasture that is suitable for sugarcane production. This estimate excludes areas with native vegetation, the Amazon and other ecologically sensitive areas, and slope over 12 degrees gives priority to degraded lands requiring minimal irrigation, and intends to avoid risk to food security and production. It is likely that not all of this land would support the energy cane yields in Figure 9.1, and some of this land may be used for food in the future (Jaiswal et al., 2017). As may be seen from Figure 9.1, key factors in reducing the footprint of biofuels include using high-yielding feedstocks that grow in tropical and semitropical climates, and utilizing lignocellulose as well as more easily processed feedstock fractions. A long time before 50 EJ of biomass are used for bioenergy, we will be able to evaluate economic and social impacts based on experience rather than speculation, affording an opportunity to reinforce what works and avoid what does not.

The shaded region is the range considered by Slade et al. (2014). Values for sugarcane and energy cane are from Leal et al. (2013), Junqueira et al. (2017), and T.L. Junqueira, personal communication. The horizontal axis may be interpreted either as primary biomass energy or as fuel energy assuming a thermodynamic process efficiency of about 50%, and biomass supply from residues equal to that from dedicated crops as assumed by Woods et al. (2015).

The growth of the biofuels industry in Brazil has been accompanied by well-documented economic and human development benefits (Moraes et al.,

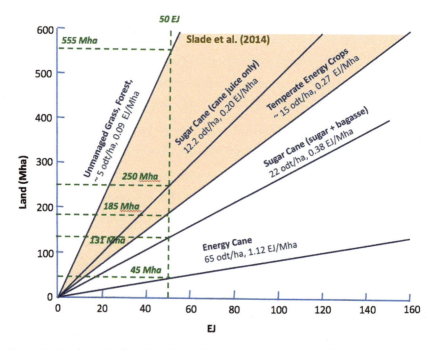

Figure 9.1 Land required as a function of energy supply as impacted by biomass yield.

2015, 2016), and bioenergy has potential to offer such benefits in Africa where land resources are similarly plentiful (Lynd et al., 2015). Incorporating perennial crops into agricultural landscapes also has well-documented positive impacts on ecosystem services (Jordan et al., 2007; Werling et al., 2014). Yet bioenergy continues to be criticized on both social and environmental grounds, and these persistent criticisms weaken the advancement and deployment of bioenergy in multiple ways. As depicted in Figure 9.2, land-based production of bioenergy feedstocks inevitably has strong linkages to food security, rural economic development, and land-based ecological services. Biofuel advocates see these linkages as opportunities, whereas bioenergy critics see these linkages as posing risks that arise to a smaller extent with other renewables. Acknowledging truth in both of these perspectives, I suggest that *intent* is a critical factor in determining which edge of the bioenergy sword is the sharpest. Simply put, if biofuels are deployed with the intent to achieve social and environmental benefits, there is a strong basis to believe that such benefits can be realized. Without this intent, there is a significant risk that negative impacts will arise. The same is true with respect to the sustainability conundrum *writ large*, which requires intent to address and will not solve itself. The collective reluctance of the world to assume and apply beneficial intent with respect to bioenergy may be related to the fact that humanity is at present more comfortable with collective responsibility for water, air,

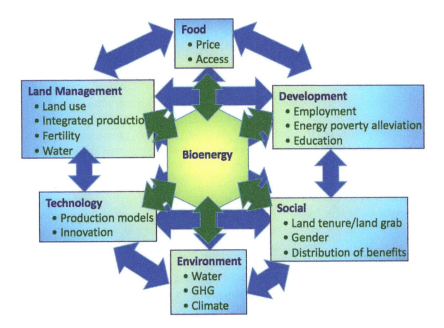

Figure 9.2 Linkages between bioenergy and other things we care about.

and climate than is the case for land. Climate-responsive land use choices are, however, required to achieve climate stabilization with or without biofuels (Prosperous living..., 2015).

2G biofuel production technology: status and prospects for non-incremental improvement

Cellulosic biomass can be converted to biofuels using either biological or nonbiological (thermochemical) processes. Although thermochemically-derived 2G biofuels merit consideration, biologically-derived 2G biofuels were largely responsible for the high expectations of a decade ago, are primarily responsible for these high expectations not being met, and are considered here. We focus on ethanol because it provides by far the shortest path to commercial production of biologically derived 2G biofuels. Once cost-competitive technology for the production of 2G ethanol is developed and deployed, the production of other molecules – including both fuels and chemicals – can confidently be expected to follow.

As reviewed elsewhere (Lynd et al., 2017a), six pre-commercial 'pioneer' cellulosic ethanol plants have come on line since 2012: the Beta Renewables plant in Crescentino, Italy; Abengoa plant in Kansas, USA; the DuPont plant in Iowa, USA; the Granbio plant in Alagoas, Brazil; the POET-DSM plant in Iowa, USA; and the Raizen plant in Piracicaba, Brazil. All of these plants use herbaceous rather than woody feedstocks, and all use a processing paradigm based on thermochemical pretreatment, added fungal cellulase, and fermentation of soluble sugars to ethanol. The capital cost for these pioneer plants, in some cases including the feedstock supply infrastructure, ranges from $100 to $500 million, and on an annualized basis exceeds the selling price of ethanol when oil was $100/barrel. The rated ethanol production capacity of the pioneer plants ranges from 40 to 120 million liters/year, although no plant is yet operating at this capacity. Operational difficulties have been reported, particularly with respect to the delivery of solids to pretreatment reactors operated at high pressure. Several plants and/or their parent companies have reported financial difficulties.

On the positive side, these pioneer 2G ethanol plants represented the first attempts to deploy complex, new technology, and provided a critically important opportunity for learning-by-doing. These plants were generally not expected to provide good investment returns on their own, but rather were motivated by the prospect of such returns once replicated at larger scale. Had oil prices and interest in cellulosic ethanol stayed at the levels seen a decade ago, the pioneer 2G ethanol plants would likely be regarded more favorably today. Still by any measure, the emergence of 2G ethanol technology into the marketplace has fallen far short of expectations. As one indication of this, cellulosic biofuel production in the USA fell short of the Renewable Fuel Standard Goal for 2016 by over 25-fold, and 98% of what was produced was biogas rather than liquid fuels as originally envisioned (Lynd, 2017).

Over the last decade, commercialization efforts aimed at 2G ethanol were based on the thermochemical pretreatment-fungal cellulase paradigm. These efforts involved both large and small companies with distinctly different assets and skill sets as summarized in Table 9.1.

Table 9.1 Features of companies involved in advanced biofuels over the last decade

Company size and capital assets	Large		Small (start-ups)
Industrial processing expertise	Great		Little or none
Microbial biotechnology expertise	Varies from a lot to a little, often enhanced by partnerships		Great
Feedstock supply chain expertise	Great	Little or none	Little or none
1G facilities available as hosts for 2G technology	Many	Few or none	None
Examples	Raizen, POET-DSM	Abengoa, GranBio, Beta Renewables, Dupont	Mascoma, Amyris, LS9, Gevo, Solazyme
Ongoing 2G activity	Substantial	Little or none	Little or none

With the benefit of hindsight, the following observations can be made:

- Although some companies and commercialization efforts proceeded assuming they could succeed largely independently, expertise across the entire supply chain – including industrial processing, microbial biotechnology, and feedstock supply – proved to be required.
- There are no examples of advanced biofuel start-ups with financial returns consistent with a successful venture investment. The expectation that start-ups would become industrial producers of biofuels was a contributing factor in some cases.
- Substantial technical and commercial challenges exist with respect to both the 'upstream' objective of accessing the carbohydrate in lignocellulose and the 'downstream' objective of producing new fuel molecules from carbohydrate. Doing both at the same time requires diverse expertise and innovation as well as increases risk. Consistent with this, companies have thus far focused on either upstream or downstream challenges but not both.
- The most successful 2G fuel producers, Raizen and POET-DSM, combined expertise in industrial processing and feedstock supply chains with substantial capital assets, partnerships with innovation-focused companies (particularly in biotech), and existing 1G processing facilities that could host smaller 'bolt-on' 2G plants.

Looking forward, it is likely that development and commercial deployment of new 2G paradigms will be most successful if they involve both producers and technology providers, often including large and small companies, each doing what they do well.

Knowledgeable people differ in their assessment of 2G biofuel production via the thermochemical pretreatment-fungal cellulase paradigm, with opinions varying from 'the technology does not work' to the view that the operational difficulties can be overcome and costs can be reduced based on experience and incremental improvement. In my view, some replication of technology based on this paradigm will likely occur as 'bolt-on' plants within larger 1G ethanol facilities processing sugarcane or maize, taking advantage of costs borne by the parent plant as well as policies that attach value to greenhouse gas emissions. For the more demanding case of plants for which 2G biofuel production is equal to or greater than 1G biofuel production and costs per GJ approach that of petroleum-derived fuels, I think we will likely need to look beyond the thermochemical pretreatment-fungal cellulase paradigm. While achieving this higher performance threshold may not be required for production facilities to be financially viable, doing so is highly desirable and arguably necessary for 2G biofuels to play their widely foreseen role in a low-carbon economy and to fully realize their considerable commercial potential.

A central point of this chapter is that there are alternatives to the conventional thermochemical pretreatment-fungal cellulase processing paradigm with potential for much lower costs and greater operational simplicity. One such alternative paradigm involves one-step consolidated bioprocessing (CBP) using engineered thermophilic bacteria in combination with milling during fermentation (cotreatment). This approach, denoted CBP/CT, builds on the following key observations reported in 2017:

- Cultures of the thermophilic, anaerobic, cellulolytic bacterium *Clostridium thermocellum* are severalfold more effective over a broad range of conditions than the fungal cellulase upon which the current industry is based (Lynd et al., 2017b);
- Enabled by recent biotechnological advances, ethanol yields and titers have been improved markedly in engineered thermophiles and are approaching levels required for commercial viability (Lynd et al., 2017b);
- Cotreatment enables *C. thermocellum* cultures to solubilize nearly all the cellulose and hemicellulose in switchgrass (Balch et al., 2017) as well as other lignocellulosic feedstocks (manuscripts in preparation) in the absence of thermochemical pretreatment.

Technoeconomic and lifecycle analysis of CBP/CT was recently reported by a team including researchers from the Lynd lab at Dartmouth, the National Renewable Energy Laboratory, Argonne National Laboratory, and the University of California at Riverside (Lynd et al., 2017a). Compared to ethanol production from corn stover via current technology based on the thermochemical pretreatment-fungal cellulase paradigm, an advanced scenario based on CBP/CT has 8-fold shorter payback period and economic feasibility at 10-fold smaller scale.

Configuring 2G 2.0 for success

About a decade ago, many voices were saying that 2G biofuel technology had reached the point that big things could be done fast with the technology in hand. Motivated by this perception, the lion's share of investment in the USA and Brazil was directed toward commercial facilities that implemented the thermochemical pretreatment-fungal cellulase paradigm, and technology development efforts were also directed toward this paradigm. In addition to far less than anticipated deployment, this course of action led to scant investment in alternative processing paradigms. There has been a tendency to assess the overall merit of 2G biofuels based on experience with the thermochemical pretreatment-fungal cellulase processing paradigm, at times suggesting a lack of awareness that other paradigms exist.

Over the same period, battery technology achieved impressive cost reductions and penetrated new markets (Lynd, 2017). These different outcomes can be traced to different strategies. Although battery development involved a succession of different chemistries – e.g. lead-acid, nickel-cadmium, lithium ion – 2G biofuels focused largely on one processing paradigm. Although it proceeded stepwise starting with the lowest barriers – e.g. from space applications to consumer electronics to hybrid cars and now looking toward grid storage – much of the effort in the 2G biofuel space embodied a 'go big or go home' approach.

To maximize the probability of developing a robust 2G biofuels industry at a scale large enough to meaningfully contribute to climate and other goals, needed technological initiatives include the following (Lynd, 2017):

1 Pursue commercial deployment in achievable, successively enabling steps, proceeding from where the industry is today. Examples of such deployment include colocation of 2G facilities at larger 1G plants, as well as '1.5G' strategies such as conversion of corn fiber, bagasse hemicellulose, or waste gasses.
2 Invest in innovation pursuant to alternative 2G processing paradigms offering potential for large cost savings.

The remainder of this article focuses on the latter, which has generally received less attention to date.

Candidate paradigms should be supported by new scientific findings and understanding and be able to demonstrate potential for large cost reductions compared to the status quo. CBP/CT meets these criteria as described above. It is likely that other technologies such as thermochemical and biological can meet as well. Government support will likely be required in light of the need to look 'upstream' within the innovation pipeline, together with the often non-monetized social and environmental benefits that motivate 2G biofuels. Industry participation, validation, and cofunding are highly desirable. Calls for proposals aimed at supporting new processing paradigms should be configured in an open way that invites new ideas while clearly specifying evaluation criteria. Consistent with research in 'Pasteur's Quadrant' as described by Stokes (1997), innovation-focused research should be inspired and justified in terms of applied impact while recognizing that innovation is often based on new fundamental understanding. Pilot-scale demonstration, potentially in multiple stages, will be necessary in order to achieve the ultimate goal of industrial deployment. Yet piloting is costly and should be undertaken only after economically attractive performance parameters, supported by technoeconomic analysis, have been achieved at small scale. Progressively greater industry involvement is desired, and should likely be required, as technologies move through pilot validation. There is a distinct possibility that the barrier posed by precommercial validation, often referred to as the 'valley of death', may be a substantially less formidable challenge for less-complex new technologies than for existing technologies.

Key challenges impeding commercial deployment of 2G biofuel production to date include high capital cost, high process complexity, and the need to mobilize feedstock supply over large catchment areas. These factors have proved to be substantial obstacles in the context of biofuel deployment in the EU, Brazil and the USA. They represent even greater challenges in less developed countries where the human development needs are greatest and a substantial fraction of the world's biomass production potential resides (Souza et al., 2015). More rapid deployment, faster learning cycles, easier integration into agricultural landscapes and material flows, and enhanced applicability in less-developed countries can be anticipated as 2G technologies become less capital-intensive, less complex, and are economically feasible on smaller scale. These features are anticipated for CBP/CT as described herein, and may well be embodied by other innovative low-cost approaches.

As pointed out elsewhere (Brown and Le Feuvre, 2017; Lynd, 2017), the window for 2G biofuels to make needed contributions to climate change mitigation by mid-century will not remain open for much longer unless deployment accelerates. Continuation of the strategies of the last decade to achieve this outcome seems risky at best. A renewed and refocused '2G 2.0' drawing from the lessons of the last decade is our best bet.

Acknowledgments

I am grateful for informative discussions with a great many colleagues in the field, and I thank Andrew Allee for editorial suggestions.

References

A better life with a healthy planet: Pathways to net-zero emissions. Shell Global. 2016.

Balch, M.L., E.K. Holwerda, M.F. Davis, R.W. Sykes, R.M. Happs, R. Kumar, C.E. Wyman, L.R. Lynd. 2017. Lignocellulose fermentation and residual solids characterization for senescent switchgrass fermentation by *Clostridium thermocellum* in the presence and absence of continuous in-situ ball-milling. *Energy Environ. Sci.* 10:1252–1261.

Brown, A., P. Le Feuvre. 2017. *Sustainable Bioenergy Roadmap*. International Energy Agency, Paris.

Dale, B.E., J.E. Anderson, R.C. Brown, S. Csonka, V.H. Dale, G. Herwick, R.D. Jackson, N. Jordan, S. Kaffka, K.L. Kline, L.R. Lynd, C. Malmstrom, R.G. Ong, T.L. Richard, C. Taylor, M.Q. Wang. 2014. Take a closer look: Biofuels can support environmental, economic and social goals. *Environ. Sci. Technol.* 48:7200–7203.

Fulton, L.M., L.R. Lynd, A. Körner, N. Greene, L. Tonachel. 2015. The need for biofuels as part of a low carbon energy future. *Biofuels Bioprod. Bioref.* doi: 10.1002/bbb.1559.

Jaiswal, D., A.P. de Souza, S. Larsen, D.S. LeBauer, F.E. Miguez, G. Sparovek, G. Bollero, M.S. Buckeridge, S.P. Long. 2017. Brazilian sugarcane ethanol as an expandable green alternative to crude oil use. *Nat. Clim. Change* 7:788–792.

Jordan, N., G. Boody, W. Broussard, J. D. Glover, D. Keeney, B. H. McCown, G. McIsaac, M. Muller, H. Murray, J. Neal, C. Pansing, R. E. Turner, K. Warner, D. Wyse. 2007. Sustainable development of the agricultural bioeconomy. *Science.* 316:1570–1571.

Junqueira, T.L., M.F. Chagas, V.L.R. Gouvela, M.C.A.F. Rezende, M.D.B Wtanabe, DC.D.F. Jesus, O. Cavalett, A.Y. Milanez, A. Bonomi. 2017. Techno-economic analysis and climate change impacts of sugar cane biorefineries considering different time horizons. *Biotechnol. Biofuels.* 10:50.

Leal, R.L.V., A.S. Walter, J.E.A. Seabra. 2013. Sugar cane as an energy source. *Biomass Conv. Bioref.* 3:17–26.

Lynd, L.R. 2017. The grand challenge of cellulosic biofuels. *Nature Biotech.* 35:912–915.

Lynd, L.R., X. Liang, M.J. Biddy, A. Allee, Hao Cai, T. Foust, M.E. Himmel, M.S. Laser, M. Wang, C.E. Wyman. 2017a. Cellulosic ethanol: Status and innovation. *Curr. Opin. Biotechnol.* 45:201–211.

Lynd, L.R., A.M. Guss, M.E. Himmel, D. Beri, C. Herring, E.K. Holwerda, S.J.L. Murphy, D.G. Olson, J. Paye, T. Rydzak, X. Shao, L. Tian, R. Worthen. 2017b. Advances in consolidated bioprocessing using *Clostridium thermocellum* and *Thermoanaerobacter saccharolyitcum*. In: C. Wittman and J.C. Liao (eds.) Industrial Biotechnology: Microorganisms (pp. 365–394). Wiley-VCH Verlag & Co., KGaA.

Lynd, L.R., M. Sow, A.F. Chimphango, L.A. Cortez, C.H. Brito Cruz, M. Elmissiry, M. Laser, I.A. Mayaki, M.A. Moraes, L.A. Nogueira, G.M. Wolfaardt, J. Woods, W.H. van Zyl. 2015. Bioenergy and African transformation. *Biotechnol. Biofuels.* 8:18–35.

Moraes, M.A.F.D., F.C.R. Oliveira, R.A. Diaz-Chavez, R.A. 2015. Socioeconomic impacts of Brazilian sugarcane industry. *Environ. Dev.* 16:31–43.

Moraes, M.A.F.D., M.R.P. Bacchi, C.E. Caldarelli. 2016. Accelerated growth of the sugarcane, sugar, and ethanol sectors in Brazil (2000–2008): Effects on municipal gross domestic product per capita in the south-central region. *Biomass Bioenergy.* 91:116–125.

Paye, J., A. Guseva, S.K. Hammer, E. Gjersing, M.F. Davis, B.H. Davison, J. Olstad, B.S. Donohoe, T.Y. Nguyen, C.E. Wyman, S. Pattathil, M.G. Hahn, L.R. Lynd. 2016. Biological lignocellulose solubilization: Comparative evaluation of biocatalysts and enhancement via cotreatment. *Biotechnol. Biofuels.* 9:1–13.

Prosperous living for the world in 2050: Insights from the Global Calculator. 2015. Globalcalculator.org.

Ruben, E.S., J.E. Davison, H.J. Herzog. 2015. The cost of CO2 capture and storage. *Int. J. Greenhouse Gas Control.* 40:378–400.

Slade, R., A. Bauen, R. Gross. 2014. Global bioenergy resources. *Nat. Clim. Change.* 4:99–105.

Smith, P. et al. 2016. Biophysical and economic limits to negative CO2 emissions. *Nat. Clim. Change.* 6:42–50.

Smith, P. et al. 2014. in *Climate Change 2014: Mitigation of Climate Change. Contribution of Working Group III to the Fifth Assessment Report of the Intergovernmental Panel on Climate Change* (eds. Edenhofer, O. *et al.*). Cambridge University Press, Cambridge, United Kingdom and New York, NY, USA.

Souza, G.M., R. Victoria, C. Joly, and L. Verdade (eds.) 2015. *Bioenergy and Sustainability: Bridging the Gaps.* SCOPE, Paris.

Stokes, D.E. 1997. *Pasteur's Quadrant: Basic Science and Technological Innovation.* Brookings.

Sugarcane agroecological zoning. 2010. http://www.unica.com.br/zoning/.

Werling, B.P., T.L. Dickson, R. Isaacs, H. Gaines, C. Gratton, K.L. Gross, H. Liere, C.M. Malstrom, T. D. Meehan, L. Ruan, B.A. Robertson, G.P. Robertson, T.M. Schmidt, A.C. Schrotenboer, T.K. Teal, J.K. Wilson, D.A. Landis. 2014. Perennial grasslands enhance diversity and multiple ecosystem services in bioenergy landscapes. *Proc. Nat. Acad. Sci.* 111:1652–1657.

Woods, J., L.R. Lynd, M. Laser, M. Batistella, D.D.C. Victoria, K. Kline, A. Faaij. 2015. Land and bioenergy. In: Souza, G.M., R. Victoria, C. Joly, L. Verdade (eds.) *Bioenergy and Sustainability: Bridging the Gaps* (pp. 258–300). SCOPE, Paris.

10 Medium- and long-term prospects for bioenergy trade

Sergio C. Trindade and Douglas Newman

Over a long period of time, energy systems dynamics studies show that a logistical substitution model correlates well the market penetration data of primary energy sources. A four-decades-old article by Marchetti and Nakicenovic depicts the energy market penetration over time by wood, coal, oil, natural gas, and nuclear sources. The data for all but the nuclear source, incipient at the time of the article publication, correlates well with the logistical substitution model. The plot suggests a market penetration trajectory for these primary sources and adds a speculative plot for nuclear, solar, and fusion energies, through the year 2050. A plot with actual data through 1998 and stretching a scenario through to 2100 also suggests that natural gas is likely to become the dominant primary energy source in the first half of the 21st century. Beyond 1979, the scenario for nuclear and fusion has not materialized by early 2018, the time of this writing. During the period through 1998, coal, oil, and natural gas were traded actively internationally and within national borders, a situation that prevails today. The cover page of *The Economist* issue of October 25–31, 2003, suggests the end of the Oil Age beyond "peak oil" at some point in the future, in agreement with the scenarios presented in the previous two references.

Bioenergy drivers and trade dimensions

In the recent past and immediate future, bioenergy trade has been and will be marginal to the total liquid fuels market picture. Globally, bioethanol has penetrated 5%–6% of the gasoline market and biodiesel about 1.5%–2% of the diesel market, based on domestic production and imports. Bioenergy trade is basically dominated by bioethanol and biodiesel, although bagasse-based electricity fed to the national grid of Brazil may be part of the power exchanges between interties with Argentina, Uruguay and Paraguay. Thus, bioenergy so far has been basically helping meet, to a very limited extent, the global mobility demand of the world, with trade playing a minor role.

Over time, production and use of bioenergy have been primed by concerns over security and the environment, farm income and rural development, and more recently over the prospect of responding to climate change. In Brazil, the fuel ethanol market got a boost from the creation of the "Proálcool" program in 1975. In the United States, the biofuels market benefitted from the establishment of the Renewable Fuel Standard (RFS). The RFS, which mandates the use of renewable fuels, was authorized under the Energy Policy Act of 2005 and expanded under the Energy Independence and Security Act of 2007. Historically, ethanol has long been employed as an octane booster, after the phasing out of tetraethyl lead and later as a source of oxygen in reformulated gasoline, especially after the banning of methyl *tert*-butyl ether (MTBE) in the United States. Of course, blended ethanol volumes have helped to extend gasoline supplies.

Global biofuels markets have grown substantially during the past decade, and trade has become an important element of these markets. This growth largely has been in response to policies in major producing and consuming countries. There are three main drivers behind these policies – farm income, energy security, and the environment. These drivers largely are the same for each major player in the global biofuels market. However, each player has different priorities based on economic history and cultural sensibilities. In the United States, the leading global biofuels market, the recent development and growth of biofuels has been driven mainly by farm income. The evolution of biofuels in Brazil, the second leading global market, has been driven mainly by energy security concerns. And the European Union (EU) market is largely driven by environmental concerns. More recently, other global markets, such as China and India, have been developing biofuels policies to incentivize demand in response to these same drivers. The priorities of these drivers have had a great impact on the nature of policies in each market, and these policies have had direct and indirect effects on biofuels trade flows.

The success of biofuels has not been without controversy. The effect of using food crops for biofuels feedstocks on food prices has been the subject of heated debate. Also, there is concern that forests and grasslands are being converted to agricultural use to compensate for the diversion of food crops to biofuels, with resulting effects on greenhouse gas emissions. And, government support for biofuels is under increasing scrutiny, as fiscal difficulties have led to reexamination of budgetary priorities in the United States and other countries.

The global biofuels market expanded rapidly during the past decade. Driven by policies focused on consumption mandates, vehicle engine technology, and trade protection, global output and consumption was more than tripled during 2003–2013 and more than doubled during 2006–2013. Trade in biofuels also responded to these drivers, although often employing esoteric policy elements such as duty drawbacks and customs rulings to take advantage of arbitrage opportunities. The tremendous growth in global biofuels markets has slowed in recent years, however, in response to a combination of market conditions, policy changes, and technological constraints.

Although the United States, Brazil, and the EU represent the greatest share of the global biofuels market, there has been substantial growth in other countries in recent years as a result of an expansion of biofuels policies. Most of this growth is occurring in Asia, albeit from a small base. These markets hold the potential for substantial future growth, as rising incomes contribute to an increase in the number of automobiles and, thus, gasoline and diesel consumption. In addition, increasing concerns regarding climate change, air pollution, and energy security in the region likely will contribute to the demand for biofuels.

Diesel fuel is the largest crude oil derivative used in the world in response to the popularity of the relatively efficient diesel engine. Biodiesel (methyl or ethyl transesterified esters of vegetable oils and animal fat) is the largest volume alternative fuel able to displace crude-based diesel fuel. Gasoline comprises the bulk of the remaining transport fuel, with ethanol the largest volume alternative fuel to gasoline. In absolute terms, ethanol is the largest biofuel volume used in transportation.

The crucial role of government policies in biofuels market growth

The inception in the mid-1970s and early 1980s, and the rapid growth of the global biofuels market since the mid-2000s resulted from government policies. The United States and Brazil are the largest producers, consumers, and exporters of fuel ethanol. The two countries overwhelmingly dominate the global fuel ethanol market, but the EU is also an important player. The industry is consolidating both in the United States and in Brazil. The global biodiesel market is more fragmented, but the United States and Brazil also play an important role in this market together with European countries, Argentina, Colombia, and Malaysia and Indonesia. There remains potential for substantial growth for biofuels in other markets, mainly in Asia and Africa, as these markets begin to establish and implement biofuels policies. This growth almost certainly will lead to a short-to-medium-term increase in trade.

Global ethanol production over 2009–2017 (estimate) ranged from 70 to 100 million m^3/a according to feedstock, with starch-based feedstock (primarily corn) dominating the picture. Since sugarcane can yield both sugar and ethanol, during the period considered, sugarcane-based ethanol volumes benefitted from weakness in sugar prices.

The world ethanol trade (exports plus imports) to contract during 2009–2018 (estimate) ranged from 9 to 12 million m^3/a, where the Americas (United States and Brazil mainly) and the EU dominate.

Figure 10.1 illustrates the world output of fatty acid methyl esters (FAMEs) or biodiesel during 2008–2017 (estimate), ranging from 13 to 28 million metric tons, where the Americas and the EU are dominant, with an important role for Asia.

Figure 10.2 indicates the largest producers of biodiesel in the Americas during 2008–2017 (estimate), in order of decreasing importance from the United States to Brazil, to Argentina, to Colombia, and to Canada.

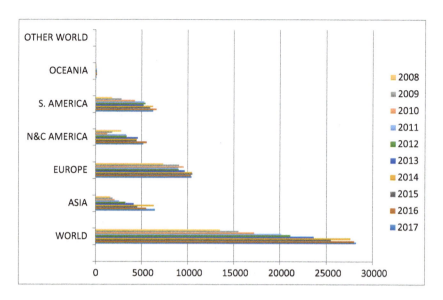

Figure 10.1 FAME world output – 1,000 tons.
Source: Personal communication from Christoph Berg (October 2017).

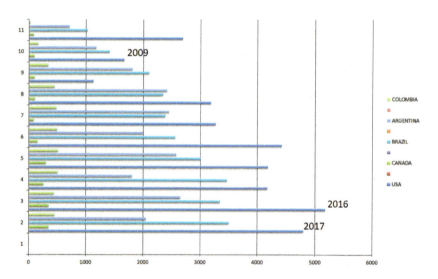

Figure 10.2 FAME output – 1,000 m. Tons – 2008–2017 (United States, Brazil, Argentina, Colombia, Canada).
Source: Personal communication from Christoph Berg (October 2017).

As in the ethanol trading market, biodiesel trading has suffered from protectionism by local producers, especially in the EU and the United States.

Biofuels trade barriers and opportunities

Global fuel ethanol trade expanded greatly in concert with consumption mandates in the mid-2000s. Prior to this period, trade consisted mainly of relatively small volumes of anhydrous ethanol shipped from Brazil and Canada or hydrous ethanol from Europe and Brazil channeled through Caribbean Basin countries benefiting from a unique U.S. dehydration origin quota, for seasonal use as an octane enhancer or oxygenate in U.S. border and coastal metropolitan areas. Global trade accelerated with the imposition of consumption mandates in the United States and the EU in the mid-2000s. Trade patterns shifted in the late 2000s as a result of weather- and financial-related supply and demand imbalances, expanding biofuels policies, carbon requirements, and customs issues.

It should be noted that there are significant data limitations regarding fuel ethanol trade. These limitations have been addressed by the creation of specific import and export categories for fuel ethanol and by increasingly well-informed estimates based on these categories and other information related to the uses of ethanol in fuel blends and products such as ethyl *tert*-butyl ether (ETBE). However, other issues, mainly related to varying customs classification practices, likely will continue to contribute to deficiencies in the comparability and accuracy of fuel ethanol trade data. Most estimates of fuel ethanol trade data are proprietary. Fuel ethanol is believed to comprise the bulk of global ethanol trade.

The United States was the leading global importer of ethanol for most years during 2003–2013, before declining in subsequent years as domestic production capacity reached saturation. Brazil became the second leading global importer of fuel ethanol in 2012, as production was adversely affected in 2011 and 2012 by weather and financial conditions and domestic demand continued a long-term rise. Brazil became the leading global ethanol importer in 2017, owing mainly to a substantial rise in domestic demand and conditions in the global sugar market affecting the production mix of sugar and ethanol.

The EU, the second leading importer for most of the period under review, slipped to third in 2012, largely the result of increasing domestic production and trade actions that adversely affected imports. EU imports rebounded slightly in 2013. EU imports remained relatively low in subsequent years. Imports by Canada generally increased during the period, as rising demand outpaced the production in response to consumption mandates. Imports by Central American countries during the period largely reflected the use of Brazilian hydrous ethanol as a feedstock for exports of anhydrous ethanol to the United States under a now-expired origin quota.

The United States, historically a minor exporter of ethanol, became the leading global exporter of ethanol in 2010. Specific data are not available for global exports of fuel ethanol. Such exports peaked in 2011 before falling in 2012 because of U.S. and EU policy changes and trade actions. Exports rebounded in 2013, as a result of diversification into a number of nontraditional markets. U.S. exports continued rising and reached record levels in 2017. Exports from Brazil, traditionally the global leader, peaked in 2008 in concert with an expansion in the production before falling through 2011 because of rising domestic demand and supply issues caused mainly by weather and financial constraints. Exports rebounded in 2012 and 2013, as production and global demand recovered. Exports then declined annually in subsequent years owing mainly to decreased demand for imports by the United States as well as generally increasing domestic demand. Exports from Central American countries followed the Brazilian trend during the period, as they were largely produced from Brazilian hydrous ethanol. After the expiration of the U.S. dehydration quota at the end of 2011, Central American exports consisted of domestic-origin product rather than ethanol dehydrated from Brazilian-origin material.

Overall, bioenergy trade has not established a firm structural foundation for a host of reasons. Trade largely has been opportunistic and part of producers' balancing acts, as every producing country wants to protect the agriculturally related domestic industry. The uncertainty introduced by weather variability and ties between food and energy markets contribute to the opportunistic nature of trade.

Government policies, such as the aforementioned RFS and the California Low Carbon Fuel Standard in the United States, the Renewable Energy Directive and the Fuel Quality Directive in the EU, and the Prorenova and RenovaBio initiatives in Brazil, have created an increased, but still modest, demand for biofuels trade during the past decade. However, this trade has been limited by a number of factors. Most countries have yet to establish stable biofuels policies, although this is changing. In addition, an extensive market for carbon credits associated with bioenergy has yet to develop firm roots. Trade regimes, including tariffs and quotas, unfair trade remedies, and duty drawbacks lend uncertainty to the development of trade. Differences in product standards pose additional costs and barriers. And, imprecise trade data obfuscate the picture.

Brazil's tariff rate responds to the country's occasional need to import ethanol. Likewise, China's tariff rates have moved up and down according to changing policies promoting market penetration of fuel ethanol and protection to domestic producers. China's ethanol import tariff used to stand at 30% for all types of ethanol. It was lowered to 5% for industrial ethanol, assumed to include fuel ethanol. This move took place after a day-long seminar on October 21, 2008, hosted by the National Information Center of the NRDC – National Reform and Development Commission, in which one of authors participated, as a consultant to BRENCO (Brazilian Renewable Energy Corporation).

The EU has imposed antidumping measures and countervailing duties on imports of biodiesel and ethanol from the United States. A 9.5% antidumping duty on ethanol imported from the United States was established in February 2013 and revoked on June 9, 2016, by the EU General Court, a measure that was appealed by the European Commission on August 26, 2016. The duty will be maintained during the appeal process on all ethanol from the United States.

Imposition of measures on biodiesel started in 2009. On September 14, 2015, the European Commission imposed a definitive antidumping duty on biodiesel originating from the United States. Antidumping measures were also in force against biodiesel from Argentina and Indonesia in the EU.

In August 2017, the US Department of Commerce issued a preliminary countervailing duty determination against biodiesel imported from Argentina and Indonesia.

Other barriers to biofuels trade include differing technical standards and blending decisions by relevant countries and groupings. In addition, most countries do not differentiate fuel ethanol in trade classification schemes. When vegetable oils are introduced in hydrocrackers in petroleum refineries and become incorporated into the propylene and diesel streams, they become hydrocarbons and lose their biodiesel identification.

A Memorandum of Understanding between Brazil and the United States on biofuels was signed on March 7, 2007. This followed a recommendation by one of the authors to the then president of Brazil and his full cabinet during a presentation in Brasilia, on July 26, 2001. This was an effort to switch from confrontation to cooperation between the two leading countries on bioenergy trade and the development of markets in select third countries. It needs to be revisited. The African and Asian markets offer the greatest prospects for growth but mostly are in their infancy. A notable exception is China, which appears to be planning for significantly expanded consumption of biofuels that would definitely require imports to bolster inadequate domestic supplies until domestic cellulosic ethanol becomes available to help meet China's expanding demand for transport fuels.

The U.S. government recently began directly promoting ethanol exports through the U.S. Department of Agriculture's Market Access Program (MAP) and the U.S. Department of Commerce's Foreign Commercial Service. In addition, biofuels are a key beneficiary of the U.S. Renewable Energy and Energy Efficiency Export Initiative, which provides multiagency support to assist U.S. exporters. Support includes funding for trade shows and trade missions. U.S. government promotion of ethanol exports largely was in response to supply overhangs, as production capacity expanded according to volume mandates under the RFS, and U.S. gasoline consumption did not increase according to projections, thus limiting the demand for ethanol blending. The 10% ethanol-gasoline blend wall in the United States opened up room for export expansion by the United States.

The statement, "The increased use of ethanol globally could provide strong and diverse export market opportunities for U.S. ethanol and ethanol by-products," such as distillers' dried grains with solubles (DDGS), encapsulates the gist of a 2017 USDA's Economic Research Service (ERS) report. In spite of the plateauing of the Brazilian and U.S. ethanol production and the 10% ethanol-in-gasoline blend wall in the United States, many countries throughout the world have introduced fuel ethanol mandates, which may expand export opportunities for the larger producers with a potential for surplus production, from traditional as well as cellulosic feedstocks.

Long-term markets and trade prospects for bioenergy

The orientation of trade varies by market and can change over time. The EU historically has relied on a relatively large proportion of imports for its consumption, as production has not kept pace with rising demand pulled by policy mandates (Figure 10.3). Brazil historically has exported a substantial share of its surplus production, which expanded rapidly in the mid- to late-2000s (Figures 10.3 and 10.4). In the United States, trade historically has played a relatively minor role, with imports slightly more prevalent than exports (Figure 10.3). However, shifts in trade orientation have occurred in recent years. Exports have become more prominent in the United States, imports more important in Brazil, and the EU reliance on imports has declined. Factors leading to this shift include recent cost and price advantages gained by the United States vis-à-vis Brazil in the EU market; carbon and greenhouse

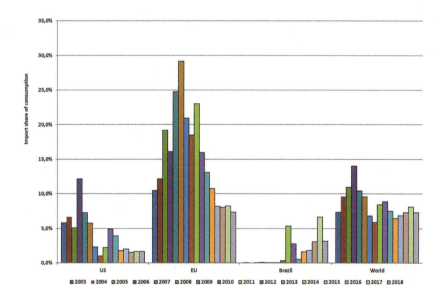

Figure 10.3 Ethanol import penetration typically is the greatest in the EU market.
Note: Includes all ethanol in HS heading 2207. Quantities include denaturants. Data for 2018 are forecast.
Source: Calculated based on data from LMC International.

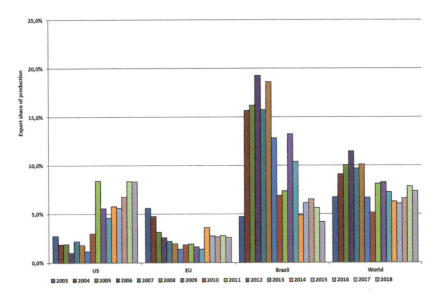

Figure 10.4 Ethanol export production share rises in the United States and falls in Brazil.
Note: Includes all ethanol in HS heading 2207. Quantities include denaturants. Data for 2018 are forecast.
Source: Calculated based on data from LMC International.

gas policy requirements and transportation economics that contributed to two-way trade between the United States and Brazil; and rising production, customs classification changes, and an antidumping action in the EU that led to higher import duties.

The long-term view for bioenergy is obviously more complex as technology innovation in many fields will shape the way people live and work and the demand for mobility. Recent trends point out the prospect of a medium- to long-term transition towards electricity-based mobility and to renewable wind and solar power generation. Major global original equipment manufacturers (OEMs) have been announcing plans to discontinue the manufacture of internal combustion engines in the future. There are also indications that the need to personally drive may abate in the future, influenced by the dissemination of working remotely patterns and by the widespread availability of on call driving services, and, in the longer term future, of driverless vehicles.

A transition towards sustainable electrical mobility that joins the value of biofuels and electrical power trains can be achieved today by using hybrid power vehicles, where the internal combustion engine is fueled by either ethanol or biodiesel, and the electric motor is charged with renewable electricity. This is possible today in the United States, Brazil, and elsewhere with a hybrid vehicle burning ethanol or biodiesel and recharging the electric motor with hydro, solar, or wind electricity.

Bioenergy trade today is based on first-generation biofuels, which are tied up to agricultural markets. Disengaging biofuels from a direct connection to agriculture would allow the prospect of significant market expansion without the hindrance of agricultural commodity restraints and associated uncertainty and volatility of supplies and prices. This development depends on the increased global demand for biofuels and the market penetration of large-scale lower cost second-generation biofuels, whose feedstock would be cellulosic in nature. Current EU policy proposals lead in this direction. Stable policies and well-established futures and options global market would help a great deal to achieve and enduring structured global biofuels market.

The relevance of climate change to biofuels development for sustainable trade

With the increasing political relevance of climate change, the competitiveness of sustainable biofuels may increase to the extent they add to sustainability. The demand for sustainability would likely increase and place a value in biofuels trade. In fact, sustainability itself could, in principle, be traded based on sustainability indices. Financial instruments that support sustainability are gradually becoming available, such as the sustainability bonds of the HSBC bank. With respect to biofuels trading, if they are produced and handled sustainably, it seems that sustainability can be promoted via international trade.

If a significant scale of biofuels can be achieved in the long-term future, supplementing the supply of first-generation biofuels, then perhaps a structural base for markets could be established to move from opportunistic trade to a more stable, structured and sustainable trade. Biofuels then would become truly energy commodities.

References

Beckman, J. and G. Nigatu. *Global Ethanol Mandates: Opportunities for U.S. Exports of Ethanol and DDGS*. BIO-05 USDA, Economic Research Service, 2017. October. www.ers.usda.gov/webdocs/publications/85450/bio-05.pdf?v=43025 (accessed on 17 February 2018).

http://biodiesel.org/docs/default-source/policy--federal/eu-notice-reinstating-anti-dumping-duties---2015.pdf?sfvrsn=2 (accessed on 17 February 2018).

www.biofuelsdigest.com/bdigest/2016/12/28/biofuels-mandates-around-the-world-2017/ (accessed on 17 February 2018).

www.biofuelsdigest.com/bdigest/2016/08/29/eu-appeals-decision-invalidating-duties-on-us-ethanol-imports/ (accessed on 17 February 2018).

www.biofuelsdigest.com/bdigest/2017/08/23/us-slaps-argentine-indonesian-biodiesel-producers-with-huge-anti-dumping-penalties/ (accessed on 17 February 2018).

https://enforcement.trade.gov/trcs/foreignadcvd/eu.html (accessed on 17 February 2018).

Larrea, S. (Sergio Trindade, Contributor). *Fiscal and Economic Incentives for Sustainable Biofuels Development: Experiences in Brazil, the United States and the European Union.* IDB Technical Note 555, August, 70p, 2013.

Macke, Y. *China - Biofuels Annual Growing Interest for Ethanol Brightens Prospects.* Global Information Agricultural Network – GAIN Report Number: CH17048, 20 October, 13p, 2017.

Marchetti, C. and N. Nakicenovic. *The Dynamics of Energy Systems and the Logistical Substitution Model.* RR-79-13, IIASA, Laxenburg, Austria, 85p, 1979.

Market Access Program (MAP), U.S. Department of Agriculture, www.fas.usda.gov/programs/market-access-program-map (accessed on 6 December 2017.) Funding for ethanol export promotion is provided to the U.S. Grains Council.

Memorandum of Understanding Between the United States and Brazil to Advance Cooperation on Biofuels, U.S. Department of State, https://2001-2009.state.gov/r/pa/prs/ps/2007/mar/81607.htm (accessed on 6 December 2017).

Nakicenovic, N., Gruübler, A., and Mcdonald A., eds. *Global Energy Perspectives, 2018.* Pure.iiasa.ac.at (accessed on 17 February 2018).

Renewable Energy and Energy Efficiency Export Initiative (RE4I), U.S. Department of Commerce, https://2016.export.gov/reee/eg_main_023036.asp (accessed on 6 December 2017).

SE^{2}T International, Ltd. *Biofuels for the Future: Strategy Briefing and Technical Solutions.* Seminar, Beijing, October 21, 2008.

Trindade, S. C. *Natural Gas and Development: The Policy Issues for Developing Countries.* In Vergara, W. et al. *Natural Gas: Its Role and Potential in Economic Development.* Westview Press, Boulder, Colorado, p. 105, 1990.

Trindade, S. C. *Mercado Internacional Sustentável De Etanol Combustível: Penetração De Etanol De Origem Brasileira E Derivados E Formulações Nos Mercados Norte-Americano E Europeu.* Report to UNICA, Abril 2001.

Part II
Why bioenergy?

11 Why promote sugarcane bioenergy production and use in Latin America, the Caribbean and Southern Africa?

Luiz A. Horta Nogueira

Introduction

Different drivers, associated with strategic, economic, social and environmental positive impacts, are fostering bioenergy development in both industrialized and developing countries. In the particular context of Latin America and Southern Africa, some of these drivers are even more relevant, making relatively straightforward to identify the reasons behind the interest to foster the development of sustainable sugarcane bioenergy schemes. Actually, these regions can be considered as especially appropriate for bioenergy production and use, as presented in the next five chapters, addressed to a very first basic question: why deploy modern sugarcane bioenergy systems in those countries?

Concisely, bioenergy sugarcane is rational and makes sense. Sugarcane is one of the most efficient solar energy converters in chemical energy, able to be used as feedstock for producing a set of energy vectors and diversified list of useful bioproducts. Sugarcane is a tropical crop largely cultivated in wet tropical countries, its culture and processing industry are already well known and practiced abroad, as well as there is enough suitable land available for expanding sugarcane production, in area large enough to increment bioenergy supply without harming other relevant uses such as food and feed production (Souza et al., 2015). In most countries in these regions, there is a large dependence of imported fuels, and a pressing demand of clean and accessible energy to attend in these regions, the situation that could be properly faced by sugarcane energy products. Regarding the environment, the lower greenhouse gas (GHG) emissions associated with bioenergy from sugarcane can contribute decisively to mitigate climate change and improve significantly the air condition in cities (BEST, 2008). Finally, it has been verified that this agro-industry can promote interesting improvements in social conditions, generating stable jobs and improving the food security (Lynd and Woods, 2011). Anyway, as all other energy technologies, sugarcane bioenergy is not a panacea and has limitations and constraints that should be taken into account. Thus, it is necessary to assess properly the effective potential for developing this energy path in those regions, as explored in the next five chapters of this section, briefly introduced in the following paragraphs, highlighting the main aspects focused in each chapter.

Why to produce and use bioenergy in Latin America, the Caribbean and Southern Africa

Chapter 12 presents an overall appraisal of sugarcane bioenergy perspective at the national level for several countries of Latin America and Southern Africa, evaluating two supply scenarios. The first one considers that ethanol is produced by processing molasses currently available in existing sugarcane industry, while another scenario is assumed the expansion of sugarcane culture in a small fraction of national land currently available for agriculture expansion. On the demand side, the automotive use of ethanol blends with gasoline is assumed, complemented by the consumption of ethanol as a clean fuel for cooking in the case of African countries. The impact on bioelectricity generation and mitigation of GHG emissions and a preliminary assessment of investment requirements for each country are estimated. The main outcome of these scenarios is that in most countries there is an interesting potential to produce sustainable bioenergy, in many of them with surpluses to trade.

Chapter 13 presents the synergic development of bioenergy production and the increase of food production and food security. Actually, the growth of the agricultural production in Brazil was essentially determined by productivity gains and a better production environment in the rural sector, associated with access to agronomic practices, availability of services and equipment and adoption of modern technology partially derived from or influenced by the sugarcane agro-industry. As remarkable indicators, during the last 35 years, the production of cereals and oil crops in Brazil increased at an annual growth rate of 3.6%, while the cultivated area expanded on average just 0.7% per year; while in cattle ranching, the pasture area was reduced by 8%, but the herd increased to 155%. Definitively, farm production depends on much more than the area cultivated (Nogueira and Capaz, 2013).

Regarding the impact of sugarcane bioenergy on the supply of energy, Chapter 14 explores two relevant dimension: (i) the national level, under the concept of energy security, and (ii) at household level, under the concept of energy poverty. In both directions, sugarcane bioenergy had played in some countries an important role, which can be replicated in other countries, reducing the dependence on imported energy and reinforcing the national supply of liquid fuels for transportation and electricity, as well as improving life conditions, increasing the electricity services coverage and, particularly in the African context, opening conditions for replacing inefficient solid biomass stoves by ethanol stoves, well accepted in pilot projects focused in low-income families.

Dealing with the socioeconomic impacts, Chapter 15 focuses more closely the situation in an African country with a bioenergy potential as large as its need to improve general living conditions and foster economic progress. Employing the national Social Account Matrix (SAM) to obtain the Input-Output matrix, it was possible to evaluate the impact of sugarcane bioenergy deployment in a broad sense, including the effects on GDP, HDI and GINI, access to electricity, jobs creation and others.

Closing this section on the main reasons and basic conditions to promote modern bioenergy in Latin America and Southern Africa, and recognizing the inherent complexity and diversity of aspects to be evaluated, introduced the Sybil-LACAf approach, a "toolbox" for integrated analysis of bioenergy systems, highlighting the relevance to include and analyze properly all data and information available as much as possible and, equally important, to consider the need of consult with stakeholders and involved people.

These chapters represent an initial approach to this question, proportioning some basic figures and facts to be detailed and closely examined in their more specific topics, such as land availability and alternative production schemes, objective of other chapters.

The ethical basis of sustainable bioenergy

To explore systematically the need of and opportunity for putting forward the programs for sustainable sugarcane bioenergy production and use in Latin America, the Caribbean and Southern Africa, it is worth to examine this question from a comprehensive point of view. With this regard, a relevant group of scholars and experts interested in judiciously recognizing and promoting sustainability in bioenergy, provided an interesting and useful framework for "evaluation on the basis of which more ethical production of current biofuels and the emerging biofuels production systems can be established" (Nuffield Council on Bioethics, 2011), which concise six principles and respective evaluation in the context of sugarcane bioenergy in those regions are summarized as follows:

Principle 1: Biofuels development should not be at the expense of people's essential rights (including access to sufficient food and water, health rights, work rights and land entitlements). This very fundamental principle should be generalized and assumed as essential for all public policy. Any energy project, including the innovative and conventional technologies, should be accepted if not following strictly this orientation. There are good examples in developed and developing countries showing that it is effectively possible to harmonize sugarcane bioenergy promotion with social and environmental requirements, reinforcing the convergence of food security, bioenergy sustainability and resource management (Kline et al., 2016). The government role is decisive in this regard, setting properly the legislation, regulation and enforcement system to take this principle into account and follow it up.

Principle 2: *Biofuels should be environmentally sustainable.* Like the principle above, this principle is clearly an obligation, a "sine qua non" condition for any energy project, not only for bioenergy. Probably as an outcome of the intense discussion on sustainability of bioenergy during the last two decades, a solid and reliable set of certification systems has been developed (Endres, 2011) to independently and consistently access and monitor the environmental sustainability indicators at production unit level, such as the Roundtable for Sustainable Biomaterials (RSB), widely accepted.

Principle 3: *Biofuels should contribute to a net reduction of total greenhouse gas emissions and not exacerbate global climate change.* In several countries, climate change mitigation and fossil carbon emissions reduction have been primary drivers for biofuels development, exactly because these energy vectors are able to reduce notably the GHG emissions, compared with the conventional fuels, as indicated by detailed life cycle assessments. It is noteworthy that, among all kinds of renewable energy vectors, sugarcane ethanol is one of the most efficient in terms of carbon emission reduction (IPCC, 2014), as endorsed by the US Environmental Protection Agency, which classified ethanol from sugarcane as an "advanced biofuel" (EPA, 2010). Indeed, sustainable biofuels can be a major factor to help many countries accomplish the pledges in their National Determined Contributions presented in the COP 21 Conference; Goldemberg (2017) estimated conservatively that if the share of bioenergy in renewable energy production in 2014 remains unchanged, an additional contribution of bioenergy will correspond to 0.8–1.0 Gtons of avoided CO_2 in 2030, in addition to about 1.7 Gtons already avoided in this year.

Principle 4: Biofuels should develop in accordance with trade principles that are fair and recognize the rights of people to just reward (including labor rights and intellectual property rights). This is an almost obvious principle, biofuels should be produced certainly under the regular labor and property rights legislation, without allowances or exceptions. However, considering the international market, the fair trade of bioenergy is still to be achieved, since protectionism and high import tariffs are relatively a common practice in many countries, imposing barriers to free trade of biofuels and hurdles to efficient producers. Regarding intellectual property rights, it is interesting to mention that although the sugarcane bioenergy has good potential for improvement, the conventional sugarcane production and processes are globally well known, basically open access. The enforcement of decent work conditions, labor rights protection and adequate payment to workers are mandatory and unconditional constraints to be followed by bioenergy producers. It should be observed that in Brazil, mainly as a result of good government oversight, the work conditions in the sugarcane agro-industry are in general better than that observed in other crops (Moraes, 2007).

Principle 5: *Equitable distribution of costs and benefits.* This principle is very general and, although desirable and rational, its implementation seems more complex, depending of course on several complementary conditions: most of them exogenous to the sugarcane production system, such as the legal framework and land property scheme, are directly dependent on the government and incumbent public policies. So far, considering that the sustainability, in a broad sense, has been an essential guideline and constant condition in all topics presented in the paragraphs above, it can be stated that hence, this principle will be consequently accomplished.

Principle 6: If the first five principles are respected and if biofuels can play a crucial role in mitigating dangerous climate change then, depending on certain key considerations, there is a duty to develop such biofuels. The key considerations are as follows:

a   Will the costs of the development be out of proportion to the benefits, compared to other major public spending priorities?
b   Are there competing energy sources that might be even better at reducing GHG emissions, while still meeting all the required ethical principles?
c   Is there is an alternative and better use of the crops needed to produce biofuels?
d   Are there areas of uncertainty in the development of a technology, and are there efforts to reduce them?
e   Will the technologies lead to irreversible harms, once they are scaled up?
f   Are the views of those directly affected by the implementation of a technology being considered?
g   Can regulations be applied in a proportionate way?

Considering the arguments, information and data presented and discussed in the following chapters, the questions above have favorable answers, based on the conditions that already exist, such as land available and proper climate, or conditions that can be achieved, depending on adequate government measures, action and enforcement.

Thus, this final principle offers a kind of answer to the initial question, "Why promote sugarcane bioenergy production and use in Latin America, the Caribbean, and Southern Africa", replicating that, as the previous principles have been accomplished, it is an obligation to implement the sustainable bioenergy path. In a few words, it is more difficult to answer, "Why not to promote sugarcane bioenergy production and use in these regions".

References

BEST, 2008. *Review of fuel ethanol impacts on local air quality, Bioethanol for Sustainable Transport*, Imperial College, London.
Endres, J.M., 2011. No free pass: putting the "Bio" in biomass. *Natural Resources & Environment.* 26, 33–38.

EPA, 2010. *Renewable Fuel Standard (RFS)*, Environmental Protection Agency, Washington. Available in www.epa.gov/otaq/fuels/renewablefuels/.

Goldemberg, J., 2017. How much could biomass contribute to reaching the Paris Agreement goals? *Biofuels, Bioproducts & Biorefining* (Editorial), 11(2), March/April 2017.

IPCC, 2014. Technical summary, In: *Climate Change 2014, Mitigation of Climate Change. Contribution of Working Group iii to the Fifth Assessment report of the Intergovernmental Panel on Climate Change* Edenhofer, O., R. Pichs-Madruga, Y. Sokona, E. Farahani, S. Kadner, K. Seyboth, A. Adler, I. Baum, S. Brunner, P. Eickemeier, B. Kriemann, J. Savolainen, S. Schlömer, C. von Stechow, T. Zwickel and J.C. Minx (eds.). Cambridge University Press, Cambridge, United Kingdom and New York, NY, USA.

Kline, K.L, Msangi, S, Dale, V.H., Woods, J., Souza, G.M., Osseweijer, P., Clancy. J.S., Hilbert, J. A., Johnson, F.X. Mcdonnell, P.C., Mugera, H.K., 2016. Reconciling food security and bioenergy: priorities for action, Global Change Biology - Bioenergy, doi:10.1111/gcbb.12366.

Lynd, L.R., Woods, J., 2011. Perspective: a new hope for Africa. Nature 474, S20–S21. doi:10.1038/474S020a.

Moraes, M.A.F.D., 2007. *O mercado de trabalho da agroindústria canavieira: desafios e oportunidades*. Economia Aplicada, v.11 (4), Ribeirão Preto.

Nogueira, L.A.H., Capaz, R.S., 2013. Biofuels in Brazil: Evolution, achievements and perspectives on food security. *Global Food Security* 07/2013; 2(2). doi:10.1016/j.gfs.2013.04.001.

Nuffield Council on Bioethics, 2011. *Biofuels: Ethical Issues*. London. Available at: http://nuffieldbioethics.org/project/biofuels-0.

Souza, G.M., Victoria, R.L., Joly C.A., Verdade, L.M. (eds), 2015. *Bioenergy & Sustainability: Bridging the Gaps*, SCOPE 72, Scientific Committee on Problems of the Environment, Paris.

12 The potential of bioenergy from sugarcane in Latin America, the Caribbean and Southern Africa

Luis Cutz and Luiz A. Horta Nogueira

Introduction

Many countries of Latin America and the Caribbean (LAC) and Southern Africa (SA) share common concerns with regard to the energy scenario such as high dependence on imported fossil fuels and limited coverage of modern energy services. In addition, they have excellent conditions for developing modern bioenergy production, such as appropriate climate and land available. This is indeed a situation deserving to be better understood and evaluated. The scope of our analysis are the wet tropical countries of LAC: Argentina, Bolivia, Brazil, Colombia, Costa Rica, Cuba, Dominican Republic, Ecuador, El Salvador, Guatemala, Honduras, Jamaica, Mexico, Nicaragua, Panama, Paraguay, Peru and Venezuela; and SA: Angola, Malawi, Mauritius, Mozambique, South Africa, Swaziland, Tanzania, Zambia and Zimbabwe. Due to its particular condition, with a large bioenergy program based on sugarcane, Brazil was not included.

During the last decades, the population and the fossil fuel consumption have been growing rapidly in these countries. Although currently there is a large dispersion in per capita consumption, as depicted in Figure 12.1, in most cases all motor fuels (gasoline and diesel) are imported, representing a heavy burden to the national economies and the main source of carbon emissions (EIA, 2012).

The type of resources and challenges in each of these countries vary widely, but during the last decades it has been observed a growing interest to increase the use of bioenergy in LAC and SA, especially from sugarcane (ECLAC, 2011; Janssen and Rutz, 2012). Sugarcane has been cultivated for centuries in these regions and used as feedstock for production of sugar, while its by-products such as bagasse and molasses are used for the production of electricity and ethanol (Goldemberg et al., 2008). Sugarcane bioenergy is a promising alternative as it can be produced at competitive costs (van den Wall Bake et al., 2009), reduce greenhouse gas (GHG) emissions compared to fossil fuels (Seabra et al., 2011) and promote social development (Diaz-Chavez et al., 2015).

It is estimated that around 40 million tons of sugarcane is grown in SA, mainly in South Africa, Swaziland, Mauritius, Zimbabwe, Zambia and Mozambique (FAO, 2012b). In LAC, sugarcane occupies less than 10% of the arable land, except in countries such as Colombia, Costa Rica, Guatemala

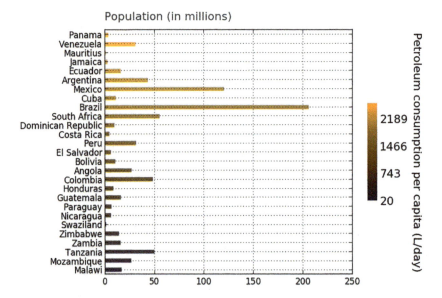

Figure 12.1 Population and petroleum consumption per capita for selected LAC and SA countries for 2014. Population data from World By Map (2016) and Petroleum consumption per capita from World By Map (2014). The countries are shown in decreasing order of their petroleum consumption.

and Mexico where sugarcane area corresponds to about one-fourth of the arable land. Particularly in Brazil, sugarcane bioenergy has a significant contribution on the energy sector, representing 16% of the national energy supply and corresponding to the second most important energy source, after oil (EPE, 2015). Brazil has the largest sugarcane area in the world, where 750 million tons of sugarcane is grown every year and shared approximately in equal amounts to sugar and ethanol production (FAO, 2012).

This chapter summarizes previous studies developed in the LACAf Project (Souza et al., 2016, 2018) and aims to foresee, from a joint perspective on both LAC and SA regions, the reasons and preliminary limits to foster sugarcane bioenergy. We estimate the potential of sugarcane for the production of electricity and ethanol for a short-term scenario (considering molasses use and the installed capacity in sugar mills) and medium- and long-term scenarios, assuming that sugarcane is cultivated on 1% of the current pastureland, electricity is produced in high-efficiency cogeneration units and ethanol is produced from molasses in existing sugar mills and direct sugarcane juice in new sugar mills.

In fact, sugarcane bioenergy has already started to participate in the energy matrix of several countries in the developing world, particularly in LAC, where sugarcane ethanol blends in gasoline is a reality in 12 countries, as indicated in Figure 12.2. But also in Africa such modern bioenergy is expanding: Angola, Malawi and Mozambique have implemented an E10 ethanol

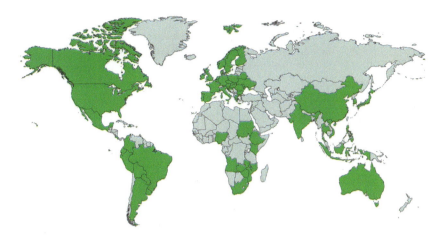

Figure 12.2 Countries adopting biofuels blending mandates.
Source: based on BiofuelsDigest (2018).

blending mandate in place and six other countries are in the initial stages to be adopted (BiofuelsDigest, 2018).

The next sections introduce the hypothesis and evaluations developed to estimate the potential impact to displace fossil fuels, generate electricity and mitigate carbon emissions, closing with a set of conclusions and recommendations.

Scenarios for sugarcane bioenergy production

The potential supply of biofuel and electricity from sugarcane in LAC and SA was estimated for two scenarios. Also, both scenarios assume that sugarcane is processed in sugar mills with a crushing capacity of 1 million tons per year and distillery self-consumption of 30 kWh/t cane (Dias et al. 2011). It is assumed that ethanol is used in two ways for using ethanol: blended with gasoline in vehicular engines and, in SA countries, also as cooking fuel. The methodology presented in this section is detailed as follows (Souza et al., 2016, 2018).

I **Business as usual (BAU):** It considers that ethanol is produced only from sugarcane molasses based on the existing sugarcane production with a yield of 10 L/ton of cane (tc). It is assumed that sugar mills operate with average efficiency cogeneration systems (boiler steam at 42 bar and 450°C), yielding 60 kWh/tc (BNDES/CGEE, 2008). Ethanol is used in 10% blends (v/v) with gasoline, E10.

II **New framework (NF):** It considers that sugarcane is cultivated on 1% of the current pastureland and assumes that this scenario is likely to be implemented over the medium to long term based on the 2030 scenario proposed by the International Energy Agency (IEA, 2014). Pasturelands are usually underutilized, and by applying better practices, such as rotational grazing and integrated crop-livestock-forestry systems, it is possible to increase the productivity without comprising the grazing activity (The Royal Society, 2009). In this scenario, it is assumed that ethanol is produced from molasses in existing sugar mills with a yield of 10 L/tc and direct juice in new sugar mills with a yield of 80 L/tc. It is also considered that cogeneration systems have higher efficiency compared to BAU scenario, operating at 65 bar and 480°C to generate 110 kWh/tc of electricity (BNDES/CGEE, 2008). The ethanol blend is set to 20% (v/v), E20.

Table 12.1 presents the sugarcane production in LAC and SA in 2015 (FAO, 2018) and the potential production assuming that 1% of pastureland in each country (FAO, 2012) is cultivated with sugarcane with an average yield of 80 t/ha. These values give the absolute relevance of the current sugarcane industry and the relative importance of land potentially available for expanding the sugarcane production, using a relatively small fraction of the national land.

Table 12.1 Current sugarcane production and 1% of the pastureland in LAC and SA

Country	Sugarcane production in 2015[a]	Sugarcane potential production on 1% of the current pastureland[b]	Potential production increase in relation to 2015
	10^3 ton/year	10^3 ton/year	%
Latin America and the Caribbean			
Argentina	22,540	86,800	385
Bolivia	7,192	26,400	367
Colombia	36,710	31,360	85
Costa Rica	4,266	1,040	24
Cuba	19,300	2,240	12
Dominican Republic	4,535	960	21
Ecuador	10,106	4,000	40
El Salvador	6,578	480	7
Guatemala	23,653	1,600	7
Honduras	5,171	1,440	28
Mexico	1,475	160	11
Nicaragua	55,396	64,720	117
Panama	2,381	2,640	111
Paraguay	6,701	1,200	18
Peru	4,186	13,600	325
Venezuela	5,569	15,040	270

(*Continued*)

The potential of bioenergy from sugarcane 145

Country	Sugarcane production in 2015[a]	Sugarcane potential production on 1% of the current pastureland[b]	Potential production increase in relation to 2015
	10^3 ton/year	10^3 ton/year	%
Southern Africa			
Angola	552	45,696	8,278
Malawi	2,894	1,568	54
Mauritius	4,009	8	0
Mozambique	3,084	37,232	1,207
South Africa	14,861	71,016	478
Swaziland	5,517	872	16
Tanzania	2,717	24	1
Zambia	4,145	16,920	408
Zimbabwe	3,348	10,240	306

a Source: Data extracted from (FAO, 2018) for year 2015.
b Assuming 80 kton/ha, pastureland from (FAO, 2012).

Values in this table span from 8 kton to 86 Mton of sugarcane per year, pointing out that different approaches would be needed to deal with the economies of scale of sugarcane projects. Although in some countries, the potential expansion is limited due to the importance of current use of land for sugarcane production, as observed in El Salvador and Guatemala, whereas in other cases, the expansion is limited because the country is relatively small, such as in Mauritius. The countries with the highest potential from the supply perspective are Argentina, Nicaragua and South Africa, where more than 60,000 kton/year could be produced annually.

Ethanol from sugarcane production and use

Figure 12.3 presents the potential ethanol supply in LAC and SA, for both production scenarios. In order to inform its relative importance of ethanol production in relation to the domestic demand, this figure shows also the level of blending (EX) which can be reached in each scenario. Details about the additional hypothesis and considerate conditions in this evaluation are available (Souza et al., 2016, 2018).

Under the BAU scenario, just using the current availability of molasses as feedstock for ethanol production, LAC countries such as Cuba, Guatemala and Nicaragua could almost immediately displace at least 10% of gasoline consumption. In SA, countries such as Malawi, Mauritius, Swaziland and Zimbabwe have the potential to easily reach E10. Noteworthy that from all the aforementioned countries, in the year 2017, only Malawi and Zimbabwe had an E10 mandate in place.

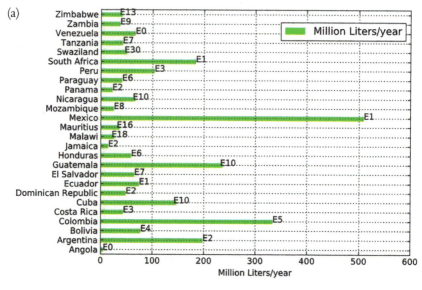

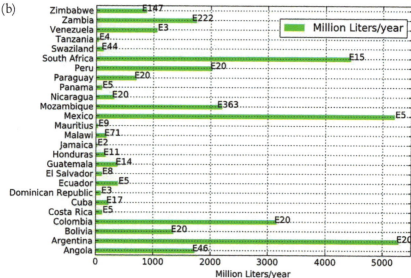

Figure 12.3 (a) BAU scenario; (b) NF scenario. Potential ethanol supply in LAC and SA. Identifications (Ex) indicate potential gasoline blend that could be achieved in each country.

For the NF scenario, most of the countries in LAC and SA could implement at least E10. Bolivia, Colombia, Nicaragua, Paraguay and Peru would be able to displace 20% of the gasoline consumption. In SA, ethanol could displace at least 15% of gasoline consumption, except in countries such as Mauritius and Tanzania due to their low availability of pastureland. On the other hand, the ethanol potential in Zimbabwe and Mozambique is larger than the current gasoline consumption in transport. It is important to mention that the projected increase in ethanol production for the NF scenario will require a significant effort beyond the BAU scenario for all countries.

For the SA countries, it was also evaluated the ethanol use as cooking fuel. In this region, for preparing food, solid biomass still is the main source of energy, typically burned in low efficiency and smoky stoves, associated with serious health problems and deforestation. For instance, charcoal and firewood supply more than 95% of the cooking energy consumed in Mozambique and Malawi (IEA, 2014). Ethanol stoves, clean and more efficient, has been promoted as an alternative technology and had good public acceptation, but initiatives to introduce them have been impaired by discontinuities of ethanol supply (REN21, 2015). To evaluate the impact of ethanol from sugarcane in this context, it was assumed that a typical household use of 1.5 kg of firewood per capita per day obtained from 0.9 ha of forest for wood gathering (IEA, 2014), which can be displaced by 360 liters of ethanol per year.

In the NF scenario, all SA countries could meet at least 50% of the cooking fuel demand with ethanol, except Malawi, Mauritius and Tanzania due to their low availability of pastureland. It is estimated that approximately 85 million tons of firewood per year (in 2030) could be saved by implementing the NF scenario in all of the SA countries, which could reduce deforestation notably. Given the promising results for ethanol as cooking fuel, further research is needed to evaluate the proper conditions to promote the use and invest on ethanol cooking stoves.

Potential bioelectricity supply

The use of sugarcane bagasse as fuel in thermal power plants has a huge potential to increase electricity production and diversify the energy mix of LAC and SA. This should induce a reduction of foreign fuel imports for electricity generation and improve the electricity supply. Figure 12.4 presents the potential bioelectricity supply in LAC and SA for the NF scenario. The impact of the BAU production scenario on electricity supply is available (Souza et al., 2016, 2018).

By using 1% of the pastureland to enlarge sugarcane cropping, in Angola, Swaziland and Zambia, bagasse could represent 22%, 23% and 15% of their total electricity generation. Sugarcane provides new opportunities for countries such as South Africa, where the power sector is highly dependent on coal-based generation. In LAC, Bolivia, Guatemala and Nicaragua, sugarcane bioelectricity could represent about 18%, 16% and 15% of their total electricity generation, respectively. In countries such as Bolivia, firing sugarcane bagasse would improve significantly the energy mix considering that currently renewable energy represents only 1.6% of the current electricity generation (IEA, 2012).

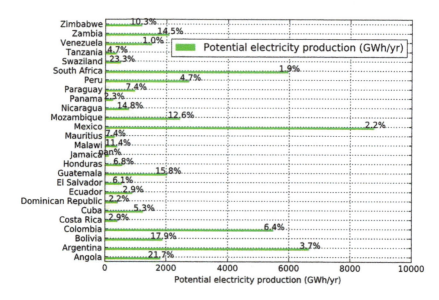

Figure 12.4 Potential bioelectricity production in the NF scenario. (x%) indicates the percentage of sugarcane bagasse in the total electricity generation.

Potential GHG mitigation in Latin America, the Caribbean and Southern Africa

The impact of sugarcane bioenergy on GHG savings was estimated for the NF scenario, assuming implicitly that the countries adopt E20 ethanol-gasoline blends and electricity generation from bagasse available, as indicated in Figure 12.5. The GHG emission factors used are available in (Wang et al., 2014). As expected, the impact is more important in those countries where electricity production depends more on fossil fuels, displaced by bagasse. A similar remark could be done as regards to the impact of ethanol on emissions, depending on the dimension of fleet, gasoline demand and ethanol availability.

In some conditions, the use of bagasse in power plants means important mitigation on GHG emissions. In South Africa, more than 90% of the electricity generation is derived from coal; the use of bagasse could help to eliminate 5,889 Mton of CO_2e emissions per year in the power sector, in addition to an annual reduction potential of 6,596 Mton CO_2e by displacing gasoline. In Swaziland, where 70% of the electricity production is based on coal, sugarcane bioelectricity could also contribute to a cleaner power sector by reducing the annual GHG emissions by more than 60%. In LAC, Cuba, Dominican Republic and Jamaica, countries where electricity generation mainly comes from fossil fuels could reduce also more than 60% of GHG emissions by switching to bagasse-based power generation, always under the scheme of NF scenario.

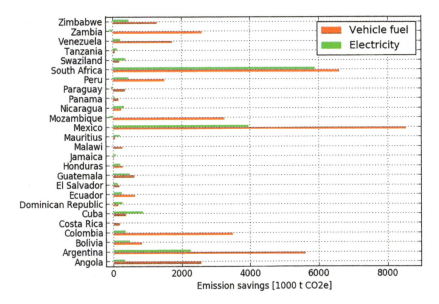

Figure 12.5 Potential GHG savings due to gasoline and electricity displacement in LAC and SA for the NF scenario.

Investment to develop sugarcane bioenergy infrastructure in SA and LAC

The investment required to expand the sugarcane cropping and build new sugar mills for the NF scenario was estimated based on a representative mill with a crushing capacity of 1 million tons of sugarcane per harvest season with 200 days. For LAC, it was considered an investment of US$ 139.7/tc (MME, 2017; PECEGE, 2015) for the agricultural sector, sugar mill and cogeneration system, while for SA it was assumed US$ 212/tc (BNDES, 2014). Investments in the distribution and transmission systems were not considered, assuming that these investments will happen regardless of the expansion of sugarcane. The investment needed to expand sugarcane cropping over 1% of the pastureland varies according to the available area of each country. Table 12.2 presents the investment required in the agricultural sector and sugar mill estates to achieve nationally the conditions of the NF scenario, related to GDP and gross capital formation, with data referred to 2015 (World Bank, 2017).

Although the numbers of Table 12.2 should be used with care, since they represent a very preliminary assessment, they could give policy maker an initial idea, to be better detailed and evaluated. Assuming that the total investment would occur over a period of 10 years, the required annual investment for all of the countries is relatively low compared to the GDP, except for Mozambique and Zimbabwe, where the investment required represents an important portion of the fixed capital formation. In LAC, the

Table 12.2 Number of new 1-Mt sugarcane mills and total investment for the NF scenario

Countries	Number of mills crushing (1 Mton/year)	Annual investment required (1,000 USD)	Investment related to GDP (%)	Investment related to gross capital formation (%)
Southern Africa				
Angola	22	470,000	0.60	1.90
Malawi	2	44,400	0.80	3.40
Mauritius	<1	115	<0.01	<1
Mozambique	28	592,000	5.80	32.60
South Africa	55	1,162,500	0.30	1.60
Swaziland	1	22,600	0.60	4.50
Tanzania	<1	440	<0.01	<1
Zambia	22	467,600	2.30	8.90
Zimbabwe	11	227,600	2.40	11.10
Latin America and the Caribbean				
Argentina	64	9,368	0.20	1.10
Bolivia	16	2,355	0.70	3.40
Colombia	35	5,206	0.10	0.50
Costa Rica	1	136	0.00	0.10
Cuba	1	152	0.02	0.30
D. Republic	1	85	0.01	0.10
Ecuador	4	576	0.10	0.20
El Salvador	1	81	0.00	0.20
Guatemala	2	273	0.00	0.30
Honduras	1	215	0.10	0.50
Mexico	59	8,652	0.10	0.30
Nicaragua	3	464	0.40	1.40
Panama	1	151	0.00	0.10
Paraguay	8	1,206	0.40	2.50
Peru	24	3,538	0.20	0.70
Venezuela	12	1,833	0.00	0.20

investment required in most of the countries would represent less than 1% of the national investment in fixed capital over a period of 10 years. Nevertheless, this level of investment has not been seen yet in any of the analyzed countries. For example, Mexico would require an investment four times higher than the one made on renewable energy projects in 2014 (BNEF,

2015). Colombia targets 6.5% of renewable energy on electricity generation by 2020, excluding large hydropower (IRENA, 2015). By investing in 35 new 1 Mt-sugarcane mills, bagasse would supply 6% of the Colombian electricity demand. Nicaragua could replace 20% of the gasoline consumption by investing US$ 464 million on three new sugarcane mills. This investment represents 10% of the country's investment target on renewable energy projects over the next 15 years (BNEF, 2015), excluding the biomass share. It is estimated that Costa Rica and Panama would require 25% of their total investment in renewable energy in 2014 (BNEF, 2015) to replace at least 5% of the gasoline consumption.

Closing remarks

This study confirms that sugarcane bioenergy is a strategic asset for developing LAC and SA, as well as to supply clean and affordable energy to the world. The initial question about the reasons to push sugarcane bioenergy in these regions has many assertive answers, some of them clearly indicated in this chapter: there are resources, the required technology is known and available, there are domestic and global needs to supply, there are environmental benefits and so on.

The current availability, in some countries, of raw material such as sugarcane molasses, easy to process for producing ethanol, a biofuel that can be blended with gasoline and used with excellent performance in conventional vehicles without any change, offers a good departure point to promote the modern bioenergy. Progressively, with an increase of sugarcane cropping, cultivating a small share of pasture areas and requiring relatively minor investment, it is possible to displace a significant amount of fossil fuel, generate electricity, and mitigate GHG emissions in the power and transportation sectors. Nevertheless, it is important to stress that to implement sustainable sugarcane bioenergy systems, a large base of natural resources, as identified in LAC and SA, and the availability of efficient and competitive technologies for producing and processing feedstock are not enough. Equally important is a detailed pre-assessment, a careful evaluation of environmental and social conditions, including restrictions.

As it has been indicated in several studies, in the global context, bioenergy can play a decisive role in the forthcoming decades. LAC an SA are effectively very well posed to play a significant role as a sustainable energy supplier and take advantage of their favorable situation as an efficient and competitive producer, promoting agro-industry and rural development. But also as an energy consumer, they develop their own domestic markets of biofuel, improving energy security and air quality in cities.

Sustainable bioenergy development in LAC and SA will depend essentially on stable policies, reinforced by long-term prospects such as international agreements on carbon emission mitigation and global trade of biofuels. These policies must integrate economic, social and environmental policies and put forward plans for land use and rural development, aligned with strategies to limit potential negative impacts of biofuels by applying sustainability guidelines based on aspects such as biodiversity, GHG emissions, land use and water use.

Acknowledgment

The authors are thankful to Dr. Simone Pereira de Souza, a former researcher in the LACAf Project for her valuable comments and contribution to the manuscript.

References

BiofuelsDigest, 2018. "Biofuels Mandates Around the World 2018." *Biofuels Digest.* www.biofuelsdigest.com/bdigest/2018/01/01/biofuels-mandates-around-the-world-2018/2/.

BNDES, 2014. "Estudo de Viabilidade de Produção de Biocombustíveis na UEMOA (União Econômica E Monetária Do Oeste Africano)" Rio de Janeiro, RJ: BNDES.

BNDES/CGEE, 2008. "Sugarcane-Based Bioethanol: Energy for Sustainable Development" Banco Nacional de Desenvolvimento Econômico e Social, Centro de Gestão e Estudos Estratégicos. Available at: file:///C:/Users/Sica/Downloads/7bioetanol_ing.pdf.

BNEF, 2015. "Global Trends in Renewable Energy Investment" Frankfurt: Bloomberg New Energy Finance (BNEF).

Dias, M.O.S., Cunha, M.P., Jesus, C.D.F., Rocha, G.J.M., Pradella, J.G.C., Rossell, C.E.V., Maciel Filho, R., Bonomi, A., 2011. "Second Generation Ethanol in Brazil: Can It Compete with Electricity Production?" *Bioresource Technology* 102 (19): 8964–8971. doi:10.1016/j.biortech.2011.06.098.

Diaz-Chavez, R., Morese, M.M., Colangeli, M., Fallot, A., Moraes, M.A., Olényi, S., Osseweijer, P., Sibanda, M.A., Mapako, M, 2015. "Social Considerations." In Souza GM, Victoria RI, Joly CA, Verdade L (Eds.), *Bioenergy & Sustainability: Bridging the Gaps*, 528–553. Scientific Committee on Problems of the Environment (SCOPE). São Paulo.

ECLAC, 2011. "Estudio regional sobre la economía de los biocombustibles en 2010: temas clave para los países de América Latina y el Caribe", by Dufey, A. and Stange, D., Economic Comission of Latin America and the Caribbean.

EIA, 2012. "International Energy Statistics." U.S. Energy Information Administration (EIA). www.eia.gov/cfapps/ipdbproject/IEDIndex3.cfm?tid=5&pid=5&aid=2.

EPE, 2015. "Brazilian Energy Balance 2015 Year 2014." Rio de Janeiro: EPE.

FAO, 2012. "Inputs, Land. FAOStat." FAO. http://faostat3.fao.org/download/R/RL/E.

FAO, 2018. "Production, Crops. FAOStat." FAO. Available at: http://faostat3.fao.org/download/Q/QC/E.

Goldemberg, J., Coelho, S., Guardabassi, P., 2008. The sustainability of ethanol production from sugarcane. *Energy Policy*, 36 (6):2086–2097.

IEA, 2014. "A Focus on Energy Prospects in Sub-Saharan Africa." In *World Energy Outlook*, 237. International Energy Agency (IEA).

IRENA, 2015. "Renewable Energy in Latin America 2015: An Overview of Policies." Abu Dhabi: IRENA.
Janssen, R., Rutz, D., 2012 (eds). Bioenergy for Sustainable Development in Africa, Springer Science+Business Media B.V. doi:10.1007/978-94-007-2181-4.
MME, 2017. "Oferta de Biocombustíveis, in: Plano Decenal de Expansão de Energia - PDE, Consulta Pública, Empresa de Pesquisa Energética, Ministério de Minas E Energia, Secretaria de Planejamento E Desenvolvimento Energético" Brasília, DF: MME.
PECEGE, 2015. "Production Costs of Sugarcane, Sugar, Ethanol and Bioelectricity in Brazil. 2015/2015 Crop Season, 2014/2016 Crop Projection" Piracicaba, SP, Brazil: University of São Paulo, Luiz de Queiroz College of Agriculture, Program of Continuing Education in Economics and Management/Department of Economics, Management and Sociology.
REN21, 2015. "Renewables 2015 Global Status Report". REN21 Secretariat. ISBN 978-3-9815934-6-4.
Seabra, J.E.A., Macedo, I.C., Chum, H.L., Faroni, C.E., Sarto, C.A., 2011. "Life Cycle Assessment of Brazilian Sugarcane Products: GHG Emissions and Energy Use" *Biofuels, Bioproducts and Biorefining* 5: 519–32.
Souza, S.P., Nogueira, L.A.H., Watson, H.K., Lynd, L.R., Elmissiry, M., Cortez, L.A.B., 2016. "Potential of Sugarcane in Modern Energy Development in Southern Africa" *Frontiers in Energy Research* 4. doi:10.3389/fenrg.2016.00039.
Souza, S.P., Nogueira, L.A.H., Martinez, J., Cortez, L.A.B., 2018. "Sugarcane Can Afford a Cleaner Energy Profile in Latin America & Caribbean" *Renewable Energy*, January. doi:10.1016/j.renene.2018.01.024.
The Royal Society, 2009. "Reaping the Benefits Science and the Sustainable Intensification of Global Agriculture". The Royal Society, London.
The World Bank, 2017. "Data of Gross Domestic Product and Gross Capital Formation from National Accounts Data." Available at: https://data.worldbank.org/indicator/
Wall Bake, J. D. van den, Junginger, M., Faaij, A., Poot, T., Walter, A., 2009. "Explaining the Experience Curve: Cost Reductions of Brazilian Ethanol from Sugarcane" *Biomass and Bioenergy* 33 (4): 644–658. doi:10.1016/j.biombioe.2008.10.006.
Wang, M., Sabbisetti, R., Elgowainy, A., Die enthaler, D., Anjum, A., Sokolov, V., et al. (2014). *GREET Model: e Greenhouse Gases, Regulated Emissions, and Energy Use in Transportation Model*. Chicago, USA: Argonne National Laboratory.
World By Map, 2014. "Petroleum Consumption" http://world.bymap.org/OilConsumption.html.
World By Map, 2016. "Population" World by Map. http://world.bymap.org/Population.html.

13 Reconciling food security, environmental preservation and biofuel production
Lessons from Brazil

João Guilherme Dal Belo Leite, Manoel Regis L. V. Leal, Luís A. B. Cortez, Lee R. Lynd and Frank Rosillo-Calle

Introduction

Over the last decade, food security has drawn attention from governments, scientists and institutions engaged in development initiatives. The food crisis of 2008, when market prices of major food commodities increased by up to 65% (nominal terms) over previous years, raised many questions about global capacity to feed an increasing population. The main drivers of the surge in food prices were as follows: (1) the increasing demand for agricultural production, pushed by, among other things, economic development and population growth (FAO, 2009a, 2009b); (2) slow growth of agricultural production, particularly in the developed regions where arable land for crop expansion is scarce and yield levels approach crop potential ceiling (Grassini et al., 2013); (3) droughts, floods and heat waves in key producing regions (FAO, 2009b); (4) rising oil prices with further consequences for agriculture production and transportation costs; (5) declining global grain stocks (Rosillo-Calle and Johnson, 2010; Trostle, 2008); (6) increased food imports, particularly in the developing countries such as China and India (Trostle, 2008), which triggered speculation on commodities market as expectations of an increase in food prices grew among investors (Lagi et al., 2011); and (7) cereal-based biofuel production, such as maize ethanol in the United States (FAO, 2008; OECD-FAO, 2008).

The contribution of biofuels to increase food prices, both past and future, is highly contentious. Driven by climate change mitigation strategies and energy security concerns, biofuels have become one of the most dynamic and rapidly growing sectors of the global energy economy (Elbehri et al., 2013; Tomes et al., 2010; UN, 2007). Liquid biofuels in particular emerged as a renewable alternative to reduce greenhouse gas (GHG) emissions through the replacement of finite fossil fuels (e.g. gasoline and diesel) and as an opportunity to increase rural income and agricultural investment (Pingali et al., 2008; Tyner, 2013). For instance, in Brazil, sugarcane ethanol employs thousands, promotes wealth creation and replaces 35% of gasoline consumption (EPE, 2013; Moraes et al., 2016). On the other hand, there are concerns as to the allocation of arable land to the production of biofuel crops, particularly if large amounts of food products, such as maize, are diverted to biofuel production (Mueller et al., 2011; Ugarte and He, 2007; Zilberman et al., 2013). Central to many critical assessments is a trade-off, often presented as inescapable, between biofuels and food security. The argument made for an inherent "food-fuel" conflict is generally based on the following logic. First, the production of biofuel crops on agricultural land displaces food crops, thus reducing food production and harming food security. Second, if not produced on agricultural lands, biofuels would expand into native forests or savannas, thus driving deforestation. Deforestation may also take place through indirect land use changes (ILUC) if, for example, biofuels push the expansion of food crops into areas of native vegetation (Plevin et al., 2010). As a result, even highly efficient fuel crops such as sugarcane may have their climate mitigation potential negated by the emissions caused by land clearing (Fargione et al., 2008; Searchinger et al., 2008). Moreover, agricultural expansion into pristine forest lands and savannas, whether directly or indirectly, drives climate change, which in the mid to long terms is likely to have negative impacts on food security as well (Kalnay and Cai, 2003; Pielke, 2005).

The underlined consequences on food security are particularly relevant in developing nations, which are expected to play a key role in meeting the world's future demand for food, animal feed and fiber, with production expected to double by 2050 (FAO, 2009a). Meeting these targets will be essential to attend the demands of a growing population, which is becoming wealthier, but also constrained by resources (e.g. available arable land in developed countries) and climate change (Hertel, 2015; Kastner et al., 2012).

The land-based logic, however, overlooks the scope for environmental-oriented policies and agreements and agricultural intensification. Increasing land output in regions with relatively low agricultural productivity, such as in developing countries, can be a sustainable way to raise food production, preserve the environment and, in some cases, pursue biofuel production (Garnett et al., 2013). And, nowhere in the developing world is this dynamic as vivid as in Brazil, where there was an explosion of agricultural production and unparalleled development of biofuels over the last decades.

In this paper, we review the simultaneous, rapid changes in biofuel production, food security and deforestation in Brazil over the past 25 years. Our objective is to explore main conditions, without concern to causal relationships, which allowed sugarcane ethanol production together with food security and the environmental preservation and to what extent the Brazilian experience may provide lessons relevant to other developing regions.

Methodological approach

In this study, we combine data on food production (i.e. grains) in Brazil with food security indicators related to availability, access, stability and utilization.

Grain production and agricultural area are particularly useful as indicators of potential impacts from cropping systems and biofuel production (i.e. sugarcane) over time. In the following sections, we explore these indicators in further detail, as well as the development of biofuels and food security over a period of 25 years in Brazil.

We also analyse how deforestation, a common externality associated with the expansion of agriculture, both for food and biofuels, develops along the same span of time. The analysis is based on a detailed assessment of available food security indicators, discussed in parallel with the expansion of sugarcane ethanol and deforestation in Brazil from 1990 to 2014. We also review major developments affecting agricultural production systems associated with intensification strategies and environmental accords aimed at reducing illegal deforestation in the Amazon.

Our approach is limited in detailing the causal effects driving food security, biofuel production and deforestation. However, it offers an exploratory and alternative perspective to the presumed consequences of agricultural and biofuel production, together with food security and environmental preservation. The Brazilian experience over these issues may also offer a unique platform to identify both opportunities and threats for policymaking and agricultural development in other developing regions.

Sugarcane ethanol in Brazil

No developing nation has invested as much as Brazil in promoting biofuel production. Biofuel production in Brazil dates back to the 1920s, but 1975 marked the first large-scale production of sugarcane ethanol in response to the national ethanol program (Proálcool). Proálcool was created under tough government regulation over fuel prices, blending strategies and subsidies that led to the production and manufacturing of 100% ethanol car engines (Rosillo-Calle and Cortez, 1998). In the early 1970s, the world was facing a notorious energy crisis with oil prices at record high levels. In this period, Brazil was primarily concerned with energy security, as the country was heavily dependent on oil imports and struggled to deal with nearly unaffordable oil expenditures.

Over the years, the ethanol sector in Brazil went through acute changes, triggered by further oil price volatility together with the removal of market regulations and government subsidies. Over time, it developed and became a modern industry, energy self-sufficient and a net producer of electricity (generated from sugarcane bagasse) to the national grid (Conab, 2008).

Nevertheless, the expansion of sugarcane over the years (more details in the following sections) has led to concerns related to the potential displacement of food crops and other environmental consequences (i.e. deforestation). Yet information on the relationships among sugarcane, food security and deforestation remains limited. Moreover, there is little discussion of how land use dynamics may be affected by agricultural management and technology adoption, such as those driving intensification, and its relationship with food production and environmental preservation.

Food security in Brazil over the last quarter century

Feeding 200 million people

The Brazilian population grew from 150 million in 1990 to over 200 million in 2015 (IBGE, 2015). This demographic expansion was followed by unprecedented developments in food security. By all metrics – availability, stability, access and utilization – Brazil has seen a tremendous increase in food security over the last 25 years, with some of the highest rates of improvement worldwide. Availability of energy (i.e. caloric intake) and protein has increased steadily since 1990 (Figure 13.1A and B). Today, Brazil has reached parity with developed nations with respect to energy supply and is rapidly approaching similar progress on protein supply, which – from a nutritional point of view – is entirely adequate (WHO, 1985).

Brazil has also achieved an extraordinary progress on the access dimension of food security (Figure 13.2). Notably, a substantial reduction in the number of hungry people was made in the early 2000s, when the prevalence of undernourishment, food deficit and food inadequacy plunged to record low levels (Figure 13.2A–C).

In the decade beginning in 2000, Brazil changed from a net importer of cereal to a net exporter (Figure 13.3A), thus reducing domestic vulnerability to shocks from international food markets. Malnutrition, the main cause of wasting and stunted growth among children, also decreased dramatically from 1990 to the present, continuing a trend initiated in the 1970s (Figure 13.3B). These outcomes in food security mirror steady progress towards the redistribution and increase in income, which enabled the country to lift millions out of extreme poverty, reduce inequality and improve human development (FAO et al., 2014; IBGE, 2015).

Among the factors enabling Brazil's success in improving food security were the economic stabilization during the mid-1990s, particularly inflation control, coupled with implementation of a suite of hunger-related policies during the 2000s (FAO et al., 2014; Sachs and Zini Jr, 1996). This period was also marked by the surge in commodities prices (i.e. the commodities boom) driven by aggressive economic growth and raising demands in emerging markets, particularly China, which fostered an extraordinary growth of the Brazilian agriculture (Helbling et al., 2008). From 2000 to 2014, grain production in Brazil is more than doubled with only a 50% increase in cultivated area (Figure 13.4A). This development was only possible due to acute changes

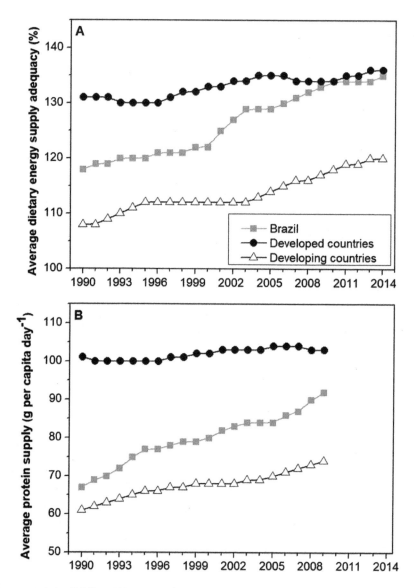

Figure 13.1 Availability: (A) average dietary energy supply adequacy (%); and (B) average protein supply (g per capita per day).
Source: FAO (2015).

in how food was produced across the country, as well as plentiful land resources. Over the years, agricultural systems went through gradual modernization, from extensive to more intensive production systems, made possible by the adoption of technical knowledge (e.g. soil conservation techniques

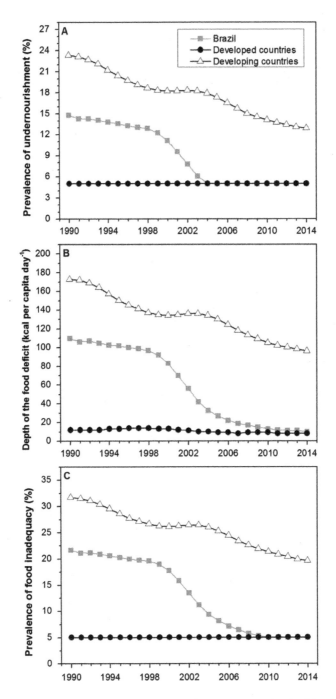

Figure 13.2 Access: (A) prevalence of undernourishment (%); (B) depth of the food deficit (kcal per capita per day); (C) prevalence of food inadequacy (%).
Source: FAO (2015).

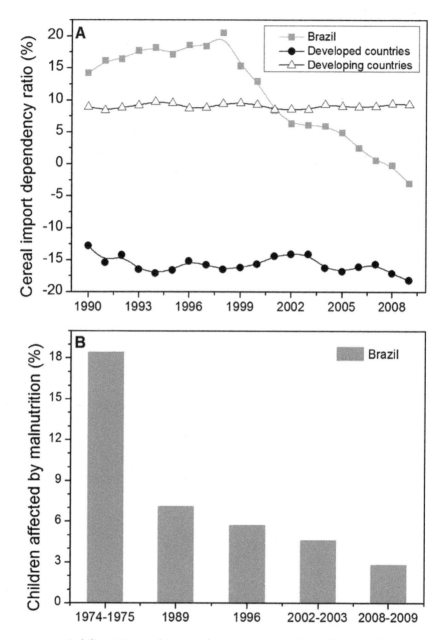

Figure 13.3 Stability: (A) cereal import dependency ratio (%); utilization: (B) children affected by malnutrition (%).
Source: FAO (2015) and IBGE (2015).

Reconciling food security, environmental preservation 161

and fertility) and innovation (e.g. improved crop genotypes, mechanization). Investments in agricultural R&D and the role of research agencies such as Embrapa propelled agricultural production to record levels. Gains in crop yield allowed the production of grains to increase significantly, from 418 kg per capita in 1980 to nearly 1000 kg per capita in 2014 (Figure 13.4B).

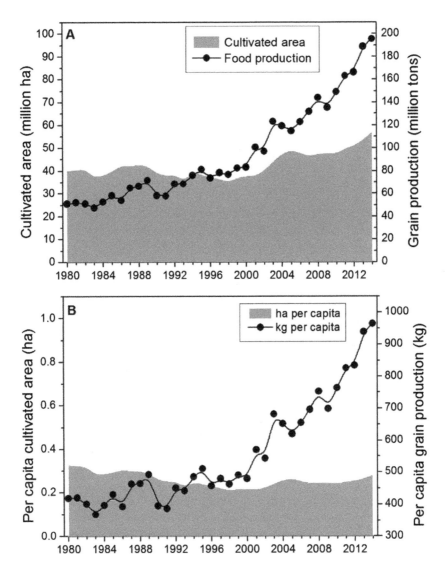

Figure 13.4 Food production and cultivated area in Brazil (A), food production and cultivated area per capita (B).
Source: Conab (2014) and IBGE (2000, 2013).

Biofuel production and environmental conservation

The growth of agricultural production in Brazil was accompanied by the expansion of biofuels. Sugarcane ethanol production, which in 2014 accounted for 90% of Brazil's biofuel output (EPE, 2014), grew from 0.6 million cubic meters in the mid-1970s to 27.6 million cubic meters in 2013 (Figure 13.5A). During the same period, the area under sugarcane grew from 1.7 million ha to 10 million ha, with roughly half allocated to ethanol and the balance to sugar production. This growth was driven both by increasing local demand, as ethanol became a viable substitute for gasoline (Goldemberg, 2007), and by growing export-oriented production of sugar for international markets. Sugarcane ethanol soon became the top renewable transport fuel for Brazilians, and sugarcane became the second most important source of energy in the Brazilian primary energy matrix (Figure 13.5B).

Biofuel initiatives have also been opposed on environmental grounds, particularly concerning to the reduction of GHG emission and impacts on biodiversity. Although a number of studies find biofuels environmentally positive (Armstrong et al., 2002; de Vries et al., 2010; Farrell et al., 2006; Gnansounou et al., 2009; Iriarte and Villalobos, 2013; Lee and Ofori-Boateng, 2013; Nogueira, 2011), such positive effects can be reversed if forests and peatlands are converted into biofuel cropping areas (Fargione et al., 2008; Lapola et al., 2010; Scharlemann and Laurance, 2008; Searchinger et al., 2008).

In Brazil, the nearly fivefold expansion of the sugarcane-planted area (Figure 13.5A) occurred primarily over the Central-South region on the former pastureland (Adami et al., 2012; Egeskog et al., 2016; Novo et al., 2012). Moreover, data on deforestation shows a consistent decline across all biomes with most of the preservation achieved in the Amazon region (Figure 13.6A and B). Recent assessments show that little if no progress was made in Cerrado from 2009 to 2011, when deforestation levels stagnated or increased (Figure 13.6A). Yet, environmental conservation advanced during the mid-2000s, which was also a dynamic period for the expansion of sugarcane (Figure 13.5A) and food crops in general (Figure 13.4A).

There are, of course, a myriad of factors underpinning land use dynamics. Nevertheless, the main reasons Brazil was able to expand agricultural production and sugarcane acreage while decreasing deforestation are rooted in (a) agricultural intensification, (b) arable land being relatively plentiful and (c) political and voluntary initiatives to stop deforestation.

Agricultural intensification is generally described as the increase of produce output per unit of area. This concept gained attention during the green revolution (Borlaug, 1971), when large efforts were made to increase food production in developing regions. In recent years, the term "sustainable intensification" has come to denote increased production per unit land accompanied by the rational and efficient use of inputs, together with cautious resource management and environmental conservation (Campbell et al., 2014; Garnett et al., 2013; Tilman et al., 2009, 2011).

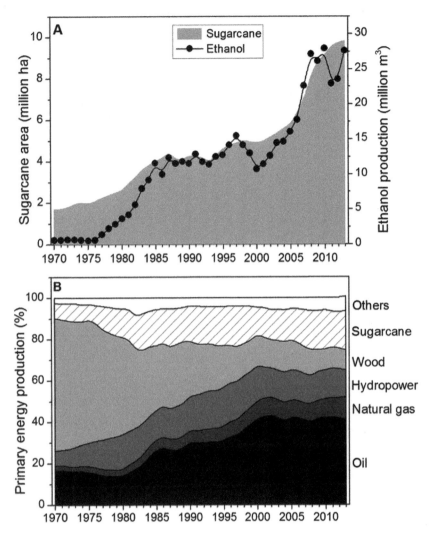

Figure 13.5 Sugarcane area and ethanol production (A), and primary energy production in Brazil (B).
Source: EPE (2014), Unicadata (2013) and FAOSTAT (2014).

Brazil has experienced intensification in both crop and livestock production systems. Cattle production in particular has been transformed by the increase of animal stocking rates, from 0.51 heads/ha in 1970 to 1.08 heads/ha in 2006 and beef productivity gains, which quadrupled, from 10 kg/ha/year to 43 kg/ha/year, over the same span of time (Martha Jr et al., 2012). Intensification continued over the last 20 years, as the cattle herd grew, the pasture area decreased simultaneously (Figure 13.7). From 1985 to 2006,

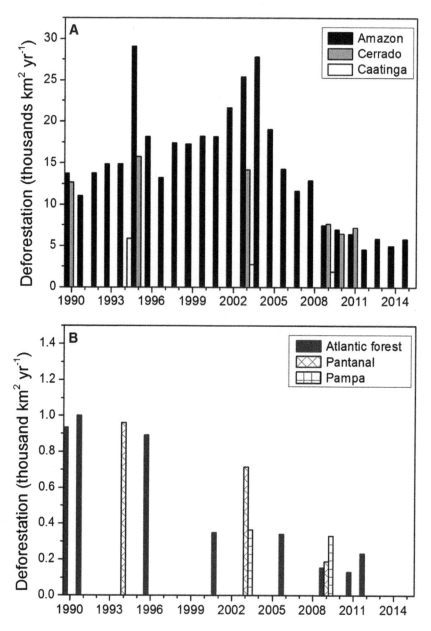

Figure 13.6 Deforested area in Brazil ((A) Amazon, Cerrado and Caatinga, and (B) Atlantic forest, Pantanal and Pampa).
Source: MMA (2016).

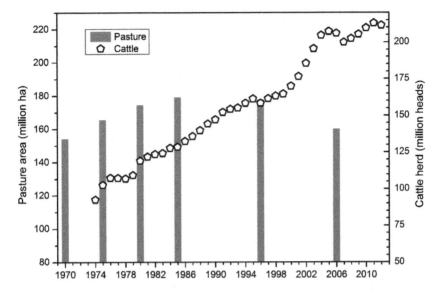

Figure 13.7 Pasture area and cattle herd in Brazil.
Source: IBGE (2006, 2012, 2014).

nearly 20 million ha of pastureland may have been freed to other uses (e.g. sugarcane production), while the cattle herd increased by over 50% during the same period (Figure 13.7).

Like livestock, cropping systems have steadily increased yield levels over the years (Figure 13.4). In this case, another form of intensification – double cropping – also took place (Stephanie et al., 2014). Maize features as the most important food crop cultivated later in the rainy season, representing 80% of the cropped area during this period (Conab, 2014). In 2013, it became a predominantly second crop, in contrast to 1990, when maize double cropping systems barely existed (Figure 13.8).

Shifting maize towards the end of the cropping season was a strategy to accommodate the large expansion of soybean, which tripled from 10 million ha in 1990 to 30 million ha in 2014 (Conab, 2014). In this period, maize double cropping systems may have freed 30% of the area used for soybean production, as soybean-maize rotations became a common practice in the Central-South region of Brazil. The widespread adoption of no-tillage production systems also contributed to reduce fossil fuel inputs and enhanced soil carbon sequestration, hence increasing soil carbon stocks (Bayer et al., 2006).

Agricultural intensification of both crop and livestock production systems likely played an important role in mitigating agriculture expansion into pristine forest areas and, to a lower extent, savannas (i.e. Cerrado). Yet, it alone may not explain the large reduction in deforestation achieved in Brazil, particularly after 2004 (Figure 13.6A and B).

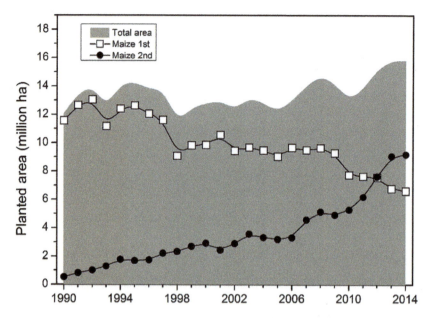

Figure 13.8 Planted area of maize (first crop, second crop and total) in Brazil from 1990 to 2014.
Source: Conab (2014).

Deforestation is not only driven by agricultural activity but also by an array of factors, including the exploitation of natural resources and land speculation. Direct impacts on deforestation levels may have also resulted from tougher environmental laws and improvements in monitoring and enforcement implemented in the early 2000s (Assunção et al., 2011). Moreover, voluntary zero-deforestation agreements, such as the soybean and beef moratoria, represented land-mark achievements in environmental preservation during the same period (Gibbs et al., 2015). These accords – implemented under stiff pressure from retailers and nongovernmental organizations (NGOs) – introduced new supply-chain governance that traces and blocks produce from newly deforested areas in the Amazon region (Nepstad et al., 2014). The sugarcane agroecological zoning is yet another policymaking tool that can guide funding and restricts sugarcane expansion towards forests and woodlands, particularly in the Cerrado region (Embrapa, 2009).

Final considerations

Over the last 25 years, Brazil has achieved remarkable progress towards food security and agricultural production. This progress was marked by improvements in a number of food security indicators, which pushed Brazil towards developed nations' standards.

Food security progress coincides with an aggressive expansion of biofuel production, which positions Brazil among the largest producers in the world. At the national level, sugarcane products (ethanol and electricity) became the second most important source of energy. Moreover, despite biofuel production and food security enhancements, deforestation levels declined over the last decade, particularly in the Amazon.

Agricultural intensification may have played a large role in limiting the potential negative impacts of sugarcane expansion on food security and the environment. The combination of double cropping and increased cattle stocking rates freed large swaths of arable land for other uses, such as sugarcane production, thus mitigating the impact of agricultural expansion on pristine forest areas. Moreover, environmental conservation improved through tougher policies and private sector agreements to curtail deforestation.

Overall, Brazil seems to offer an opportunity to rethink the presumed mandatory interplay between food security and biofuel production. If the right conditions are in place, such as effective institutions to protect native vegetation and scope for crop intensification, particularly pasture, then agricultural and biofuel production can help both food security and environmental preservation. These conditions are not present everywhere. Therefore, a cautious assessment – aware of the complex and dynamic changes that can unfold on land, including intensification – is necessary to guide site-specific policies and strategies that, in some regions, may include biofuels among the alternatives to sustainably meet global demands today and in the future. Land-abundant countries, such as Brazil and other developing nations, have an edge over more constraint – often over exploited – regions, and may well lead this process.

References

Adami M, Rudorff BFT, Freitas RM, et al. (2012) Remote sensing time series to evaluate direct land use change of recent expanded sugarcane crop in Brazil. *Sustainability* 4(4): 574–585. Available from: www.mdpi.com/2071-1050/4/4/574.

Armstrong AP, Baro J, Dartoy J, et al. (2002) *Energy and greenhouse gas balance of biofuels for Europe - An update*. Brussels: CONCAWE Reports. Available from: www.scopus.com/inward/record.url?eid=2-s2.0-4544224277&partnerID=40&md5=c068a6ac35d4bfda21ebf9161afe2d4c.

Assunção J, Gandour CC and Rocha R (2011) *Deforestation slowdown in the legal Amazon: prices or policies?* Rio de Janeiro: Climate Policy Initiative (CPI).

Bayer C, Martin-Neto L, Mielniczuk J, et al. (2006) Carbon sequestration in two Brazilian Cerrado soils under no-till. *Soil and Tillage Research* 86(2): 237–245. Available from: www.sciencedirect.com/science/article/pii/S0167198705000723.

Borlaug NE (1971) The green revolution: for bread and peace. *Bulletin of the Atomic Scientists* 27(6): 6–48.

Campbell BM, Thornton P, Zougmoré R, et al. (2014) Sustainable intensification: What is its role in climate smart agriculture? *Current Opinion in Environmental Sustainability* 8: 39–43. Available from: www.sciencedirect.com/science/article/pii/S1877343514000359.

Conab (2008) *Perfil do setor do açúcar e do álcool no Brasil*. Available at: www.conab.gov.br/conabweb/download/safra/perfil.pdf Accessed in November 07th, 2015.

Conab (2014) Safras - Séries Históricas. Available at: http://conab.gov.br/conteudos.php?a=1028&t=2 Accessed in November 07th, 2014.

de Vries SC, van de Ven GWJ, van Ittersum MK, et al. (2010) Resource use efficiency and environmental performance of nine major biofuel crops, processed by first-generation conversion techniques. *Biomass and Bioenergy* 34(5): 588–601. Available from: www.sciencedirect.com/science/article/pii/S0961953410000024.

Egeskog A, Barretto A, Berndes G, et al. (2016) Actions and opinions of Brazilian farmers who shift to sugarcane - an interview-based assessment with discussion of implications for land-use change. *Land Use Policy* 57: 594–604. Available from: www.sciencedirect.com/science/article/pii/S0264837715302593.

Elbehri A, Segerstedt A and Liu P (2013) *Biofuels and the sustainability challenge: a global assessment of sustainability issues, trends and policies for biofuels and related feedstocks*. Rome: FAO.

Embrapa (2009) *Zoneamento agroecológico da cana-de-açúcar. Expandir a produção, preservar a vida, garantir o futuro*. Manzatto CV, Assad ED, Bacca JFM, et al. (eds), Rio de Janeiro: Embrapa Solos.

EPE (2013) *Balanço energético nacional*. Available at: https://ben.epe.gov.br/BENSeries Completas.aspx Accessed in November 2014.

EPE (2014) *Brazilian energy balance*. Rio de Janeiro: Empresa de Pesquisa Energética.

FAO (2008) *The state of food and agriculture 2008- Biofuels: prospects, risks and opportunities*. Rome: FAO.

FAO (2009a) *High level expert forum - How to feed the world in 2050*. Rome: FAO.

FAO (2009b) *The state of agricultural commodity markets: high food prices and the food crisis - experience and lessons learned*. Rome: Food and Agriculture Organization of the United Nations (FAO).

FAO (2015) Statistics: food security indicators. Available at: www.fao.org/economic/ess/ess-fs/ess-fadata/en/#.VkSPTeIzs30 Accessed in November 11th, 2015.

FAO, IFAD and WFP (2014) *The State of Food Insecurity in the World 2014. Strengthening the enabling environment for food security and nutrition*. Rome: FAO.

FAOSTAT (2014) Production - Crops. Available at http://faostat3.fao.org/download/Q/QC/E Accessed in January 04th, 2015.

Fargione J, Hill J, Tilman D, et al. (2008) Land clearing and the biofuel carbon debt. *Science* 319(5867): 1235–1238. Available from: www.sciencemag.org/content/319/5867/1235.abstract.

Farrell AE, Plevin RJ, Turner BT, et al. (2006) Ethanol can contribute to energy and environmental goals. *Science* 311(5760): 506–508. Available from: www.sciencemag.org/content/311/5760/506.abstract.

Garnett T, Appleby MC, Balmford A, et al. (2013) Sustainable intensification in agriculture: premises and policies. *Science* 341(6141): 33–34.

Gibbs HK, Rausch L, Munger J, et al. (2015) Brazil's soy moratorium. *Science* 347(6220): 377–378. Available from: www.sciencemag.org/content/347/6220/377.short.

Gnansounou E, Dauriat A, Villegas J, et al. (2009) Life cycle assessment of biofuels: energy and greenhouse gas balances. *Bioresource Technology* 100(21): 4919–4930. Available from: www.sciencedirect.com/science/article/pii/S0960852409006245.

Goldemberg J (2007) Ethanol for a sustainable energy future. *Science* 315(5813): 808–810. Available from: www.sciencemag.org/content/315/5813/808.abstract.

Grassini P, Eskridge KM and Cassman KG (2013) Distinguishing between yield advances and yield plateaus in historical crop production trends. *Nat Commun*, Nature Publishing Group, a division of Macmillan Publishers Limited. All Rights Reserved. 4. Available from: doi:10.1038/ncomms3918.

Helbling T, Mercer-Blackman V and Cheng K (2008) Riding a wave. *Finance and Development* 45(1): 10–15.

Hertel TW (2015) The challenges of sustainably feeding a growing planet. *Food Security* 7(2): 185–198. Available from: www.scopus.com/inward/record.url?eid=2-s2.0-84924238314&partnerID=40&md5=5651bc878ce8b8f1bbad5e4ed-59cebc9.

IBGE (2000) *Projeção da população do Brasil por sexo e idade para o período de 1980–2050*. Instituto Brasileiro de Geografia e Estatística.

IBGE (2006) *Censo Agropecuário*. Rio de Janeiro: Instituto Brasileiro de Geografiae Estatística. Available from: www.ibge.gov.br/servidor_arquivos_est/.

IBGE (2012) Pesquisa pecuária municipal. Available at: www.sidra.ibge.gov.br/bda/tabela/listabl.asp?c=73&z=t&o=24 Accessed in November 11th, 2014.

IBGE (2013) *Projeção da população do Brasil por sexo e idade para o período 2000–2060*. Instituto Brasileiro de Geografia e Estatística.

IBGE (2014) Estatísticas do Século XX. Estabelecimentos agropecuários recenseados e área dos estabelecimentos, segundo a propriedade das terras, a condição do responsável e grupos de area total - (1920–85) - (1985–1995–96). Available at: http://seculoxx.ibge.gov.br/e.

IBGE (2015) *Indicadores de desenvolvimento sustentável - Brasil 2015*. Rio de Janeiro: Instituto Brasileiro de Geografia e Estatística (IBGE).

Iriarte A and Villalobos P (2013) Greenhouse gas emissions and energy balance of sunflower biodiesel: Identification of its key factors in the supply chain. *Resources, Conservation and Recycling* 73(0): 46–52. Available from: www.sciencedirect.com/science/article/pii/S0921344913000232.

Kalnay E and Cai M (2003) Impact of urbanization and land-use change on climate. *Nature* 423(6939): 528–531. Available from: www.nature.com/nature/journal/v423/n6939/suppinfo/nature01675_S1.html.

Kastner T, Rivas MJI, Koch W, et al. (2012) Global changes in diets and the consequences for land requirements for food. *Proceedings of the National Academy of Sciences* 109(18): 6868–6872. Available from: www.pnas.org/content/109/18/6868.abstract.

Lagi M, Bar-Yam Yavni, Bertrand KZ, et al. (2011) The food crises: a quantitative model of food prices including speculators and ethanol conversion. *arXiv: 1109.4859v1*: 56. Available from: http://arxiv.org/abs/1109.4859.

Lapola DM, Schaldach R, Alcamo J, et al. (2010) Indirect land-use changes can overcome carbon savings from biofuels in Brazil. *Proceedings of the National Academy of Sciences* 107(8): 3388–3393. Available from: www.pnas.org/content/107/8/3388.abstract.

Lee K and Ofori-Boateng C (2013) Environmental sustainability assessment of biofuel production from oil palm biomass. In: *Sustainability of Biofuel Production from Oil Palm Biomass*, Springer Singapore, pp. 149–187. Available from: doi:10.1007/978-981-4451-70-3_5.

Martha Jr GB, Alves E and Contini E (2012) Land-saving approaches and beef production growth in Brazil. *Agricultural Systems* 110: 173–177. Available from: www.scopus.com/inward/record.url?eid=2-s2.0-84861921853&partnerID=40&md5=b551bf9e4f2d7d5fe01ca71ff5476120.

MMA (2016) Projeto de Monitoramento do Desmatamento dos Biomas Brasileiros por Satélite - PMDBBS. Ministério do Meio Ambiente (MMA), Brasília. Available at http://siscom.ibama.gov.br/monitora_biomas/index.htm Accessed in May 24th, 2016.

Moraes MAFD de, Bacchi MRP and Caldarelli CE (2016) Accelerated growth of the sugarcane, sugar, and ethanol sectors in Brazil (2000–2008): effects on municipal gross domestic product per capita in the south-central region. *Biomass and Bioenergy* 91: 116–125. Available from: www.sciencedirect.com/science/article/pii/S0961953416301489.

Mueller SA, Anderson JE and Wallington TJ (2011) Impact of biofuel production and other supply and demand factors on food price increases in 2008. *Biomass and Bioenergy* 35(5): 1623–1632. Available from: www.scopus.com/inward/record.url?eid=2-s2.0-79953289600&partnerID=40&md5=fecad4b985b-fac2ef5407446147d1bff.

Nepstad D, McGrath D, Stickler C, et al. (2014) Slowing Amazon deforestation through public policy and interventions in beef and soy supply chains. *Science* 344(6188): 1118–1123.

Nogueira LAH (2011) Does biodiesel make sense? *Energy* 36(6): 3659–3666. Available from: www.sciencedirect.com/science/article/pii/S0360544210004718.

Novo A, Jansen K and Slingerland M (2012) The sugarcane-biofuel expansion and dairy farmers' responses in Brazil. *Journal of Rural Studies* 28(4): 640–649. Available from: www.sciencedirect.com/science/article/pii/S0743016712000733.

OECD-FAO (2008) *Agricultural outlook 2009–2018*. Paris: OECD.

Pielke RA (2005) Land use and climate change. *Science* 310(5754): 1625–1626. Available from: http://science.sciencemag.org/content/sci/310/5754/1625.full.pdf.

Pingali P, Raney T and Wiebe K (2008) Biofuels and food security: missing the point. *Applied Economic Perspectives and Policy* 30(3): 506–516. Available from: http://aepp.oxfordjournals.org/content/30/3/506.short.

Plevin RJ, Jones AD, Torn MS, et al. (2010) Greenhouse gas emissions from biofuels' indirect land use change are uncertain but may be much greater than previously estimated. *Environmental Science & Technology* 44(21): 8015–8021.

Rosillo-Calle F and Cortez LAB (1998) Towards proalcool II-A review of the Brazilian bioethanol programme. *Biomass and Bioenergy* 14(2): 115–124. Available from: www.scopus.com/inward/record.url?eid=2-s2.0-0031780695&partnerID=40&md5=07090de7a5e5977c3af33ed3aaef19b9.

Rosillo-Calle F and Johnson FX (2010) *Food versus fuel: an informed introduction*. London: Zed Books.

Sachs J and Zini Jr AA (1996) Brazilian inflation and the plano real. *World Economy* 19(1): 13–37. Available from: www.scopus.com/inward/record.url?eid=2-s2.0-0029773725&partnerID=40&md5=e0008d81d1a1d24e92a76e811cf278b7.

Scharlemann JPW and Laurance WF (2008) How green are biofuels? *Science* 319(5859): 43–44. Available from: www.sciencemag.org/content/319/5859/43.short.

Searchinger T, Heimlich R, Houghton RA, et al. (2008) Use of U.S. croplands for biofuels increases greenhouse gases through emissions from land-use change. *Science* 319(5867): 1238–1240. Available from: www.sciencemag.org/content/319/5867/1238.abstract.

Stephanie AS, Avery SC, Leah KV, et al. (2014) Recent cropping frequency, expansion, and abandonment in Mato Grosso, Brazil had selective land characteristics. *Environmental Research Letters* 9(6): 64010. Available from: http://stacks.iop.org/1748-9326/9/i=6/a=064010.

Tilman D, Socolow R, Foley JA, et al. (2009) Beneficial biofuels – The food, energy, and environment trilemma. *Science* 325(5938): 270–271. Available from: www.sciencemag.org/content/325/5938/270.short.

Tilman D, Balzer C, Hill J, et al. (2011) Global food demand and the sustainable intensification of agriculture. *Proceedings of the National Academy of Sciences* 108(50): 20260–20264.

Tomes D, Lakshmanan P and Songstad D (2010) *Biofuels: global impact on renewable energy, production agriculture, and technological advancements.* New York: Springer.

Trostle R (2008) *Global agricultural supply and demand: factors contributing to the recent increase in food commodity prices.* USDA (ed.), *A Report from the Economic Research Service*, Economic Research Service (ERS) USDA.

Tyner WE (2013) Biofuels and food prices: Separating wheat from chaff. *Global Food Security* 2(2): 126–130. Available from: www.scopus.com/inward/record.url?eid=2-s2.0-84881088696&partnerID=40&md5=b80deee399db725fb3546b-642ffdd5f5.

Ugarte DDLT and He L (2007) Is the expansion of biofuels at odds with the food security of developing countries? *Biofuels, Bioproducts and Biorefining* 1(2): 92–102. Available from: doi:10.1002/bbb.16.

UN (2007) *Sustainable Bioenergy: A Framework for Decision Makers.* United Nations.

Unicadata (2013) Área plantada com cana-de-açúcar. Available at: www.unicadata.com.br/historico-de-area-ibge.php?idMn=33&tipoHistorico=5 Accessed in November 04th, 2014.

WHO (1985) *Energy and Protein Requirements.* Geneva: World Health Organization.

Zilberman D, Hochman G, Rajagopal D, et al. (2013) The impact of biofuels on commodity food prices: assessment of findings. *American Journal of Agricultural Economics* 95(2): 275–281. Available from: http://ajae.oxfordjournals.org/content/95/2/275.short.

14 Energy security and energy poverty in Latin America and the Caribbean and sub-Saharan Africa

Manoel Regis L.V. Leal, João Guilherme Dal Belo Leite and Luiz A. Horta Nogueira

Introduction

Energy security (ES) is a concept that accounts for nation's ability to successfully meet their energy needs. The first concerns about ES emerged in the 20th century when a number of industrialized countries realized the risks of depending on a small number of oil suppliers; the 1973 oil shock increased these concerns. According to the International Energy Agency (IEA), ES is "the uninterrupted availability of energy sources at an affordable price" (IEA, 2016). Reliable and continuous access to modern sources of energy is well recognized as an essential item for economic development and progress of any nation. The former United Nations (UN) Secretary-General Ban Ki-moon said that "energy is the golden thread connecting economic growth, social equity and environmental sustainability" (UN, 2014).

Energy poverty (EP) complements the national approach of ES as EP focuses on household access to electricity and clean cooking fuels (Vandeweerd, 2012). EP is a serious obstacle to the creation of opportunities for economic growth and a real threat to health and well-being in several countries, particularly for low-income families. When the Millennium Development Goals (MDGs) were enacted in 2000, with the strong support of the UN, the aim was to half extreme poverty by 2015 (UN, 2015), yet ES was not explicit among the eight MDGs. From 1990 to 2015, the population living below the poverty line decreased from 1.9 billion to 836 million, which means that more than 1 billion people got out of the extreme poverty (UN, 2015). Despite this extraordinary achievement, tackling barriers such as EP became a must to guarantee any further progress in fighting poverty. Hence, in 2012, the UN Secretary General launched the Sustainable Energy for All (SE4ALL) initiative (IEA/WB, 2017). The SE4ALL foresees three interlinked objectives to be achieved by 2030: (1) ensure universal access to modern energy services, (2) double the rate of improvement in energy efficiency, and (3) double the share of renewable energy in the global mix (UN, 2012). Around 80 countries joined the initiative, including several developing ones, involving public and private sectors such as the World Bank and IEA. These two institutions became responsible for tracking the progress towards the achievement of SE4ALL objectives through the implementation of a Global Tracking Framework, with the assistance of other 15 international organizations to produce biannual reports containing energy baseline data updates. Another product of the SE4ALL is the identification of the High Impact Opportunities (HIOs), which are priority areas for action with potential to produce important progress in the three objectives. Among the HIOs are clean cooking alternatives, renewable energy procurement, sustainable biofuels, off-grid lighting and appliance efficiency (UN, 2017).

In 2015, the MDGs were replaced by the Sustainable Development Goals (SDGs). SDG 7 Affordable and Clean Energy recognizes the lack of access to modern energy as a constraint to human and economic development and aims to ensure access to affordable and modern energy for all by 2030.

From 2010 to 2016, SE4ALL achieved encouraging progress regarding global electrification (i.e. 86%), yet the results on access to clean cooking fuels and facilities remain poor (38% globally) (IEA, 2017). In the energy efficiency side, there was a reduction in the annual energy intensity indicator (MJ/PPP US$) of about by 2.1% from 2012 to 2014, thus close to the target of 2.6%. As to the participation of renewable energy in the global energy mix, the overall progress remains below expectations with encouraging advances in electricity generation but disappointing on heat and transportation (IEA/WB, 2017).

With the adoption of the SE4ALL and the SDGs, EP was clearly introduced in the agenda to promote the development and fight extreme poverty. For developing countries, it means promoting the switch from traditional biomass (fuel wood, agricultural residues and animal dung) to modern and sustainable forms of energy. In this chapter, our aim is to provide an overview on global ES and explore opportunities associated with sugarcane, particularly in developing nations.

ES: a global view

Energy poor countries commonly have large rural populations that are highly depend on traditional biomass to satisfy their energy needs. The IEA/WB (2017) shows that there are a strong correlation between EP and gross domestic product (GDP) per capita. When GDP per capita crosses the US$ 500 to 1,000 threshold, there is a sharp increase in electricity access. EP is a chronic problem in Africa and, to a lesser extent, in smaller Asia-Pacific countries. In general, it is not a serious problem in Latin America and Caribbean (Table 14.1).

It is quite clear that sub-Saharan Africa (SSA) is the most critical region in terms of electricity access; it is the only region where the population growth exceeded the increase in electrification rate, meaning that despite a relative increase in electricity access the total number of people without electricity increased. Developing Asia is not doing so badly in relative terms, but it accounts for a large number of people without access. These two regions hold more than 96% of the global population without access to electricity (Table 14.1).

The variation across countries is quite high in SSA, from 10% in Burundi to 86% in South Africa. It is interesting to note that energy resource-rich countries (due to recent discoveries os oil and gas), such as Nigeria and Mozambique, remain energy poor. In Central and South America, access to electricity, both in relative and absolute tems, is not a serious problem and is progressing satisfactorily towards the SE4ALL 2030 goals (Table 14.2).

Table 14.1 Global energy access 2016

Region	Electrification rate population access (%)	Population access (million)
World	86	1,060
Developing countries	86	1,060
Africa	52	588
SSA	43	588
Developing Asia	89	439
Central and South America	95	17
Middle East	93	17

Source: IEA (2017).

Table 14.2 Energy access in selected countries in SSA and LA&C

Country	Electrification rate population with access (%)	Population without access (million)
Africa		
Botswana	55	1
Burundi	10	10
Ghana	84	4
Malawi	11	16
Mozambique	40	16
Nigeria	61	74
South Africa	86	8
Central and South America		
Argentina	100	<1
Bolivia	92	1
Brazil	100	1
Colombia	98	1
Guatemala	94	1
Nicaragua	89	1

Source: IEA (2017).

Table 14.3 Global access to clean cooking fuels in 2015

Region	Population without access (%)	Population without access (million)
World	38	2,792
Developing countries	49	2,792
Africa	71	848
SSA	84	846
Developing Asia	49	1,874
China	33	457
India	64	834
Central and South America	12	59

Source: IEA (2017).

Access to clean cookfuels, both globally and regionally, presents more causes for concern. The present situation and the progress toward the 2030 goal are quite sluggish when compared with electicity (Table 14.3). The GDP per capita treshold from traditional to modern cooking fuels is somewhat blurred, in the range of US$ 12,000 (IEA/WB, 2017). This is due to cultural reasons, economic and fisical access to fuels such as wood, agricultural residues and dung. Hence, increasing fears of missing the 2030 goals by a significant margin (IEA/WB, 2017).

The indoor cooking with traditional biomass in ineffcent stoves causes serious health problems, specially to women and children, being responsible for 4.3 million premature deaths per year (WHO, 2014).

Of the 2.8 billion people without access to clean cooking, 2.5 billion use traditional biomass, 120 million kerosene and 170 million coal (IEA, 2017). SSA accounts for 846 million, India for 834 million and China for 457 million. It is important to notice that around 85% of the people without access to clean cooking fules lives in only 20 countries (IEA/WB, 2017).

Globally, there is a trend to use LPG to replace traditional biomass. This means replacing traditional biomass by fossil alternatives, certainly with great social benefits, but negative environmental impacts. Therefore, efforts to offer viable renewable alternatives of clean cooking fuels, such as ethanol and pellets, should be welcome.

Lack of access to clean cooking fuels in SSA is a very complicated issue. A total of 846 million energy poor people are scattered in the whole region, with very few exceptions such as South Africa, many countries are close to full dependence on traditional biomass (Table 14.4). In Central and South America, the situation is quite different and presents hopes for achieving the 2030 goals. Nevertheless, there are 59 million (12%) people without clean cooking facilities and fuels in the region (Table 14.4).

Table 14.4 Access to clean cooking fuels in selected countries of SSA and Central and South America in 2014

Country	Population without access (%)	Population without access (million)
Africa		
Botswana	42	<1
Burundi	>95	11
Ghana	71	20
Malawi	>95	17
Mozambique	95	27
Nigeria	94	171
South Africa	17	10
Central and South America		
Argentina	–	–
Bolivia	17	2
Brazil	5	10
Colombia	13	6
Guatemala	30	5
Nicaragua	52	3

Source: IEA (2017).

Sugarcane as an improver of ES and a reducer of EP

Sugarcane is essentially a food crop that responds for around 80% of the sugar produced globally (USDA, 2017). It is produced in more than 100 countries (FAO, 2015), in some of them for centuries, with mature technologies and well-known problems. However, since the dawn of the 20th century, sugarcane became an energy crop – Brazil issued the first ethanol blending with gasoline mandate in 1931, after several years of field trials (Walter et al., 2013). Today, sugarcane is the feedstock for around one-third of the ethanol produced around the world. It also accounts for a significant amount of surplus electricity generated by sugar/ethanol mills in several countries. Therefore, it has become an important feedstock for energy. On average conditions, the total primary energy content of sugarcane in the field is around 7,400 MJ t^{-1} (Leal, 2007), representing almost 600 GJ ha^{-1} for an 80 t ha^{-1} yield. Assuming ethanol production at 86 L t^{-1} and electricity surplus at 60 kWh t^{-1}, the total energy output is 2,220 MJ t^{-1} or 178 GJ ha^{-1} (30% conversion efficiency) of useful energy. These estimates show that sugarcane is an excellent energy crop in terms of primary and useful energy production per unit area, with significant potential for efficiency improvement, especially through straw recovery (green cane harvest) to boost power generation.

Latin America and Caribbean

LA&C is a traditional region for sugarcane with 939 Mt produced in 2012 (FAO, 2013 in Cruz et al., 2016), representing approximately half of the global sugarcane production and harvested area. Brazil was the largest producer with 721 Mt, remaining 218 Mt for the other LA&C producers, a still significant figure. Table 14.5 shows the production and consumption of sugarcane and sugar for selected countries in the region, amounting 93% of the regional production, indicating a high concentration of the total sugarcane production in only seven countries, particularly Brazil.

LA&C is an important sugar-exporting region, featuring Brazil, Guatemala and Cuba as main players in the international market. In general terms, ethanol can be produced without threatening the regional supply of sugar. Moreover, some conversion flexibility between sugar and ethanol provides mixed mills (for more details see Chapter 22 in this book) and economic edge to deal with high price volatility of the international sugar market. Table 14.6 shows the main global ethanol producers.

Ethanol production is highly concentrated, with the USA and Brazil responding for 86% of the total production and the main five producers making 92% (Table 14.6). Roughly, two-thirds of ethanol-producing countries use grains (mostly maze) as feedstock and the other third uses sugarcane. A few countries such as Brazil and Argentina use both feedstocks, grains and sugarcane. Other crops, such as cassava, have negligible contribution to ethanol production and are restricted to Thailand.

Table 14.5 LA&C sugarcane and sugar statistics for selected countries

Countries	Sugarcane		Raw centrifugal sugar	
	Production (Mt)	Harvested area (Mha)	Production (Mt)	Consumption (Mt)
World	1,839	26.1	179.1	176.6
LA&C	938.6	13.0	60.47	38.63
Argentina	19.8	0.360	2.19	2.09
Brazil	721.1	9.7	40.22	12.57
Colombia	33.4	0.409	2.08	1.93
Cuba	14.7	0.361	1.47	0.64
Guatemala	23.7	0.256	2.46	0.95
Mexico	50.9	0.735	5.05	4.87
Peru	10.4	0.081	1.10	1.11

Source: FAO (2015).

Table 14.6 Global ethanol production in selected countries

Country	Ethanol production (billion liters/year)	Accumulated production (%)
The United States	58.0	58.8
Brazil	27.0	86.2
China	3.2	88.2
Canada	1.7	91.2
Argentina	0.9	92.1
Germany	0.9	93.0
India	0.9	93.9
Rest of the World	6.0	100.0
World total	98.6	100.0
EU-28	3.4	3.4

Source: REN21 (2017).

Sub-Saharan Africa

In 2012, Africa produced 94.3 Mt of sugarcane and 10.6 Mt of sugar (also from sugar beet) to balance a consumption of 14.7 Mt, what makes the region a net sugar importer. A number of countries in the region belong to the Africa, Caribbean and Pacific (ACP) group of countries, and many are considered least developed countries (LDCs). LDCs are entitled to preferential status in the European Sugar Regime, with sugar export quotas and minimum prices above the international sugar market. This condition allowed some countries to export sugar to EU at higher prices and buy from the international market with a profit. With the end of the EU Sugar Regime, on September 30, 2017, the EU will likely increase its sugar production with consequences to the sugar international prices. In an increasingly competitive market, LDCs are left with two alternatives: claim government subsidies or improve competitiveness. One way to achieve the latter is to diversify by including coproducts along with the conventional products (USAID, 2015). Several countries have already done that like Brazil, Colombia, Argentina, Guatemala, Mauritius, Reunion, India, Thailand and others by adding ethanol, electricity, specialty sugars and other products to their production. In Chapter 23, diversification of a sugar mill in the African context is evaluated in terms of its contribution to improve ES at different levels. This assessment shows that an average size sugarcane mill can supply electricity to thousands of households as well as enabling ethanol blending policies of up to 7% (E7) countrywide. Malawi is a successful case that produces ethanol to meet blending mandates with the gasoline since the 1980s, in two distilleries with a total capacity around 36 ML/year. The production is approximately half of the installed capacity and is enough to make a 10% ethanol blend (E10).

Table 14.7 Mozambique and South Africa energy profiles: electricity and gasoline consumption

Country	Population (1,000 people)	Electricity (GWh/year)	Per capita electricity (kWh/capita/year)	Gasoline (ML/year)
Mozambique	25,203	19,521	775	240
South Africa	52,386	311,012	5,937	11,470

Source: Souza et al. (2016).

Considering that southern African countries are important sugarcane producers, Souza et al. (2016) explored the implementation of sugarcane ethanol and electricity production in Mozambique and South Africa (Table 14.7).

The energy profile of the two countries are quite different. In Mozambique, average per capita electricity consumption remains low for global standards. While gasoline consumption in Mozambique remains very modest and dependent on the international market, South Africa meet its domestic demand from coal in coal-to-liquid (CTL) process, making the fossil fuel even more environmentally unfriendly. Coal is also used to generate 94% of the electricity in South Africa. In Mozambique, nearly 100% of the electricity comes from hydro power.

As to the sugarcane sector, South Africa features as the main sugar producer in SSA. Mozambique remains a promise, with large availability of land suitable for sugarcane expansion and an active industry in operation in the South and Central provinces of the country.

Closing remarks

The energy scenario of Central and South America is quite different from that of SSA. Central and South America have achieved 97% of electricity access, with several countries already at full universal access and only three with less than 94% access (Haiti, Honduras and Nicaragua). In SSA, on the other hand, only 43% of the population have access to electricity, accounting for almost 600 million people.

With respect to access to clean cooking fuels, the situation is even more favorable to Central and South America with only 12% of the population without access, representing 59 million people, still a large number in absolute terms. On the other side, SSA has 84% of its population without access to clean cooking, or 846 million, and 783 million of those totally dependent on traditional biomass; there are several countries where the share of the population without clean cooking is above 95%.

Therefore, SSA desperately needs alternatives to come out of this noxious EP condition, or otherwise it will not be able to improve its economy and other social indicators. Sugarcane could be an important contributor in this endeavor mainly because important lessons could be learned from other regions where it proves to be a sustainable energy alternative. Moreover, it is a well-known crop in many countries of SSA where an incipient but active industry awaits the right incentives and policies to boost food (i.e. sugar) and renewable energy production, dragging with it local and regional development.

References

Cruz, C.B.; Cortez, L.A.B.; Nogueira, L.A.H.; Baldassin Jr, R.; Rincón, J.M. 2016. *Bioenergy.* Current status and perspectives in Latin America & Caribbean: addressing sugarcane ethanol, Chapter 6, pp. 156–189. Guide Towards Sustainable Energy: Future for the Americas, Editors John Millhone, Claudio Estrada, Adriana de la Cruz Molina, Cuernavaca, Mexico, IANAS/IAP 2016, 206p.

FAO, 2015. *FAOStat: Production, Crops.* FAOStat 2013. http://faostat3.fao.org/download/Q/QC/E (access January 1, 2015).

IEA, 2016. *Energy Security.* International Energy Agency. Paris. Available at: www.iea.org/topics/energysecurity/subtopics/whatisenergysecurity/.

IEA, 2017. *Africa Energy Outlook.* International Energy Agency. Paris.

IEA/WB, 2017. International Energy Agency (IEA) and the World Bank. Sustainable Energy for All 2017. Global Tracking Framework: Progress toward Sustainable Energy 2017. International Bank for Reconstruction and Development/World Bank and the International Energy Agency (IEA), Washington, DC. 208p.

Leal, M.R.L.V. 2007. The potential of sugarcane as an energy source. *Proceedings of the XXVI International Society of Sugar Cane Technologists (ISSCT) Congress*, Durban, South Africa, August 31st–September 3, Vol. 26, 23–34.

REN21, 2017. Renewables Global Status Report (GSR 2017). Renewable Energy Policy Network for the 21st Century.

Souza, S.P.; Nogueira, L.A.H.; Watson, H.K.; Lynd, L.R.; Elmissiry, M.; Cortez, L.A.B. 2016. Potential of sugarcane in modern energy development in Southern Africa. *Front. Energy Res.* 4:39.

Vandeweerd, V. 2012. Foreword, *Energy Policy* 47:1.

UN. 2012. *Sustainable energy for all: a framework for action.* The Secretary-General's Sustainable Energy for All. UN, Geneva.

UN. 2014. UN Secretary-General Ban Ki-moon's message to the Clean Energy Ministerial Meeting, Seoul, May 2014.

UN. 2015. The Millennium Development Goals Report.

UN, 2017. *Sustainable energy for all: an overview.* 2p. Available at www.un.org/millenniumgoals/pdf/SEFA.pdf.

USAID, 2015. *Sugar in Mozambique: balancing competitiveness with protection.* United States Agency for International Development. 30p.

USDA, 2017. *Commodity basis: sugar prices.* United States Department of Agriculture. Available at www.commoditybasis.com.

Walter, A.S.; Galdos, M.V.; Scarpare, F.V.; Leal, M.R.L.V.; Seabra, J.E.A.; Cunha, M.P.; Picoli, M.C.A.; Oliveira, C.O.; 2013. Brazilian sugarcane ethanol: developments so far and challenges for the future. *WIREs Energy Environ*, 3:70–92. doi:101002/wene.87.

WHO, 2014. *Burden of disease from household air pollution for 2012.* Geneva: World Health Organization.

15 Sustainable bioenergy production in Mozambique as a vector for economic development

Marcelo Pereira da Cunha, Luiz Gustavo Antônio de Souza, André de Teive E Argollo, João Vinícius Napolitano da Cunha and Luiz A. Horta Nogueira

Introduction

Behind the search process to reach feasible renewable energies, the special case of the bioethanol has been considered as one of the alternative routes to decrease the dependency on fossil fuel and a way to mitigate the greenhouse gases (GHGs) emissions. The use of this kind of bioenergy could as well be a way to improve the energy security and sustainability of countries and to promote the rural economic sector, especially in tropical developing countries [1, 2].

In order to improve the production of bioenergy, the first requirement is to know where and how such production will be allocated in the world. Actually, 60% of the total world's potential available land could be used to produce bioenergy by 2050 (440 Mha), of which around 60% (250 Mha) will be in Latin America and Caribbean, and 180 Mha in Africa [3].

The availability of fertile land with good climatic conditions will determine the overall scale of bioenergy production in next decades, mainly in developing countries. However, it has widely been assumed that in order to increase or to introduce the production of bioenergy, a sustainable (social, economic and environmental) overview will be required, and during the process, the perception is that a growth in biofuels could sacrifice the food security, particularly for the world's poor [4].

The Mozambican agriculture has considerable agricultural potential, since it is a country with vast reserves of land suitable for cultivation. Thus, the Mozambican government hopes to take advantage of this aspect to encourage agriculture in view of the economic development. The economic impact of bioenergy sector can bring positive social impacts, as the majority of the Mozambican population lives in rural areas.

Considering liquid fuels, sugarcane bioethanol is recognized as the best current option for sustainable biofuel. In the context, the purpose of this study is to quantify the socioeconomic impacts of a sustainable sugarcane bioethanol production in Mozambique, including all direct, indirect and induced effects along the production chain – depending on the socioeconomic variable, the indirect and induced effects can be the most important one.

Taking these factors into account, this study aims to determine the direct, indirect and induced impacts created by a starting-level bioethanol and bioelectricity production in Mozambique. The available area suitable for sugarcane production in the country is around 21 Mha; this analysis focused on a conservative scenario of 600 kha or 2.86% of the total sugarcane available land.

Of course, such a production would affect many socioeconomic variables and indicators. Therefore, this study focused on studying the ones among the most critical for Mozambique, namely the output level, gross domestic product (GDP), jobs creation, government income and the Human Development Index (HDI).

In order to investigate the socioeconomic impacts, this chapter addresses the use of the national Social Accounting Matrix (SAM) for Mozambique published by Global Trade Analysis Project (GTAP) considering the data of 2011, built using the supply-use tables from national accounts, government budgets and balance of payments for separated 57 activities and commodities to obtain the input-output (I-O) matrix. In addition, a sugarcane bioenergy sector was included in the I-O model totalizing 58 activities.

This chapter follows the I-O approach as the methodology for the socioeconomic impacts evaluation, which presents the description and the relationship among all productive activities of a country or a region – in this case, Mozambique.

To evaluate the impacts, it was considered a scenario with the following assumptions: (i) adoption of sugarcane manual harvesting and (ii) distillery and bioelectricity production. The scenario considers the estimation of suitable sugarcane expansion area as well as the commercial available technologies in agricultural and industrial phases.

This chapter is structured in four sections, in which this introduction is included. The next section discusses the I-O methodology and SAM methods. The third section shows the main results and discussions, and the final section shows the main conclusions and final remarks.

Methods and materials

I-O analysis

I-O analysis is widely applied to conduct national economic analyses and structural research, and is used to assess macroeconomic impacts of bioenergy production [5, 6]. The methodology allows evaluating the impacts of new economic activities on a regional or national economy by using I-O tables, which represent annual monetary flows of goods and services among different sectors in the economy. In this study, I-O analysis is used to determine the impacts of sustainable sugarcane ethanol and bioelectricity production in Mozambique.

Sustainable bioenergy production in Mozambique

This methodology enables one to view a "snapshot" of the economy, representing from one side input suppliers – sectors that produce intermediate inputs for other sectors –, and from another side buyers – sectors that purchase inputs from other sectors to meet their need. In this sense, it shows the interdependence between flows [5, 7].

One essential set of data for an I-O model are monetary values of the transactions between pairs of sectors (from sector i to sector j) representing its origin and destination, with the variable z_{ij}. Moreover, in any country, there are sales to purchasers who are more external or exogenous to the industrial sectors that constitute the producers in the economy – for example, households, government and foreign trade. The demand of these external units is generally referred to as final demand [5].

Assume that the economy can be categorized into n sectors. If we denote by x_i, the total output (production) of sector i, and by f_i, the total final demand for sectors i's product, we may write a simple equation accounting for the way sector i distributes its product through sales to other sectors and to final demand:

$$x_i = z_{i1} + \cdots + z_{ij} + \cdots + z_{in} + f_i = \sum_{j=1}^{n} z_{ij} + f_i \qquad (15.1)$$

The z_{ij} terms represent *interindustry* sales by sector i (also known as *intermediate* sales) to all sectors j (including itself, when $j = i$). Equation (15.1) represents the distribution of sector i *output*. There will be an equation like this that identifies sales of the output of each of the n sectors, and arranging in a matrix[1] notation:

$$\mathbf{x} = \begin{bmatrix} x_1 \\ \vdots \\ x_n \end{bmatrix}, \quad \mathbf{Z} = \begin{bmatrix} z_{11} & \cdots & z_{i1} \\ \vdots & \ddots & \vdots \\ z_{n1} & \cdots & z_{nn} \end{bmatrix} \text{ and } \mathbf{f} = \begin{bmatrix} f_1 \\ \vdots \\ f_n \end{bmatrix} \qquad (15.2)$$

Or in algebraic matrix view,

$$\mathbf{x} = \mathbf{Z}\mathbf{i} + \mathbf{f} \qquad (15.3)$$

In I-O approach, a fundamental assumption is that the interindustry flows from i to j – recall that these are for a given period, say a year – depend entirely on the total output of sector j for that same time period. Given z_{ij} and x_j, the ratio between them for a specific year will be called a technical coefficient a_{ij}, which represents the amount of inputs from sector i required to produce a unit of final product of sector j. The a_{ij} are viewed as measuring fixed relationships between a sector's output and its inputs. Economies of scale in production are thus ignored; the production in a Leontief system operates under what is known as constant returns to scale [5].

Using the definition of technical coefficients, we can see that in the Leontief model, this becomes

$$x_j = \frac{z_{1j}}{a_{1j}} = \frac{z_{2j}}{a_{2j}} = \cdots = \frac{z_{nj}}{a_{nj}} \tag{15.4}$$

Thus, the more usual specification of the kind of production that is embodied in the I-O model is

$$x_j = \min\left(\frac{z_{1j}}{a_{1j}}, \frac{z_{2j}}{a_{2j}}, \cdots, \frac{z_{nj}}{a_{nj}}\right) \tag{15.5}$$

where min $(x, y, \ldots, z)$ denotes the smallest of the numbers $x, y, \ldots,$ and z. In the I-O model, for those a_{ij} coefficients that are not zero, these ratios will all be the same and equal to x_j. For those a_{ij} coefficients that are zero, the ratio will be infinitely large and hence will be overlooked in the process of searching the smallest ratios. This specification of the production function reflects the assumption of constant returns to scale; multiplication of $z_{1j}, z_{2j}, \ldots, z_{nj}$ by any constant will multiply x_j by the same constant [5].

If we change z_{ij} values in Eq. (15.1) for the correspondent $a_{ij}x_j$, Eq. (15.3) will become

$$\mathbf{x} = \mathbf{A}\mathbf{x} + \mathbf{f} \tag{15.6}$$

where $\mathbf{x}$ is a vector ($n \times 1$) representing the total output production of each sector, $\mathbf{f}$ is a vector ($n \times 1$) with the values of final demand and $\mathbf{A}$ is a matrix ($n \times n$) with the technical coefficients of production. Solving the $n \times n$ equation (15.6), we obtain

$$\mathbf{x} = (\mathbf{I} - \mathbf{A})^{-1}\mathbf{f} = \mathbf{L}\mathbf{f} \tag{15.7}$$

where $\mathbf{L} = (\mathbf{I} - \mathbf{A})^{-1} = [l_{ij}]$ is known as the Leontief inverse or the total requirements matrix (which measures the direct and indirect effects). In other words, the element l_{ij} is the total output of sector i needed to produce one unit of the final demand in sector j.

The model described in Eq. (15.7) depends on the existence of an exogenous sector, disconnected from the technologically interrelated productive sector, since it is here that the important final demands for outputs originate. The basic kinds of transactions that constitute the activity of this sector, as we have seen, are consumption purchases by households, sales to government, gross private domestic investment and shipments in foreign trade. In the case of households, especially, this "exogenous" categorization is something of strain on basic economic theory.

Households (consumers) earn incomes (at least in part) in payment for their labor inputs to production processes, and as consumer, they spend this in a rather well-patterned way. And, in particular, a change in the amount of labor needed for production in one or more sectors – say an increase in labor inputs due to increased output – will lead to a change (in this scenario, an increase) in the amounts spent by households as a group for consumption. Although households tend to purchase goods for "final" consumption, the amount of their purchases is related to their income, which depends on the outputs of each of the sectors [5].

One could thus move the household sector from the final-demand column and labor and capital owned by households-input row to the technically interrelated table, making it one of the *endogenous* sectors. This is known as closing the model with respect to households, allowing to include the assessment of the induced (or income) effects on the economy, for example, considering the expansion of exports. So, the *i*th equation, as shown in Eq. (15.1), would now be modified to

$$x_i = z_{i1} + \cdots + z_{ij} + \cdots + z_{in} + z_{i,n+1} + f_i^* \tag{15.8}$$

where f^* represents the remaining final demand for sector *i* output – exclusive of that from households, which is now captured in $z_{i,n+1}$. In addition to this kind of modification on each of the equations, there would be one new equation for the total "output" of the household sector, defined as the total value of its sale of labor services to the various sectors – total earnings. Thus,

$$x_{n+1} = z_{n+1,1} + \cdots + z_{n+1,j} + \cdots + z_{n+1,n} + z_{n+1,n+1} + f_{n+1}^* \tag{15.9}$$

In other terms, the matrix of technical coefficients **A** would change, so denote by $\overline{\mathbf{A}}$ the $(n + 1) \times (n + 1)$ technical coefficients matrix with households included. Using partitioning to separate the old **A** matrix from the new sector,

$$\overline{\mathbf{A}} = \begin{bmatrix} \mathbf{A} & \mathbf{h}_C \\ \mathbf{h}_R & 0 \end{bmatrix} \tag{15.10}$$

where $\mathbf{h}_C$ is the column vector of household consumption coefficients, and $\mathbf{h}_R$ is the row vector of labor and capital input coefficients owned by households. Let $\overline{\mathbf{x}}$ denote the $(n + 1)$th element column vector of gross outputs, $\mathbf{f}^*$ be the *n*th element vector of remaining final demands for output of the original *n* sectors and $\overline{\mathbf{f}}$ be the $(n + 1)$th element vector of final demands, including that for the output of households:

$$\overline{\mathbf{x}} = \begin{bmatrix} x_1 \\ \vdots \\ x_n \\ x_{n+1} \end{bmatrix} = \begin{bmatrix} \mathbf{x} \\ x_{n+1} \end{bmatrix}, \quad \text{and } \overline{\mathbf{f}} = \begin{bmatrix} f_1^* \\ \vdots \\ f_n^* \\ f_{n+1}^* \end{bmatrix} = \begin{bmatrix} \mathbf{f}^* \\ f_{n+1}^* \end{bmatrix} \tag{15.11}$$

Then, the new system of $(n + 1)$ equations, with households as an endogenous variable, can be represented as

$$(\mathbf{I} - \overline{\mathbf{A}})\overline{\mathbf{x}} = \overline{\mathbf{f}} \tag{15.12}$$

or

$$\overline{\mathbf{x}} = (\mathbf{I} - \overline{\mathbf{A}})^{-1}\overline{\mathbf{f}} = \overline{\mathbf{L}}\overline{\mathbf{f}} \tag{15.13}$$

The next subsection will present the principles of SAM and its application in a complementary way over I-O approach.

Social Accounting Matrix

The SAM seeks to expand the I-O analysis to incorporate these other elements of national accounts that usually are not present in the I-O analysis [5].

There is no standard definition of what SAM is. Following Pyatt [8], a SAM is a simple and efficient way to represent the fundamental law of economics that for each revenue, there must be a corresponding expense.

According to Robinson and Roland-Holst [9], despite the variations on definitions for the entries in SAM, there are some basic properties that it shall comply with:

a   it is a square matrix where the total of lines and columns representing the incomes and expenditures of the various agents should always be the same;
b   there is a double entry convention that ensures there will be no leaks or injections of resources in the system and that each flow should go from one agent to another;
c   by convention, revenues are recorded in lines and spending in columns.

A general structure of a SAM is presented in Table 15.1.

Taking into account the use of SAM approach to evaluate the structural characteristics of the economy of Mozambique, Arndt et al. [10–12] provided an analysis, as well, of biofuels and poverty in this country.

The work of Arndt and Tarp [13] is exceptional in showing how the various scenarios established production of biofuels (ethanol and biodiesel, produced by sugarcane and Jatropha, respectively) cause economic and social impacts.

Among the works in the social science, Jelsma et al. [14] is one of the most in-depth. The effort was done to register the dynamics of one of the few areas where the sugarcane cultivation took place with the participation of smallholders. This report is important because the use of hand labor of small farmers, instead of large estates, has a greater effect on poverty reduction.

As noted by Arndt and Tarp [13], there is little quantitative work that deals with the impacts of biofuels. A topic related to sustainability of biofuels is about the issue of GHG pollutants, to investigate how green are biofuels compared to fossil fuels.

Results

In some low-income countries, a very poor population lives in a rich natural endowment. However, it is not an easy task to convert untapped resources in welfare and development. The case of Mozambique is emblematic: holder of one of the ten lowest HDIs (0.418 in 2015), about 60.7% of the population

Table 15.1 SAM structure

	Activities	Commodities	Factors	Enterprises	Households	Government	Investment	Rest of the World	Total
Activities		Marketed output			Home consumption	Government consumption		Exports	Activity income
Commodities	Intermediate inputs	Transaction costs			Marketed consumption	Government consumption	Investment, change in stocks		Total demand
Factors	Value added							Foreign factor earnings	Factor earnings
Enterprises			Factor income to enterprises			Transfers to enterprises		Foreign enterprise receipts	Enterprise earnings
Households			Factor income to households	Indirect capital payments	Inter-household transfers	Transfers to households		Foreign remittances received	Household income
Government	Producer taxes	Sales taxes, import tariffs	Factor taxes	Corporate taxes	Personal taxes			Government transfers from rest of	Government income
Savings				Enterprise savings	Household savings	Government savings		Foreign savings	Savings
Rest of the World		Imports		Repatriated earnings	Foreign remittances paid	Government transfers to rest of world			Foreign exchange outflow
Total	Gross output	Total supply	Factor expenditure	Enterprise expenditure	Household expenditure	Government expenditure	Investment	Foreign exchange inflow	

Source: Based in Miller and Blair [5].

lives below the poverty line, with adult literacy rate at 50.6%, and average life expectancy just 55.1 years, with alarming lacks on health and education services [15]. Nevertheless, this country has a large reserve of natural resources, and its economy is taking off, expanding the national infrastructure and demanding more energy, posing challenges on improving social conditions and protecting the environment.

In this context, the relevant and recurrent question about how to increase the modern energy supply arises, either exploring the promising offshore oil reserves or fostering renewable energy sources, such as bioenergy. Actually, sugarcane bioenergy development has been presented as a possible way to mobilize resources and expand the economy, creating thousands of jobs to unskilled workers, improving domestic energy supply and generating tradable surpluses, in sustainable coexistence with the expansion of food production. Sugarcane is particularly favored as solar energy vector, able to produce liquid biofuels and electricity by efficient and competitive processes.

In the simulation proposed by this chapter, the production is just at a starting level – it could be much larger – even though the results would bring a new dynamic to Mozambique. In order to process the sugarcane production from 600,000 ha, 24 distilleries would be needed, which account for an investment of US$ 10.2 billion. Annually, about 48.0 million tons of sugarcane would be harvested in this area to produce 4.11 billion liters of fuel-grade ethanol and 2.7 TWh of electricity (the total electricity consumption in Mozambique was 11.6 TWh in 2013), representing a total investment of US$ 10.2 billion.

Output multipliers

Output multipliers show the total output (production of all activities in an economy) to meet a final demand in one monetary value (for example, US$ 1) for a specific sector. Type-I multipliers consider all combined direct and indirect effects over the production chain, and Type-II multipliers include the induced (or income) effect.

Multipliers are very important in I-O analysis because they can be easily applied when one wants to evaluate socioeconomic impacts considering some exogenous change in final demand.

Of course, the magnitude of the multipliers for each sector will depend on the structure of the economy at the region or country studied. In this chapter, we have introduced a sugarcane-based bioenergy sector in the Mozambican economy, taking into account its structure in 2011. The output multipliers are shown in Table 15.2 (Type-I multipliers) and Table 15.3 (Type-II multipliers).

The Type-I output multipliers in Table 15.2 show that sugarcane bioenergy sector and sugarcane production are ranked, respectively, at 37th and 51st positions among 58 activities. Usually, sectors of the secondary sector (industries) present the highest values due to their strong linkage with other

Table 15.2 Type-I output multipliers

Rank	Sector	Multiplier	Rank	Sector	Multiplier
26	Paddy rice	1.577	25	Wood products	1.603
12	Wheat	1.834	28	Paper products, publishing	1.529
46	Cereal grains nec	1.322	3	Petroleum, coal products	2.204
52	Vegetables, fruit, nuts	1.234	8	Chemical, rubber, plastic products	1.958
58	Oil seeds	1.011	17	Mineral products nec	1.750
51	Sugarcane, sugar beet	1.248	31	Ferrous metals	1.489
23	Plant-based fibers	1.614	20	Metals nec	1.666
43	Crops nec	1.376	22	Metal products	1.619
41	Cattle, sheep, goats, horses	1.405	7	Motor vehicles and parts	1.999
54	Animal products nec	1.207	16	Transport equipment nec	1.810
27	Raw milk	1.576	6	Electronic equipment	2.050
2	Wool, silk-worm cocoons	2.578	14	Machinery and equipment nec	1.828
44	Forestry	1.346	10	Manufactures nec	1.928
33	Fishing	1.473	57	Electricity	1.101
32	Coal	1.477	42	Gas manufacture, distribution	1.405
50	Oil	1.251	21	Water	1.637
48	Gas	1.283	30	Construction	1.502
34	Minerals nec	1.467	53	Trade	1.222
15	Meat: cattle, sheep, goats, horse	1.811	45	Transport nec	1.343
35	Meat products nec	1.466	55	Sea transport	1.179
39	Vegetable oils and fats	1.431	49	Air transport	1.253
18	Dairy products	1.725	40	Communication	1.426
5	Processed rice	2.061	9	Financial services nec	1.945
11	Sugar	1.887	1	Insurance	2.600
24	Food products nec	1.612	38	Business services nec	1.450
36	Beverages and tobacco products	1.455	29	Recreation and other services	1.519
13	Textiles	1.832	47	PubAdmin/Defence/Health/Education	1.318
19	Wearing apparel	1.680	56	Dwellings	1.128
4	Leather products	2.100	37	Sugarcane bioenergy	1.450

Table 15.3 Type-II output multipliers

Rank	Sector	Multiplier	Rank	Sector	Multiplier
13	Paddy rice	2.948	21	Wood products	2.801
45	Wheat	2.295	29	Paper products, publishing	2.629
27	Cereal grains nec	2.646	2	Petroleum, coal products	3.544
33	Vegetables, fruit, nuts	2.586	10	Chemical, rubber, plastic products	3.042
42	Oil seeds	2.424	40	Mineral products nec	2.456
30	Sugarcane, sugar beet	2.629	49	Ferrous metals	2.235
12	Plant-based fibers	2.959	37	Metals nec	2.536
28	Crops nec	2.637	41	Metal products	2.431
24	Cattle, sheep, goats, horses	2.737	7	Motor vehicles and parts	3.172
35	Animal products nec	2.547	20	Transport equipment nec	2.844
15	Raw milk	2.904	5	Electronic equipment	3.286
1	Wool, silk-worm cocoons	3.901	17	Machinery and equipment nec	2.886
25	Forestry	2.727	11	Manufactures nec	3.007
38	Fishing	2.527	51	Electricity	2.218
50	Coal	2.227	52	Gas manufacture, distribution	2.216
56	Oil	1.851	44	Water	2.295
54	Gas	2.076	48	Construction	2.256
43	Minerals nec	2.348	32	Trade	2.601
9	Meat: cattle, sheep, goats, horse	3.131	53	Transport nec	2.100
16	Meat products nec	2.893	58	Sea transport	1.611
22	Vegetable oils and fats	2.767	57	Air transport	1.759
18	Dairy products	2.854	46	Communication	2.284
6	Processed rice	3.219	23	Financial services nec	2.758
8	Sugar	3.144	3	Insurance	3.437
39	Food products nec	2.478	26	Business services nec	2.704
47	Beverages and tobacco products	2.282	31	Recreation and other services	2.610
14	Textiles	2.947	36	PubAdmin/Defence/Health/Education	2.540
19	Wearing apparel	2.845	55	Dwellings	2.063
4	Leather products	3.315	34	Sugarcane bioenergy	2.573

domestic activities; the result of 1.450 for bioenergy means that to meet one monetary value (US$ 1 , for example) by bioenergy in final demand, the whole economy needs to produce 1.450 monetary value, taking into consideration all direct and indirect effects at the production chain.

Including the induced effect, the Type-II output multipliers in Table 15.3 show that sugarcane bioenergy sector and sugarcane production are ranked, respectively, at 34th and 30th positions among 58 activities, i.e., they moved up relatively to the other industries. The result of 2.573 for bioenergy means that to meet one monetary value (US$ 1 , for example) by bioenergy in final demand (excluding the household consumption), the whole economy needs to produce 2.573 monetary values, taking into consideration all direct, indirect and income effects at the production chain.

Comparing the results between Tables 15.2 and 15.3, it is possible to calculate the increase in each activity; they vary on the range of 25.1% (wheat) to 139.7% (oil seeds). In reality, these sectors were not important for the Mozambican economy in 2011. The increases in sugarcane bioenergy and sugarcane production are 77.4% and 110.7%, respectively.

Socioeconomic impacts – output level, GDP, and taxes over products and jobs

The main objective of this research is to provide a quantitative assessment of the socioeconomic impacts over the Mozambican economy in 2011, considering the hypothetical presence of a sugarcane-based bioenergy sector in that year, producing 4.11 billion liters of fuel ethanol and 2.7 TWh surplus of electricity, requiring a production, as well, of 48.0 million tons of sugarcane on 600,000 ha. It was assumed that these amounts of bioethanol and bioelectricity would be exported to external markets.

Considering all direct and indirect effects, Table 15.4 shows the main results: there would be increases in output level, GDP and taxes over products of, respectively, US$ 3,069 million, US$ 2,019 million and US$ 69 million, and would be necessarily close to 823,000 more jobs. In relative terms, these impacts represent increases of 13.7%, 16.6%, 10.8% and 6.9% on these variables, compared to the 2011 levels. From Table 15.4, it is possible to see that the major impacts are very concentrated in bioenergy and sugarcane sectors.

These results are very representative, mainly considering their relative size to the Mozambican economy. Of course, the increase in household income due to the presence of bioenergy sector and its whole production chain would give another incentive to the domestic economy, as the household consumption would experience an additional increase (the so-called income or induced effect); in Table 15.5, all the impacts on the socioeconomic variables selected are estimated taking into account the direct, indirect and induced effects.

Taking into consideration all direct, indirect and induced effects, Table 15.5 shows the main results: the output level would increase at US$ 5,445 million (24.3% higher than 2011 level), GDP at US$ 3,378 million (27.7% higher

Table 15.4 Socioeconomic impacts considering direct and indirect effects

Direct and indirect effects

Sector	Output (million US$ 2011)	GDP (million US$ 2011)	Taxes (million US$ 2011)	Jobs
Bioenergy from sugarcane	2,117	1,395	57	20,826
Sugarcane	504	395	8	785,281
Other agricultural products and forestry	2	1	0	5,750
Animal production and fishing	1	1	0	2,597
Oil, gas and coal	0	0	0	3
Other mineral products	0	0	0	18
Food and beverages products	8	2	0	987
Textiles, wearing apparel and leather products	4	1	0	556
Wood and paper products and publishing	2	1	0	653
Petroleum, coal products	0	0	0	0
Chemical, rubber, plastic products	21	2	0	508
Mineral products	1	0	0	28
Ferrous metals and metal products	24	5	0	1,255
Motor vehicles and parts and Transport equipment	1	0	0	22
Electronic equipment, machinery and other equipment	25	3	0	1,085
Other manufactures	0	0	0	4
Electricity, gas manufacture and distribution and water supply	15	12	0	50
Construction	1	0	0	0
Trade	153	123	2	0
Transport	66	25	1	1,186
Communication	11	6	0	124
Financial services and insurance	72	21	0	294
Business services	35	20	0	1,333
PubAdmin/Defence/Health/Education	3	2	0	138
Other services	3	1	0	94
Total	3,069	2,019	69	822,793

Table 15.5 Socioeconomic impacts considering direct, indirect and induced effects

Direct, indirect and induced effects

Sector	Output (million US$ 2011)	GDP (million US$ 2011)	Taxes (million US$ 2011)	Jobs
Bioenergy from sugarcane	2,117	1,395	57	20,826
Sugarcane	507	398	8	790,964
Other agricultural products and forestry	704	523	10	2,079,680
Animal production and fishing	121	76	1	283,856
Oil, gas and coal	0	0	0	12
Other mineral products	5	2	0	250
Food and beverages products	293	96	2	46,209
Textiles, wearing apparel and leather products	54	14	0	9,216
Wood and paper products and publishing	25	10	0	7,406
Petroleum, coal products	0	0	0	0
Chemical, rubber, plastic products	50	5	0	1,212
Mineral products	9	1	0	166
Ferrous metals and metal products	25	5	0	1,269
Motor vehicles and parts and Transport equipment	13	1	0	405
Electronic equipment, machinery and other equipment	37	5	0	1,508
Other manufactures	3	0	0	98
Electricity, gas manufacture and distribution and water supply	85	48	1	180
Construction	6	2	0	0
Trade	431	347	7	0
Transport	370	140	3	6,501
Communication	66	35	1	750
Financial services and insurance	199	54	1	781
Business services	83	48	1	3,178
PubAdmin/Defence/Health/Education	55	34	1	2,447
Other services	188	136	3	2,614
Total	5,445	3,378	96	3,259,531

than 2011 level) and taxes over products at US$ 96 million (15.0% higher than 2011 level) and 3.26 million more jobs (15.0% higher than 2011 level) would be necessary to meet the growth of the entire economy. The results figure as very impressive at relative basis.

A more accurate observation on Table 15.5 shows that in this case, the relative importance of bioenergy and sugarcane sector is lower when compared with the situation where only direct and indirect effects are assessed. For example, in terms of jobs creation, the other activities are responsible for 75.1% of the total; in particular, other agricultural products and animal production account for 72.5% of all new jobs, showing the importance of these sectors on household consumption in Mozambique.

It is worth to note, as well, that all sectors in agriculture and animal production sum 96.8% of all increase in jobs (3.155 million), which represents 26.6% more employed people in Mozambican economy in 2011. As unemployment rate in this country was 23.3% and a majority of that is concentrated in unskilled workers, a bioenergy sector supported by manual harvesting sugarcane activity and family farming agriculture would be a real solution to alleviate the socioeconomic implications for million people with no education in a poor country – at least, until the economic growth reaches a level that allows the introduction of other policies.

This means that direct, indirect and induced impacts play an important role here, showing that the agricultural sector of Mozambique can create a strong final demand across different sectors.

Returning to the total increase of US$ 3,378 million in GDP, 46.9% are due to the sugarcane and bioenergy production. That addition of 27.7% to the GDP, in comparison to 2011, would create another impulse for a country that has seen an impressive average of 9% per year growth in the last decade (2004–2014).

Moreover, by creating more job opportunities, the families could count on more income stability, since despite the great numbers, the Mozambican economy still faces one of the highest unemployment rates, standing at 24.4% in 2016.

In terms of the increase of US$ 96 million on taxes over products, this amount is 6.8% of the Mozambican government expenditure, in 2011, on public education, health, defense and public administration services; this amount would offer, as well, an opportunity to reinforce the benefits of such a bioenergy sector in this country.

Influence over HDI

Another indicator that would see its numbers surge is the HDI. In the last decade (2005–2015), it has grown by 0.065, whilst if the bioenergy based on sugarcane sector were to be installed, this number would climb another 0.030, reaching 0.448 from the 2015 level. This would mean the last six years of development in the country.

To estimate this impact, the HDI and GDP of the 178 countries, in 2011, were collected to calculate how the last explained the first. The empirical relation between these two variables is shown in Figure 15.1. In this figure, an adjusted curve was set, where the parameter R^2 is equal to 0.8737. As one can notice, the variables are strongly correlated, which allowed to infer what would be the impact on the HDI brought by the increase of 27.7% on the GDP (and in GDP per capita) of Mozambique in 2011.

The same adjusted curve was used to estimate the evolution of the HDI in Mozambique from 1990; further, the results were compared to the HDIs observed in this country, as shown in Figure 15.2.

In Figure 15.2, it is clear that the difference between both estimated and observed HDIs for Mozambique is very tiny from 2009 onward, i.e., always

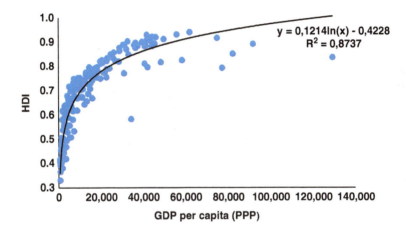

Figure 15.1 Relationship between HDI and GDP for 178 countries in 2011.

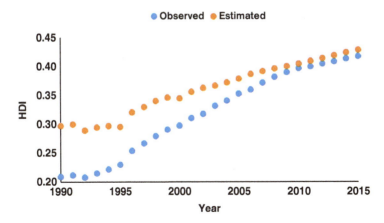

Figure 15.2 Comparison between estimated and observed HDIs for Mozambique.

less than 3%. The expression relating both variables, at the adjusted curve in Figure 15.1, is

$$y = 0.121373 \ln(x) - 0.422828 \qquad (15.14)$$

where y is the estimated HDI, and x is the GDP per capita.

To calculate the increase on HDI, the adjusted curve was applied considering an increase of that 27.7% on GDP per capita, resulting in an increase of 0.030 on the HDI in 2011.

Final remarks and conclusions

Mozambique is a country with vast reserves of land suitable for cultivation; if a sugarcane-based bioenergy sector producing 4.11 billion liters of fuel ethanol and 2.7 TWh surplus of electricity per year, requiring a production, as well, of 48.0 million tons of sugarcane on 600,000 ha (only 2.86% of the total sugarcane available land in the country), was hypothetically presented in Mozambican economy in 2011, the domestic GDP would increase 27.7%, generating around more 3.3 million jobs – most of which (96.8%) in agriculture and animal production due to the increase in household consumption explained by income effect.

The increase in GDP per capita would bring, as well, a gain of 0.030 point at HDI, which represents the development of the country in last six years. As the unemployment rate in Mozambique was at 23.3% in 2011 and a majority of that is concentrated in unskilled workers, a bioenergy sector supported by manual harvesting sugarcane activity and family farming agriculture would be a real solution to alleviate the socioeconomic implications for million people with no education in a poor country – at least, until the economic growth reaches a level allowing the introduction of other policies.

As suggestions to improve this research, the authors consider (i) a regional assessment over socioeconomic implications of stronger bioenergy sugarcane sector in Mozambique and (ii) applying a General Equilibrium framework, which brings a more detailed relationship among other elements of the economy: government consumption, investment and external sector.

Acknowledgments

The authors are grateful to FAPESP for providing financial support to the LACAf project (2012/00282-3).

Note

1 Here and throughout this text, we use lowercase bold letters for (column) vectors, as in **f** and **x** (so **x'** is the corresponding row vector) and uppercase bold letters for matrices, as in **Z**. In addition, we use **i** to represent a column vector of 1's of appropriate dimension.

References

[1] Cuvilas CA, Jirjis R, Lucas C. Energy situation in Mozambique: A review. *Renewable and Sustainable Energy Reviews.* 2010;14:2139–2146.

[2] de Souza LGA, de Moraes MAFD, Dal Poz MES, da Silveira JMFJ. Collaborative networks as a measure of the innovation systems in second-generation ethanol. *Scientometrics.* 2015;103:355–372.

[3] Doornbosch R, Steenblik R. *Biofuels: Is the Cure Worse Than the Disease? Round Table on Sustainable Development.* Paris: OECD; 2007.

[4] Lynd LR, Woods J. Perspective: A new hope for Africa. *Nature.* 2011;474:S20–S21.

[5] Miller RE, Blair PD. *Input-output analysis: Foundations and extensions.* New York: Cambridge University Press; 2009.

[6] Herreras Martínez S, van Eijck J, Pereira da Cunha M, Guilhoto JJM, Walter A, Faaij A. Analysis of socio-economic impacts of sustainable sugarcane–ethanol production by means of inter-regional input–output analysis: Demonstrated for Northeast Brazil. *Renewable and Sustainable Energy Reviews.* 2013;28:290–316.

[7] Guilhoto JJM. *Análise de insumo-produto: teoria e fundamentos.* 2011.

[8] Pyatt G. A SAM approach to modeling. *Journal of Policy Modeling.* 1988;10:327–352.

[9] Robinson S, Roland-Holst DW. Macroeconomic structure and computable general equilibrium models. *Journal of Policy Modeling.* 1988;10:353–375.

[10] Arndt C, Thurlow J. *A 2007 Social Accounting Matrix (SAM) for Mozambique.* Washington, DC: International Food Policy Research Institute (IFPRI); 2014.

[11] Arndt C, Jensen HT, Tarp F. Structural characteristics of the economy of Mozambique: A SAM-based analysis. *Review of Development Economics.* 2000;4:292–306.

[12] Arndt C, Benfica R, Tarp F, Thurlow J, Uaiene R. Biofuels, poverty, and growth: A computable general equilibrium analysis of Mozambique. *Environment and Development Economics.* 2010;15:81–105.

[13] Arndt C, Tarp F. Agricultural technology, risk, and gender: A CGE analysis of Mozambique. *World Development.* 2000;28:1307–26.

[14] Jelsma I, Bolding A, Slingerland M. *Smallholder Sugarcane Production Systems in Xinavane, Mozambique-Report from the Field.* Wageningen: Wageningen University; 2010.

[15] United Nations Development Programme (UNDP). Human Development Report 2015. United Nations Development Programme, New York, USA, 2015.

16 Integrated analysis of bioenergy systems
The path from initial prospects to feasibility and acceptability

Luiz A. Horta Nogueira

Introduction

Sustainable bioenergy has been recognized as a valuable energy alternative, fostering socioeconomic development and allowing to deal with climate change and environmental concerns. According to a detailed evaluation of its global potential, bioenergy annual production could grow to about 138 EJ by 2050, reaching 40% of the projected world energy supply. Depending on the biomass productivity, between 50 and 200 million rainfed hectares would be required, corresponding to 0.4%–1.5% of total global land (Souza and Vitoria et al., 2015). Just considering the land for rainfed agriculture expansion, without forests or areas to be protected, FAO estimated an availability of 1.4 billion ha of prime and good land and an additional 1.5 billion ha of spare and usable marginal land, several times greater the area required for food production and bioenergy production in that time horizon. About 960 Mha of this land, mostly pasture/rangeland, are in sub-Saharan Africa (47%) and Latin America and the Caribbean (38%), wet tropical regions, favorable to biomass production (Souza and Vitoria et al., 2015).

However, even when disposing of proper technologies for developing this bioenergy potential, especially in developing countries, modern bioenergy systems are still in discussion, without actions towards their effective deployment. In some countries, proposals of modern bioenergy systems receive an initial positive evaluation, but are not implemented due to restrictions from the local community perspective or exaggerated risk perception by society and decision makers, generally based on limited information and superficial assessment. For instance, can be mentioned the long debate, still unsolved in some countries, about the convenience of introducing gasoline blends with ethanol, despite several independent studies confirming the economic and technical consistence of this alternative, and the associated effects of energy security, cleaner environment, and social benefits, including food safety (Rosillo-Calle and Walter, 2006; Lynd and Woods, 2011).

Possibly, this situation is a consequence of the intrinsic complexity of modern bioenergy systems. Among all other energy resources and technologies, bioenergy is the most diversified in scale, conversion processes and final energy product, presenting a large set of relationships with the environment, economic sectors, public institutions, and society; imposing careful analysis of each case, able to consider the inherent relationship and the overlap of areas and aspects related to agricultural activity and energy production; and, as much as possible, adopting an integrated evaluation. Assessment methods of agricultural and bioenergy systems are currently available, however emphasizing particular dimensions, and usually do not allowing an integrated perspective, essential for understanding and evaluating the system's sustainability. Besides, those assessment methods do not consider the public perspective and interest, which deserves attention to put forward robust bioenergy projects.

The aim of this chapter is to propose a pathway to support policymakers in the assessment process of deploying sustainable bioenergy systems or evaluating existing ones, particularly for developing countries. Considering the usual constraints in the data and information required, a set of evaluation methods is compiled in a logical and sequential structure to draw a set of indicators used for assessing a given project. These indicators are evaluated in a Strengths, Weaknesses, Opportunities, and Threats (SWOT) matrix in order to obtain a set of options for evaluating bioenergy projects. If the outcome of this analysis indicates a feasible project and acceptability is recognized as an issue, this process should be followed by a Public Consultation and Communication (PC&C) scheme. This approach was named Sustainable and Integrated Bioenergy Assessment for Latin America and Africa (SIByl-LACAf[1]), which is expected to constitute a comprehensive framework for addressing the inherent complexity of bioenergy systems, able to make explicit the sustainability and acceptability as part of the evaluation process of bioenergy systems and projects. A more complete and detailed version is available (Nogueira et al., 2017).

The next section sets the problem of evaluating the feasibility of inherently complex bioenergy systems and, in some cases, with lack of data and information, as well as the frequent wrong perception about bioenergy impacts. The third section introduces the SIByl-LACAf framework, the essentials of the selected methods and the steps in following such approach. A final section presents the main remarks and final considerations.

Identifying the problem and proposing an approach

The referential framework for this work were the integrated analysis (IA) studies, which can be defined as a reflective and iterative participatory process that links knowledge (science) and action (policy) regarding complex global issues such as bioenergy production and climate change (Kloprogge and Van Der Sluijs, 2006). Dale et al. (2013) reported that fewer studies used IA for bioenergy system approaches than those using isolated approaches for qualitative analysis of indicators used for understanding the socioeconomic factors in such a system.

To integrate the different perspectives of analysis and address the relationships among them, a network approach can be adopted, in which all the data and information, models, and methods are part of a full network, connecting areas and subareas (Souza and Moraes et al., 2015). By this approach, less time and effort are required than evaluating each one separately. For example, to understand the sustainability of a local process, it is more important to understand the connections among environmental, economic, and social aspects than to examine each part individually.

The implementation of a sustainable bioenergy system, given a feedstock, technology, and final use, in a specific country, involves several direct and indirect factors, with an obvious complexity of situations and data. Hence, the crucial elements to support consistent decisions are the clear definition of the scope to be evaluated and the data and information collection to feed correctly the valuation processes, allowing to produce trusty indicators, to be finally used as decision criteria, as presented in the following section. External sustainable guidance from international agencies such as the Global Bioenergy Partnership (GBEP, 2016) and the Biofuels Sustainability Scorecard (IDB, 2016) can be useful complements in this process.

The SIByl-LACAf approach

The SIByl-LACAf approach is developed in seven steps, as indicated in Figure 16.1: (i) definition of objectives, (ii) recognizing the complexity of

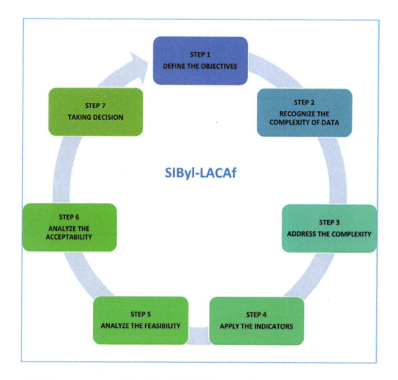

Figure 16.1 The structure of the SIByl-LACAf approach.
Source: based on Nogueira et al. (2017).

data, (iii) addressing the complexity, (iv) applying indicators, (v) analysis of feasibility, (vi) analysis of acceptability, and (vii) taking decision.

Step 1. Definition of objectives

In a general context, the following issues arise: (i) the purpose, (ii) the extent, and (iii) the methods used to implement a sustainable bioenergy system, guiding the definition of the objectives of analysis. Basically, two categories of projects can be evaluated: (i) Greenfield and (ii) Brownfield projects. A Greenfield project offers opportunities to an investor such as creating an entirely new organization with unique requirements, but implies a gradual market entry owing to the barriers and sunk costs already paid by other firms. Otherwise, an acquisition facilitates quick entry and immediate access to local resources, although the acquired company may require deep restructuring to overcome a lack of fit between the two organizations.

Step 2. Recognizing the complexity of data

As well known, bioenergy systems involve necessarily several areas of knowledge, which highlights the importance of conducting IA from informational and methodological perspectives. As examples of areas to be explored to provide information that may be sought to facilitate analysis by the proposed integrated model, one could mention the following:

i Environmental: CO_2 emissions, natural resources use, soil and pastureland management, climate change;
ii Economic: market clearing, feasibility, productivity, economies of scale, input-output relationship;
iii Social: skills, work conditions, wages, unions;
iv Institutional: laws, bureaucracy, government, research institutes;
v Technological: techniques, innovations, patents, knowledge;
vi Market for factors: capital, land, labor use;
vii Market for inputs: acquiring inputs for agriculture and industry (imports or local, regional, and national);
viii Market for outputs: selling outputs for external or national markets (exports or local, regional, and national);
ix Logistics: distribution logistics of inputs, outputs, and infrastructure.

Step 3. Addressing the complexity

The main methods and models used to draw quantitative and qualitative indicators for an IA of bioenergy systems are briefly presented in Table 16.1, with some examples of studies in this context. Although there is no preferred method, it is desirable to use a set of methods since every method processes different information and dependent variables that can be used in

Table 16.1 Analytic methods used for evaluating bioenergy systems by IA

Method	Synthetic description and reference of application on bioenergy
Agent-based model (ABM)	ABM provides a simulation approach for local-level assessment and considers important microlevel constraints such as environmental externalities, limited adaptive capacity, and behavioral barriers (Davis et al., 2009; Berger and Troost 2014)
Econometric model	Econometric models approach develop, estimate, and evaluates models that relate economic or financial variables, allowing to make diagnostic tests, ex-post forecasting, and simulations (Taylor and Taylor, 2009; Seyffarth, 2015)
General equilibrium model (GEM)	GEM looks for simulating the whole economy with several or many interacting markets, assuming an overall general equilibrium to explain the behavior of supply, demand, and prices (Dandres, 2012)
Input-output analysis (I-O)	I-O enables a snapshot of the economy under analysis by exposing the intra- and inter-sectorial factors representing goods and services suppliers and direct and indirect consumers, indicating the interdependence between economic flows (Herreras Martínez et al., 2013)
Landscape design	Landscape Design considers the effects of interventions and conditions at different spatial scales of a given project, taking into account physical, biological, and anthropic factors emphasizing the importance of self-organization and local action in the environment and allowing to identify the susceptible points (Dale et al., 2016)
Life cycle assessment (LCA)	LCA assesses the explicit and implicit impacts of a given process or product, "from the cradle to the grave", reflecting the environmental consequences of a choose set of inputs and outputs. LCA also follows internationally accepted methods and practices used to evaluate the requirements and impacts of technologies, processes, and products to determine their propensity to consume resources and generate pollution (Cherubini and Strømman, 2011; McKone et al., 2011)
Multi-criteria analysis (MCA)	MCA explicitly considers multiple criteria in helping individuals and groups to explore important decisions. MCA stands in contrast to single goal optimization and approaches that use "unifying units" to offset poor performances of one criterion by relying of good performances of another criterion, therefore exposing the actual trade-offs in each case (Buchholz and Rametsteiner et al., 2009; Scott et al., 2012)
Social network analysis (SNA)	Emerged from the graph theory, SNA models socioeconomic system by a graph or network is composed of nodes (people or groups of people with a common goal), edges (interactions or links between two or more nodes) and flow (direction of interaction, that may be unidirectional or bidirectional), interpreted using parameters such as geodesic distance, density and centrality degree to evaluate relevance, cohesion and other indicators (Souza and Moraes et al., 2015)

Method	Synthetic description and reference of application on bioenergy
Survey approach	Survey Approach provides a means of measuring a population's characteristics, self-reported and observed behavior, awareness of programs, attitudes, opinions, and needs. Repeating surveys at regular intervals can assist in the measurement of changes over time. Its reliability depends basically of the consistence of statistical sampling and surveying procedure (Buchholz and Luzadis et al., 2009)
System dynamics (SD)	SD models the interrelationships between or among subsystems that are linked by variables and aids in determining how such interlinkages will produce specific overall system behavior. SD adopts causal loop diagrams, which is a visual representation of the feedback loops in a system whereby the stocks and flows involving different variables, parameters, and indicators are connected by either positive or negative loops (Barisa et al., 2015; Martinez-Hernandez et al., 2015)

an integrated framework, offering supplementary perspectives. The methods presented in this table, commented in more detail by Nogueira et al. (2017), were applied in these references in a single or integrated way, in some applications using two or three methods, although not in the sense suggested here. Indeed, the literature review highlights the scarcity of integration among methods to support decision process.

Considering the period 1970–2016, Figure 16.2 shows the distribution of papers related to bioenergy for each selected method, indicating that the more

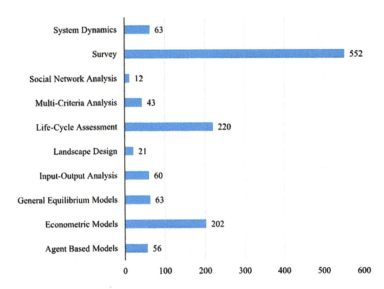

Figure 16.2 Number of papers (published between 1970 and 2016) applying evaluation method in bioenergy systems.

explored methods are related to surveys, life cycle assessment (LCA), and econometric models. The number of scientific papers indicated in this figure was determined using the WEB of Science database, adopting a search in topics (title, abstracts, keywords). To a detailed explanation of the procedure, see Souza and Moraes et al. (2015).

Addressing correctly the complexity in the SIByl-LACAf scheme (Steps 2 and 3) is crucial because their output are the indicators, which should support properly the decision process. The next step focuses on defining and selecting these indicators.

Step 4. Applying indicators

Indicators that measure parameters occurring before or after an event are essential for IA of bioenergy systems. Lying at the intersection of energy and agricultural activity, bioenergy systems are related to activities having important socioeconomic and environmental sustainability repercussions, thus the temporal dimension is relevant. To select and identify the more adequate indicators of sustainability, Dale and Beyeler (2001) suggested the following criteria:

i Practicality;
ii Sensitivity and responsiveness to both natural and anthropogenic stresses to the system;
iii Clarity with respect to what is measured, how measurements are made and how response is measured;
iv The ability to anticipate impending changes;
v The ability to predict changes that can be averted with management action;
vi Estimation capacity with known variability in response to changes;
vii Sufficiency when considered collectively.

A good example for indicators to assess bioenergy systems is the set of 24 structured indicators proposed by the GBEP Task Force on Sustainability (GBEP, 2016), presented in Table 16.2, to foster the sustainable development of bioenergy.

In the framework of SIByl-LACAf, it is suggested the use of qualitative and quantitative indicators from the methods presented in Step 3 as a main source of data to determine the appropriate indicators for a consistent feasibility analysis, as discussed in the following section.

Steps 5 and 6. Analysis of feasibility and acceptability

Of course that key point in the conception, planning and deployment of sustainable bioenergy projects is the feasibility analysis, which imposes

Table 16.2 GBEP sustainability indicators for bioenergy

Sustainability dimension

Environmental	Social	Economic
Life cycle GHG emissions	Allocation and tenure of land for new bioenergy production	Productivity
Soil quality	Price and supply of a national food basket	Net energy balance
Harvest levels of wood resources	Change in income	Gross value added
Emissions of non-GHG air pollutants, including air toxics	Jobs in the bioenergy sector	Change in consumption of fossil fuels and traditional use of biomass
Water use and efficiency	Change in unpaid time spent by women and children collecting biomass	Training and re-qualification of the workforce
Water quality	Bioenergy used to expand access to modern energy services	Energy diversity
Biological diversity in the landscape	Change in mortality and burden of disease attributable to indoor smoke	Infrastructure and logistics for distribution of bioenergy
Land use and land-use change related to bioenergy feedstock production	Incidence of occupational injury, illness and fatalities	Capacity and flexibility of use of bioenergy

Source: GBEP (2016).

understanding the indicators, models, and metrics to be applied. Such analysis determines, if the project is worth and guide, in principle, the final decision of the policymaker, investors, and other stakeholders.

Indeed, good results in sustainability and feasibility assessment can be not enough. To go ahead, it is important to evaluate also the project's acceptability (or desirability). Sometimes a generally positive condition for bioenergy development can be translated into a negative perception of relevant stakeholders and public opinion. In addressing this issue, it is advisable to recognize two separate factors that are not automatically observed and can often be in disagreement: inherent context (IC) and perceived context (PC).

We consider IC as the overall objective in a fact-based environment in which bioenergy projects already exist or will exist. This context is defined by a series of quantifiable indicators classified into thematic categories such as economic, social, environmental, agricultural, technological and legal factors. In the SIByl-LACAf framework, the previously discussed steps guide the policymaker in achieving feasibility. These aspects are part of a specific context following scientific procedures with indicators as main results.

In contrast, we consider PC as a subjective environment in which highly diversified perceptions and opinions of bioenergy interact and affect the potential development of sustainable bioenergy systems. Unlike IC, PC is based entirely on public perception rather than fact. External forces are present that can directly influence the actors and the decisions. The power of public perception should never be underestimated. It can often result in stronger arguments in favor or against bioenergy development compared with scientific facts that underpin the objective feasibility of bioenergy projects. Indeed, public perception is critical in determining the acceptability of a sustainable bioenergy project regardless of its feasibility.

As public opinion and perception are so diverse and are often not based on scientific fact and evidence, these aspects are difficult to evaluate. This challenge is not limited to academic analysis, however. The private industry actors that implement bioenergy projects are sometimes affected to an even greater extent by the opportunities and threats presented by varied and volatile public perception on bioenergy-related issues.

This is precisely why PC&C (Osseweijer, 2006) mechanisms are important steps in the implementation of any bioenergy project, as a systematic process that seeks the public's input on civil matters. The basic rationale underlying PC&C is the right of the public to be informed and consulted and to express opinion on matters of relevance. Its main objective is to improve the efficiency, transparency, and public involvement in large-scale projects or laws and policies. This process usually involves public notification to introduce the matter under consideration, consultation including a two-way flow of information and opinion exchange, and participation of interest groups in the drafting of policy or legislation. The PC&C process should lead to better decisions and can lead to improve relations between a project developer and the public. Where PC&C processes have been implemented in sustainable bioenergy projects, the minutes and findings of the reports on such processes are a rich source of information that can be used to analyze and evaluate PC in the development of bioenergy.

In SIByl-LACAf, the analytical framework used to evaluate all of the factors stemming from IC and PC is included in IA, resulting a slightly modified version of the traditional SWOT analysis model used in the structured planning of a project or business venture. Traditional SWOT analysis aims to identify the key internal and external factors deemed important for achieving an objective (see Okello et al. (2014) for a bioenergy application). Strengths and weaknesses are grouped as internal factors included in a business or organization, and opportunities and threats are grouped as external environmental factors of the business or organization. Strengths and opportunities are considered as helpful in the achievement of objectives, whereas weaknesses and threats are viewed as harmful.

The relevant difference between the traditional SWOT analysis model and SIByl-LACAf is that the latter substitutes the internal factors with IC. In the former method, the scientific fact-based characteristics that determine the practical feasibility of sustainable bioenergy projects are considered in the external environment with PC. That is, highly varied perception of sustainable bioenergy systems, real or imagined, is held by a variety of stakeholders external to the project.

Figure 16.3 shows the decision flow in the SIByl-LACAf approach to address the feasibility-acceptability problem. The analysis of feasibility in Step 5 is developed on the basis of information determined from the previous steps. Here, the policymaker can determine whether a project is feasible in terms of previously reported information. If the consultant verifies a "no" as the answer, there is no reason to continue the analysis and the project does not present interest. In fact, a review of the objectives and reconsideration of the problem can be developed in such cases, depending on the context. However, if a "yes" answer is verified, Step 6 should be done. In this step, the desirability of the project perceived by the community is addressed.

As previously mentioned, the public consultation is an important tool for understanding the perception of the actors. The Delphi method (Sackman, 1974) is also suggested to verify the opinion of more specialized actors of the community. As a group, both procedures feed Step 6, analysis of the desirability, and a new question emerges: Is the sustainable bioenergy project acceptable and desired by the actors?

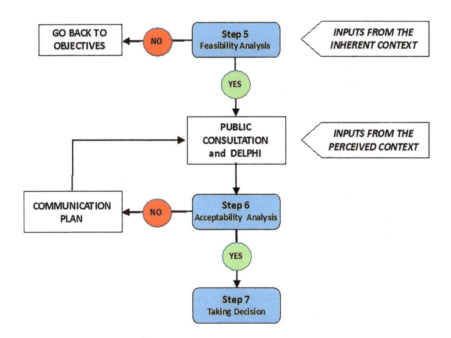

Figure 16.3 The decision flow in the feasibility and acceptability evaluation in the SIByl-LACAf.

If the consultant verifies a "no" answer, differently from Step 5, there is no reason to review the objectives. Here, the communication plan can be directly used to inform the community of the feasibility analysis results to avoid misunderstandings and incorrect preconceptions. However, we stress that this process is not intended to force the opinions in achieving a desirable result.

In Latin America, the Caribbean, and Africa, regions focused by LACAf Project, the acceptability issues of bioenergy projects appears more frequently than expected in the literature. Obviously, after implementing the communication plan, returning to the consultation stage and reanalysis of Step 6 is suggested to reevaluate if a positive answer is obtained, to go ahead or abandon the project.

Step 7. Taking decision

When the decision makers achieve Step 7, all of the procedures, information, indicators, and limitations must be known. Then a SWOT matrix can be used to translate the results into positive and negative categories to facilitate the decision. Hence, in this final step of SIByl-LACAf procedure, this matrix is constructed using the information of the feasibility analysis in the first line, the IC line, to identify the strengths and the weaknesses of the project, and in the second line, the PC line, to elucidate the opportunities and threats to the project, which included PC&C results.

Obviously, comparing different projects that are both feasible and acceptable is easier if the strengths and opportunities outnumber the weakness and threats. In summary, SIByl-LACAf provides guidelines for decision makers by expertly answering four questions that are relevant to similar sustainable bioenergy projects in the future: (i) How can the strengths in the IC be used to take advantage of the opportunities presented in the PC? (ii) How can the strengths in our IC be used to reduce the likelihood and impact of the threats present in the PC? (iii) How can the weaknesses in the IC that translate into threats in the PC be overcome? (iv) How can such weaknesses be addressed?

Educated and carefully evaluated answers to these questions can form the basis of guidelines and recommendations for policymakers and decision makers in future sustainable bioenergy projects with similar conditions and characteristics.

Final remarks

Effective indicators can help to identify and quantify the multivariate attributes of bioenergy options. However, in the process of developing and using criteria and indicators, the limitations of data and modeling deserve careful attention. Even exhaustive and comprehensive analytical frameworks that account for factors within all of these spheres have demonstrated limitations. This is evidenced by the fact that, even in cases in which all or most of the evaluated contexts indicate positive results and the potential for bioenergy development, such projects can face overwhelming resistance by a wide variety of stakeholders and public opinion. In other words, while such projects are deemed objectively feasible, they are not subjectively acceptable.

After the execution of the previous steps for analysis, the stakeholder will face different questions with different developments. If a taking decision process will be used only for feasibility, there will be three scenarios: (i) not feasible, but deserving changes and reevaluation of objectives are reevaluated; (ii) not feasible, and the project is denied; and (iii) feasible. However, as suggested, the stakeholder might want to understand how worth and acceptable the project is and how to implement the project. In this context, we suggest additional steps for achieving feasibility and acceptability under the SIByl-LACAf approach. A problem emerges because relevant stakeholders and public opinion can sometimes translate a generally positive condition for bioenergy development into negative perception. Thus, as already observed, we must first recognize that there are two separate contexts at work here, which are not automatically causal and can often be at odds with each other. We identified these contexts as IC and PC. The IC is immediately related to the overall objective (fact-based) environment in which bioenergy projects already exist or will exist, and PC as the external and public vision, not necessarily directly affected by the project.

The SIByl-LACAf framework intends to contribute to the state of the art of decision-making, integrating and combining analytic methods to reinforce the feasibility assessment and adding an acceptability evaluation, considering communications and consultation processes with local communities across the countries that can deploy modern bioenergy systems in the next years. In Nogueira et al. (2017), a preliminary application of this approach for Mozambique is summarized, indicating a route to evaluating actual contexts and handle with the acceptability issues for sustainable bioenergy development.

Note

1 The acronym SIByl-LACAf is a tribute to the legend of Greek oracle named Sibyl, represented as an old woman able to make clever predictions.

References

Barisa A, Romagnoli F, Blumberga A, Blumberga D. Future biodiesel policy designs and consumption patterns in Latvia: a system dynamics model. *Journal of Cleaner Production.* 2015;88:71–82.

Berger T, Troost C. Agent-based Modelling of Climate Adaptation and Mitigation Options in Agriculture. *Journal of Agricultural Economics.* 2014;65:323–348.

Buchholz T, Luzadis VA, Volk TA. Sustainability Criteria for Bioenergy Systems: Results from an Expert Survey. *Journal of Cleaner Production.* 2009;17, Supplement 1:S86–S98.

Buchholz T, Rametsteiner E, Volk TA, Luzadis VA. Multi Criteria Analysis for Bioenergy Systems Assessments. *Energy Policy.* 2009;37:484–495.

Cherubini F, Strømman AH. Life Cycle Assessment of Bioenergy Systems: State of the Art and Future Challenges. *Bioresource Technology.* 2011;102:437–451.

Dale VH, Beyeler SC. Challenges in the Development and Use of Ecological Indicators. *Ecological Indicators.* 2001;1:3–10.

Dale VH, Efroymson RA, Kline KL, Langholtz MH, Leiby PN, Oladosu GA, et al. Indicators for Assessing Socioeconomic Sustainability of Bioenergy Systems: A Short List of Practical Measures. *Ecological Indicators.* 2013;26:87–102.

Dale VH, Kline KL, Buford MA, Volk TA, Tattersall Smith C, Stupak I. Incorporating Bioenergy into Sustainable Landscape Designs. *Renewable and Sustainable Energy Reviews.* 2016;56:1158–1171.

Dandres T, Gaudreault C, Tirado-Seco P, Samson R. Macroanalysis of the Economic and Environmental Impacts of a 2005–2025 European Union Bioenergy Policy using the GTAP Model and Life Cycle Assessment. *Renewable and Sustainable Energy.* 2012;16:1180–92.

Davis C, Nikolić I, Dijkema GP. Integration of Life Cycle Assessment into Agent-based Modeling. *Journal of Industrial Ecology.* 2009;13:306–325.

GBEP. The GBEP Sustainability Indicators for Bioenergy. Global Bioenergy Partnership, Food and Agriculture Organization of the United Nations, Rome, 2016.

Herreras Martínez S, van Eijck J, Pereira da Cunha M, Guilhoto JJM, Walter A, Faaij A. Analysis of Socio-Economic Impacts of Sustainable Sugarcane–Ethanol Production by Means of Inter-Regional Input–Output Analysis: Demonstrated for Northeast Brazil. *Renewable and Sustainable Energy Reviews.* 2013;28:290–316.

IDB. Biofuels Sustainability Scorecard. Inter-American Development Bank, available at www.iadb.org/biofuelsscorecard/, consulted in October 2016.

Kloprogge P, Van Der Sluijs JP. The Inclusion of Stakeholder Knowledge and Perspectives in Integrated Assessment of Climate Change. *Climatic Change.* 2006;75:359–389.

Lynd L R, Woods J. Perspective: A New Hope for Africa. *Nature.* 2011;474:S20–S21. doi:10.1038/474S020a

Martinez-Hernandez E, Leach M, Yang A. Impact of Bioenergy Production on Ecosystem Dynamics and Services. A Case Study on UK Heathlands. Environ Sci Technol. 2015;49:5805-12.

McKone TE, Nazaroff WW, Berck P, Auffhammer M, Lipman T, Torn MS, et al. Grand Challenges for Life-Cycle Assessment of Biofuels. *Environmental and Science Technology.* 2011;45:1751–1756.

Nogueira LAH, Souza LGA, Cortez LAB, Leal LV. Sustainable and Integrated Bioenergy Assessment for Latin America, Caribbean and Africa (SIByl-LACAf): The Path from Feasibility to Acceptability. *Renewable and Sustainable Energy Reviews* 2017;76:292–308.

Okello C, Pindozzi S, Faugno S, Boccia L. Appraising Bioenergy Alternatives in Uganda Using Strengths, Weaknesses, Opportunities and Threats (SWOT)-Analytical Hierarchy Process (AHP) and a Desirability Functions Approach. *Energies.* 2014;7:1171–1192.

Osseweijer P. A Short History of Talking Biotech: Fifteen years of iterative action research in institutionalising scientists' engagement in public communication. 2006.

Rosillo-Calle F, Walter A. Global Market for Bioethanol: Historical Trends and Future Prospects. *Energy for Sustainable Development.* 2006;10:20–32.

Sackman H. Delphi assessment: Expert opinion, forecasting, and group process. DTIC Document. 1974.

Scott JA, Ho W, Dey PK. A Review of Multi-Criteria Decision-Making Methods for Bioenergy Systems. *Energy*. 2012;42:146–156.

Seyffarth A. The Impact of Rising Ethanol Production on the Brazilian Market for Basic Food Commodities: An Econometric Assessment. *Environmental and Resource Economics*. 2015:1–26.

Souza GM, Victoria RL, Verdade LM, Joly CA, Netto PEA, Cruz CHB, et al. Technical Summary. In: Souza GM, Victoria RL, Joly CA, Verdade LM, editors. *Bioenergy & Sustainability: Bridging the Gaps*. São Paulo: Scientific Committee on Problems of the Environment (SCOPE) 2015.

Souza LGA, Moraes MAFD, Dal Poz MES, Silveira JMFJ. Collaborative Networks as a measure of the Innovation Systems in second-generation ethanol. *Scientometrics*. 2015:1–18.

Taylor CR, Taylor MM. A brief description of AGSIM: an econometric-simulation model of the agricultural economy used for biofuel evaluation. *BioEnergy Policy Brief*, BPB. 2009;70109.

Part III
Availability of land and sugarcane potential production

17 Initial considerations to estimate sugarcane potential in Mozambique and Colombia

*Marcelo Melo Ramalho Moreira and
Felipe Haenel Gomes*

Introduction

Land is one of the most important resources for biofuels production. It has other fundamental uses such as food and fiber production, protection of natural environment, and biodiversity. It is an important source of income for farmers, among others. Greenhouse gas (GHG) emissions from land use change can be significant if particular types of land cover are converted (directly or indirectly) for biofuels production.

Biofuels are often called "bad" or "good" for the environment, local population, etc. The reality is that they can be either. They can be made using diverse crops, based on very different technological routes, respecting or not existing social infrastructure. The benefits for local population and environment are highly dependent on planning, selecting, and implementing the production patterns to fit local reality and objectives.

Land use planning varies widely between countries, or even within regions of the same country. Launched in 2009, the Brazilian Sugarcane Zoning restricted the expansion of sugarcane over several land use classes including areas covered by natural vegetation, the Upper Paraguay Basin, the Amazon and Pantanal Biomes, and indigenous areas, among others. According to the new criteria, 92.5% of the national territory was considered not suitable for sugarcane expansion having several restrictions for its expansion (Manzatto et al., 2009). The other 7.5% were found to have no environmental or social conflict and has good production potential, so it could receive specific credit lines. The sugarcane zoning in Brazil is recognized as a good example on how proper planning can conciliate production and protection objectives.

Launched in 2009, the Global Sustainable Bioenergy Project (GSB) is an initiative aimed at contributing to the sustainability of large-scale bioenergy production, while also ensuring the sustainability of food production and of the environment. The project's working hypotheses are (i) that it is possible to "make room" for bioenergy while honoring other land use priorities, and (ii) that a systematic approach to food and bioenergy production could positively and synergistically impact multiple challenges facing humanity (GSB, 2017).

On a regional level, the project named Bioenergy contribution of Latin America, Caribbean and Africa (LACAf) seeks to contribute to the GSB project through local studies related to the production of sugarcane-derived bioenergy. The LACAf project is aimed at proving an updated perspective of the potential for sustainable production of bioenergy from sugarcane in the abovementioned regions. In order to achieve this objective, the LACAf project involves different research centers and teams, assigning different activities according to the expertise of each group involved in the project.

This chapter provides information for evaluating the suitability of areas for sustainable expansion of sugarcane-based ethanol, having two countries as case studies. Mozambique and Colombia were chosen, once they have shown significant potential for biofuels expansion (see Chapters 19 and 20). The database provides a foundation for in-depth study of the potential for sugarcane production in the country, impact evaluation studies, as well as for the selection of production models that maximize the benefits and minimize possible risks of sugarcane ethanol production. It uses the GSB–LACAf infrastructure that provides international and multidisciplinary collaboration while providing results that others can build on.

Target audiences include researchers, governments, NGOs, and investment and development banks (among others) intended to further understanding possibilities and impacts of biofuels development. The final maps will be available on LACAf's webpage in order to allow easy access to all people who want to contribute to the land use-biofuels development.

References

GSB – Global Sustainable Bioenergy Initiative. Feasibility and Implementation Paths. Available in: http://bioenfapesp.org/gsb/index.php. Access in October 2017.

Manzatto, C. V. et al. Zoneamento agroecológico da cana-de-açúcar. Rio de Janeiro: Embrapa Solos, 2009.

18 Sustainability aspects
Restrictions and potential production

*Marcelo Melo Ramalho Moreira and
Felipe Haenel Gomes*

Strong restrictions

Strong physical restrictions

The physical restrictions are those that are considered mandatory because they are strongly restrictive. They can be related to soil and climate conditions.

Three sources of data were considered in the selection of areas with soils strongly restricted: (i) soil classification, prepared by the CTC sugarcane experts; (ii) the updated land use and land cover maps; (iii) climatic maps, such as Köppen classification, precipitation and temperature maps.

The CTC team classified soils into five levels according to their production potential: "high", "medium", "low", "restricted" and "inappropriate". The high, medium and low degrees correspond to the areas that are suitable for sugarcane production, with no restriction, and will be treated in the sections below. The "inappropriate" level corresponds to the permanently flooded, shallow and/or rocky soils or, in other words, those with some kind of strong restriction for plant growth. The "restricted" level corresponds to the areas where production is possible but requires significant interventions, such as drainage projects. Therefore, along with the "inappropriate" level, the "restricted" level indicates areas that are not being considered for expansion of sugarcane production.

The classification of unsuitable soils prepared by the CTC team was complemented by additional information from the land use and land cover map. Areas corresponding to water bodies, areas saturated by water, urban areas and bare soils, dunes or rocky outcrops were considered unsuitable. Even though they were not considered as improper, areas with sugarcane were excluded at this stage of the study, since the objective of the study is to seek new frontiers of expansion.

According to Scarpari (2002), temperature is probably the most significant factor for sugarcane performance. The sugarcane crop fits very well in regions of tropical, hot and humid climate, where the predominant temperature is between 19°C and 32°C (Marin, 2016). According to Humbert (1968) and Alexander (1973), temperatures below 20°C reduce sprouting rate and plant development. The areas with average annual temperature (AAT) lower than 20°C or higher than 32°C are therefore considered strongly restricted as they are improper to produce sugarcane.

The water requirement of sugarcane varies according to the periods of growth. The crop needs, on average, 1,500–2,500 mm of rain distributed evenly during the cycle (Salassier, 2006), and 1,500–2,000 mm of rain per year to reach a productivity of 100–150 Mg ha^{-1} (Doorenbos and Kassan, 1979). According to Marin (2016), sugarcane adapts very well where rains are well distributed, with cumulative rainfall above 1000 mm. Lack of rainfall can be remedied by using irrigation (in different intensities), which is commonly the case for Mozambican sugarcane plantations. In this chapter, it is treated as moderated restrictions. On the other hand, excess of rainfall brings significant restrictions, such as strong limitation of operational activities, lack of sucrose concentration or even plant death. Different from irrigation, high levels of excess of rainfall is very hard to remediate, so they are treated as strong restrictions.

Strong climate restrictions – low and high temperatures, as well as excess of rainfall – is a particular issue for the Colombian territory and quite rare in Mozambican territory (although it can happen in very specific places). For that reason, they were studied in depth in Colombian case and disregarded for Mozambique.

Strong legal restrictions

It is essential that sugarcane expansion areas comply with current legislation, particularly with regard to areas designated by law for environmental conservation and traditional communities. The legal framework of are locally defined and used to estimate the available land for sugarcane production.

In Mozambique, there are several legal restrictions for agricultural expansion, as well as a series of incentives for economic activities including mining, logging, agriculture and hunting grounds, among others. Additionally, the country recognizes the rights of society on the land which, in theory, prohibits corporate agricultural expansion on areas with community rights over the land (Moçambique, 1997, 1998). This multiplicity of incentives and restrictions makes the job of mapping legal restrictions subject to interpretations.

This study focused on Mozambican legislation and, particularly, on the 2008 Agricultural Zoning (Moçambique, 2008) seeking to identify the actual legal restrictions, which should necessarily be respected. The conservative alternative of not excluding areas that may come to be considered available by the Mozambican government and society was the chosen alternative in order to avoid a decision based solely upon one's point of view. The other restrictions are treated in a second phase, which is focused on moderate restrictions. This study considered as strong legal restrictions for agricultural expansion: Conservation Areas, Reserves and Coutadas (private reserves), urban uses and activities, industrial and trade activities, aquaculture, different types of tourism and social services and sports (Mozambique, 2008; Batistella and Bolfe, 2010).

In Colombia, the legally protected areas are formed by natural parks, indigenous lands and black communities. The country has an entity called National Natural Parks (PNN) that manages 56 natural parks, which are part of the Natural National Parks System (SPNN), and coordinates the National System of Protected Areas. This large natural area has 12.6 million ha, which represents approximately 10% of the country's territory (PNNC, 2016).

According to the General Census of the National Administrative Department of Statistics (DANE, 2005), the indigenous population occupies an area of 32 million ha, while 5.4 million belongs to the black communities. Considering the overlap of indigenous lands or black communities in national parks, the total legally protected area is estimated at 46.2 million ha, representing 40.4% of Colombian territory.

Moderate restrictions

Non-mechanizable Areas

Sugarcane harvesting can be manual or mechanized. This choice depends on the availability of labor, the physical and social configuration of the region, the viable modes of transportation and the amount of sugarcane to be harvested. Manual harvesting requires a large amount of labor, so it tends to generate a significant quantity of jobs. However, due to the risks to the workers and the low yield of manual harvesting in a "green" (non-burned) sugarcane field, pre-burning technique is recommended, which leads to the emission of particulates and greenhouse gases (GHGs) (Balsalobre, 1999). Mechanized harvesting uses specific machinery and requires less but skilled labor with higher remuneration. It also decreases environmental impact (Braunbeck and Magalhães, 2010). In a prospective view, mechanization will be essential for the utilization of sugarcane straw and tip, an important source of fuel to be used in cogeneration and second-generation ethanol production (Leal et al, 2013).

Non-mechanizable areas are identified when a slope greater than 13% is observed, where mechanical harvest becomes difficult due to the risk of harvester overturn. The maximum slope limit of 12% for mechanizable areas was considered, based on the criteria adopted in Law 11,241/2002 of the State of São Paulo, which envisages for the elimination of the sugarcane preburned harvesting techniques. It is noteworthy that, like mechanized harvesting, manual harvesting can occur in areas of lower slope.

Areas with high carbon stocks

High-carbon stock protection areas are internationally recognized as a priority for preservation. Depending on the carbon content of the area in question, emissions from land use change for biofuel production can be quite significant. In some cases, like areas with high carbon stocks and processes with low area conversion in biofuels, the expansion of biofuels can generate a debt (Tilman et al., 2006; Riguelato and Spracklen, 2007; Fargione et al., 2008). Although sugarcane is one of the most efficient technologies for converting agricultural crops into biofuels, it is advisable that sugarcane expansion does not occur in areas of high carbon stock.

The Bonsucro,[1] the production standard does not allow sugarcane production in high conservation value areas (CVA). The CVAs must protect areas with different ecosystem services (Bonsucro, 2011), but it does not establish a cutting line for carbon content. The Global Bioenergy Partnership suggests some indicators for sustainable production of sugarcane, such as life cycle analysis, land use change and soil and water quality (GBEP, 2011). On the other hand, the Renewable Energy Directive (RED) of the European Commission does not allow sugarcane production in primary forest areas, reserves and high conservation value pastures. However, the definition of these areas is not currently consolidated.

On the other hand, there is a widespread understanding that forests should not be converted for biofuel production in many parts of Africa. Therefore, the lowest amount of forest carbon stock (soil carbon + biomass carbon) was used as a criterion for the selection of environmentally sensitive areas. It was assumed that areas with a carbon stock equal or greater to the forest should not be used to produce biofuels. In brief, the adopted criterion expands the preservation coverage for areas with non-forest coverage that have high carbon content in the soil, resulting in a total carbon stock that exceeds 286 tCOe/ha. In comparison to the Mozambican agricultural zoning, the criterion adopted updates and expands the following restrictions: productive forests, arboreal crops (cashew, tea and coconut), forest plantations (current reforestation) and concessions: forestry, mining.

A stricter value was selected for the Colombian case 136 tons/ha. This limit was established considering direct emissions from land use change would be arbitrary lower 12 g CO_2/MJ of ethanol. The biomass values for the forests were extracted from a biomass map from the IDEAM (2012a, 2012b), and for the other land uses, data from Harris et al. (2009) were used. These values were associated with the 2012–2014 land use map. The biomass map was reclassified and areas with a cut above the estimated value were considered as moderately restricted.

Areas with potential land use conflicts

Social aspects are treated as a priority in current analysis. Most relevant issues are land rights, land grabbing and food security. In some cases, land rights are recognized through the delimitation of socially protected areas, such as community areas. However, formal recognition of such rights is far from simple and is not systematically organized in maps. Subsistence agriculture may indicate fragile social organization in that particular region, more subjected to land grabbing threats and loss of purchasing power. It does not mean that biofuels should be avoided in these areas. Rather, it urges that planned production systems should take such reality into consideration. Additionally, the present study has a prerogative to avoid conflicts between biofuels and food production, together with economic capacity to acquire food.

Two land use classes were therefore used as proxy of areas with potential land use conflicts. The land use class cropland/natural vegetation mosaic can be used as proxy for fragile social organization. The cropland class, characterized by annual and continuous crops, is a proxy of intensive food production areas.

For Mozambique, the abovementioned classes can be considered an update of the *cultivated fields* restrictions (annual crops, sugarcane plantations); itinerant agriculture with forests; Forests with itinerant agriculture. It can also serve as a proxy for an update of the *community areas* (which are not properly mapped and updated in the Land Registry), considering the presence of communities in the areas of *cropland/natural vegetation mosaic*.

In Colombia, the land grabbing conflict seems to be less intense; however, so as to ensure a feasible outcome, the analysis was realized ensuring that the expansion of sugarcane does not conflict with the guarantee of food security in the country. For this, the areas already occupied by agriculture and by mosaics of agriculture and natural vegetation were excluded.

Areas with climatic restrictions

Although the climatic restriction is mostly considered a strong physical restriction, the possibility of producing with irrigation is a constant reality in some regions. Different sorts of irrigation are present in almost all sugarcane areas in the country. If this restriction is ignored, attention should be paid to the possible impact of crop irrigation on the availability of water resources in the region, in addition to irrigation costs, which are often prohibitive. Technologies with higher water use efficiency would therefore be welcome. Due to availability of good climate data, different approaches were considered for the Mozambican and Colombian cases.

For the climate, the Köppen classification, adapted by the CTC (CGEE, 2009), was used for the Mozambican case. It takes into account seasonality and the annual and monthly averages of rainfall. The following climatic types were deemed suitable for sugarcane cultivation, with the use of, at the most, salvation irrigation:[2] Am, Aw, Cwb, Cwa and Cfa. On the other hand, climate types BSh and BWh were considered unfit due to the high water deficit, where cultivation would only be possible using full irrigation.

For the Colombian case, the climate was defined by the variables: temperature, rainfall and water deficit. Each variable was classified according to sugarcane needs, generating the suitable and unsuitable classes. The unsuitable classes (low and high average temperatures and excess of rainfall) were considered to be strong restrictions. The areas classified as suitable may require salvation irrigation or not. Areas with annual precipitation lower than 1000 mm were considered to require full irrigation.

Potential production of sugarcane

Classification of suitable areas in different suitability degrees (high, medium and low) was carried out according to the sugarcane production potential system of CTC used in Brazil (CGEE, 2009). This system, what determines the best or worst production potential is a more generalized way of grouping production environments (A to E), used in detailed maps (Joaquim et al., 1994, 1997), in so-called production potentials, used in large area maps, with smaller scales. It involves a combination of characteristics, among which are as follows: depth, texture, layout and thickness of soil horizons; presence of impediments for root penetration; cation exchange capacity (CEC); acidity and alkalinity; base saturation and toxicity by salts or exchangeable aluminum. Figure 18.1 summarizes the main characteristics used in the CTC potential production classification.

The production potential of both countries was obtained through the use of available soil data. In the Mozambican case, the data used comes from the parallel project (Batistella and Bolfe, 2010), obtained through soil maps of the Institute of Agricultural Research of Mozambique (IIAM) carried out in the 1:1,000,000 scale. The soils in those maps were classified in the international systems – Soil Taxonomy (Soil Survey Staff, 1999) and World Reference Base for Soil Resources (IUSS Working Group, 2014), as well local

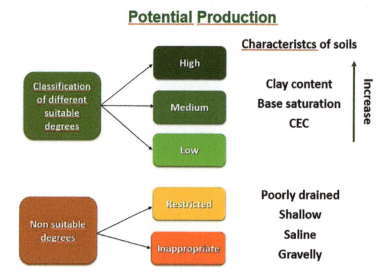

Figure 18.1 Summary of the main characteristics used in the CTC potential production classification.

classification considering mainly color, texture and parent material. Then, a correlation between the soils data and the Brazilian Soil Classification System (Embrapa, 2013) was performed. The description of the main reference soil groups that are suitable to plant sugarcane in both countries, as well as those that are restricted for sugarcane crops, is given in Table 1. It is noteworthy that the Colombian map, besides presenting a larger scale leading to greater data precision, presents the highest quality information of each soil spot, determining the percentage of associated soils, as in this level of detail the spots correspond to associations of more than one class of soil.

Soil with high potential has a productivity estimated at 81.4 t ha^{-1} is considered for Brazil, according to CGEE (2009). This potential comprises soils with good water storage capacity and few limitations to the sugarcane production, being easily remedied, through liming and maintenance fertilization. They correspond to the production environments A and B according to the CTC Production Environment System (Joaquim et al., 1994, 1997). The medium potential is represented by soils with some limitations, mainly related to water storage capacity and/or problems related to low CEC and base saturation, and may also present exchangeable aluminum toxicity. These soils have the expected average productivity around 73.1 t ha^{-1} considering data from Brazil and correspond to C and D production environments (Joaquim et al., 1994, 1997). The production potential considered as low presents strong limitations in water storage capacity, low fertility or even limitations in effective soil depth with or without the presence of surface stoniness. They correspond to the production environment E, with an expected productivity of 64.8 t ha^{-1} for Brazil (Joaquim et al., 1994, 1997). Although they are suitable for sugarcane, the recommendation of the CTC is that not exceed 20% of the total area of a sugarcane mill with these types of soils, unless it is in a full-time irrigated area.

Table 18.1 WRB Reference Soil Group description and correlation with Brazil and US systems

Reference Soil Group	Description and correlation
Suitable soil orders	
Ferralsols	Ferralsols represent the classical, deeply weathered, red or yellow soils of the humid tropics. These soils have diffuse horizon boundaries, a clay assemblage dominated by low-activity clays, mainly kaolinite and sesquioxides. Local names usually refer to the color of the soil. Many Ferralsols are known as Oxisols (USA) and Latossolos (Brazil)
Nitisols	Nitisols are deep, well-drained, red tropical soils with diffuse horizon boundaries and a subsurface horizon with at least 30% clay and moderate to strong angular blocky structure with, in moist state, shiny aggregate faces. Weathering is relatively advanced but Nitisols are far more productive than most other red tropical soils. Many Nitisols correlate with Nitossolos (Brazil), Kandic Great Groups of Alfisols and Ultisols and different Great Groups of Inceptisols and Oxisols (USA).

(Continued)

Reference Soil Group	Description and correlation
Suitable soil orders	
Acrisols	Acrisols have a higher clay content in the subsoil than in the topsoil, as a result of pedogenetic processes (especially clay migration). Acrisols have low-activity clays in the subsoil horizon and a low base saturation in the 50–100 cm depth. Many Acrisols correlate with Argissolos (Brazil) and Ultisols with low-activity clays (USA).
Lixisols	Lixisols have a higher clay content in the subsoil than in the topsoil, as a result of pedogenetic processes (especially clay migration). Lixisols have low-activity clays in the subsoil horizon and a high base saturation in the 50–100 cm depth. Many Lixisols are included in Argissolos (Brazil) and Alfisols with low-activity clays (USA).
Luvisols	Luvisols have a higher clay content in the subsoil than in the topsoil, as a result of pedogenetic processes (especially clay migration). Luvisols have high-activity clays throughout the subsoil horizon and a high base saturation in the 50–100 cm depth. Many Luvisols are known as Luvissolos (Brazil). In the USA, they belong to Alfisols with high-activity clays.
Fluvisols	Fluvisols accommodate genetically young soils in fluvial, lacustrine or marine deposits. Despite their name, Fluvisols are not restricted to river sediments (Latin fluvius, river); they also occur in lacustrine and marine deposits. Many Fluvisols correlate with Fluvents (USA) and Neossolos Flúvicos (Brazil). Some Fluvisols can be classified as restricted because of the drainage.
Nonsuitable soil orders (restricted and inappropriate)	
Gleysols	Gleysols comprise soils saturated with groundwater for long enough periods, underwater and tidal soils to develop reducing conditions. This pattern is essentially made up of reddish, brownish or yellowish colors, in combination with greyish/bluish colors in the soil. Many underwater soils have only the latter. Gleysols with a thionic horizon or hypersulfidic material (acid sulfate soils) are common. In the USA, many Gleysols belong to Aquic Suborders and Endoaquic Great Groups of various Orders (Aqualfs, Aquents, Aquepts, Aquolls, etc.) or to the Wassents. In Brazil, they belong mainly to Gleissolos.
Histosols	Histosols comprise soils formed in organic material accumulating as groundwater peat (fen), rainwater peat (raised bog) or mangroves or without water saturation in cool mountain areas. Histosols are found at all altitudes, but the vast majority occurs in lowlands. In Brazil, they belong to Organossolos Order and Histosols and Histels in the USA.

Source: Adapted from IUSS Working Group (2014).

Notes

1 Bonsucro is an international not-for-profit, multi-stakeholder organization established in 2008 to promote sustainable sugarcane production. It does this through setting sustainability standards and certifying sugarcane products.
2 Salvation irrigation is a special type of irrigation that consist in applying small amounts of water, between 30 and 60 mm, in order to keep the productivity of the sugar cane, using portable irrigation systems.

References

Alexander, A.G. *Sugarcane physiology: a comprehensive study of the Saccharum source-to-sink system.* Amsterdam: Elsevier, 1973. 752p.

Balsalobre, M. A. A. Cana-de-açúcar: quando e como cortar para o consumo animal. Balde Branco, n. 421, pp. 19–13, 1999.

Batistella, M.; Bolfe, E. L. (Org.). *Paralelos: Corredor de Nacala.* Campinas, SP: Embrapa Monitoramento por Satélite, 2010. 80p.

Bonsucro. Bonsucro Production Standard Including Bonsucro EU Bonsucro Production Standard. Version 3.0 March 2011.

Braunbeck, Oscar A.; Magalhães, Paulo Sérgio Graziano. Avaliação tecnológica da mecanização da cana-de-açúcar. Bioetanol de Cana-de-Açúcar: P&D para Produtividade e Sustentabilidade, v. 5, p. 14, 2010.

Centro de Gestão de Estudos e Estratégias - CGEE. *Bioetanol combustível: uma oportunidade para o Brasil.* Brasília: CGEE, 2009. 536p.

DANE - Departamento Administrativo Nacional de. Estadística. Censo de población y vivienda de Colombia. Año 2005. Análisis de la estructura y composición de las principales variables demográficas y socioeconómicas del Censo 2005. Informe final. Bogotá, D.C. 30 de September de 2008.

Doorenbos, J.; Kassam, A. H. *Efectos del agua sobre el rendimiento de los cultivos.* Roma: FAO, 1979. 212p.

Embrapa - Empresa Brasileira de pesquisa Agropecuária. Sistema brasileiro de classificação de solos. 3rd ed. Brasília, 2013. 353p.

Fargione, J., Hill, J., Tilman, D., Polasky, S. and Hawthorne, P. 2008. Land clearing and the biofuel carbon debt. *Science* 319: 1235–1238.

GBEP – Global Bioenergy Partnership Sustainability Indicators for Bioenergy. Rome: FAO, 2011. 223p. Available in: www.globalbioenergy.org/fileadmin/user_upload/gbep/docs/Indicators/The_GBEP_Sustainability_Indicators_for_Bioenergy_FINAL.pdf. Access in Oct. 2017.

Harris, N.L., Grimland S. and Brown, S. 2009. Land Use Change and Emission Factors: Updates since Proposed RFS Rule. Report submitted to EPA.

Humbert, R. P. *The growing of sugar cane.* Amsterdam: Elsevier, 1968. 779p.

IDEAM – Instituto de Hidrología, Meteorología y Estudios Ambientales. Mapa de Coberturas de la Tierra. Metodología CORINE Land Cover adaptada para Colombia. Período 2000–2002. Escala 1:100.000. 2012a.

IDEAM – Instituto de Hidrología, Meteorología y Estudios Ambientales. Mapa de Coberturas de la Tierra. Metodología CORINE Land Cover adaptada para Colombia. Period 2005–2009. Scale 1:100.000. 2012b.

IUSS WORKING GROUP WRB. 2015. World Reference Base for Soil Resources 2014, update 2015 International soil classification system for naming soils and creating legends for soil maps. World Soil Resources Reports No. 106. FAO, Rome.

Joaquim, A.C.; Donzelli, J.L.; Bellinaso, I.F.; Quadros, A.D.; Barata, M.Q.S. Potencial e manejo de solos cultivados com cana-de-açúcar. In: *Seminário Copersucar de Tecnologia Agronômica*, vol. 6, 1994, Piracicaba. Anais. Piracicaba: Centro de Tecnologia Copersucar, 1994. pp. 1–9.

Joaquim, A.C.; Donzelli, J.L.; Quadros, A.D.; Sarto, L.F. Potencial de Produção de cana-de-açúcar. In: Seminário Copersucar de Tecnologia Agronômica, vol. 7, 1997. Anais. Piracicaba: Centro de Tecnologia Copersucar, 1997. pp. 68–76.

Leal, M. R. L., Galdos, M. V., Scarpare, F. V., Seabra, J. E., Walter, A., E Oliveira, C. O. (2013). Sugarcane straw availability, quality, recovery and energy use: a literature review. *Biomass and Bioenergy*, 53, 11–19.

Marin, F. R. Características – Cana de Açúcar. Available in: www.agencia.cnptia. embrapa.br/gestor/cana-de-acucar/arvore/CONTAG01_20_3112006152934. html. Access in Sept. 2016.

Moçambique. Regulamento da Lei de Terras. Decreto n° 66/98 de 8 de dezembro Boletím da República, Maputo. 1998.

Moçambique. Lei n° 19/97 de 1 de outubro – Lei de Terras. Maputo: Boletím da República, 1997.

Moçambique. Zoneamento agrário a nível nacional. Grupo de Trabalho Técnico. Moçambique, 2008.

Parques Nacionales Naturales de Colombia (PNNC). Disponível em: www.parquesnacionales.gov.co/portal/es/. Acesso em: 06 de dez. 2016.

Riguelato, R.; Spracklen, D.V., 2007. Carbon Mitigation by Biofuels or by Saving and Restoring Forest. *Science* 317: 902.

Salassier, B. Manual de Irrigação. Viçosa: UFV, 2006.

Scarpari, M. S. Modelos para a previsão da produtividade da cana-de-açúcar (Saccharum spp.) através de parâmetros climáticos. Dissertação (Mestrado) - Escola Superior de Agricultura Luiz de Queiroz. Piracicaba, 2002, 202p.

Soil Survey Staff. 1999. Soil taxonomy: A basic system of soil classification for making and interpreting soil surveys. 2nd ed. Natural Resources Conservation Service. U.S. Department of Agriculture Handbook 436.

Tilman, D.; Reich, P. B.; Knops, J. M.H. Biodiversity and ecosystem stability in a decade-long grassland experiment. 2006.

19 Case study

Potential of sugarcane production for Mozambique

Marcelo Melo Ramalho Moreira,
Felipe Haenel Gomes and Karine Machado Costa

Mozambique context

Mozambique is a country in the east coast of sub-Saharan Africa, with a territorial extension of about 80 million hectares divided into 11 provinces. The population, estimated at just over 24.4 million people, is mostly young (45% of habitants are less than 15 years old) and concentrated in rural areas (68%) (Moçambique, 2016).

A Civil War that followed the Mozambican independence war and lasted until 1994 destroyed the existing much of the country's agricultural production infrastructure and initiated large migratory movements. Although the economy of Mozambique is growing at rates above 7% per year, still it is one of the ten lowest human development indexes in the world (UNDP, 2016). As in other least developed countries, foreign currency availability is a major restriction to development objectives.

Energy security is a particular challenge. Almost all electricity produced in Mozambique is hydro-based and sold to international markets. Only 18% of the electricity needs are fulfilled, so the great majority of the population relies on traditional biomass (mostly from deforestation) to source minimal energy needs. The country also imports 100% of its oil from international sources (IEA, 2016). The energy security is so fragile that the country imports gasoil, gasoline and even jet fuels to compensate for systematic cuts in power generation using small-scale generators (KPMG, 2013). Along with new gas discoveries in the northern part of the country, the production of agricultural-based biofuels is a potential means to address Mozambique's energy insecurity.

Mozambique has a Biofuels Policy and Strategy, as part of the National Fight against Poverty agenda. The main objectives of the Biofuels Policy are as follows: valuing agricultural production and rural development, reducing energy dependency on foreign oil and improving the foreign trade balance. Sugarcane and sweet sorghum have been identified as the preferred raw materials for ethanol production (Moçambique, 2009).

Although the potential for bioenergy, the sugarcane industry in Mozambique in 2014 was still restricted to four mills built before the Civil War. All four are focused solely on the production of raw sugar. The dominant production system is that of large-scale agriculture in plain and continuous areas near watercourses. Thus, there is still time and need for a territorial planning exercise in Mozambique involving detailed characteristics related to social, economic and environmental sustainability.

The main objective of this study is to provide information for evaluating the suitability of areas for sustainable expansion of biofuels in Mozambique, focusing on sugarcane ethanol.

Material and methods

The Agricultural Zoning of 2008 is the main reference for land use planning in order to guide agricultural expansion (Moçambique, 2008). Drawn up in two stages (preliminary qualification and local verification), and excluding areas according to 24 criteria, the zoning resulted in identification of lands available for agricultural expansion. Although significant efforts have been involved in the development of the Agricultural Zoning, it needs to be updated and adjusted to guide expansion of sugarcane-based biofuels through a sustainability lens. The final document admits that some exclusions should be transitional. Other exclusions would not be complete or require constant updates. Finally, there is some difficulty in applying certain exclusion criteria. This is the case, for example, for the areas occupied or intended for agricultural use (classes: cultivated fields, arboreal crops, itinerant agriculture with forests and agricultural land use rights – direito de uso e aproveitamento da terra (DUATs)). While these areas require special attention, it is possible, for example, to find productive systems that protect and enhance agricultural production (particularly food production systems and smallholders).

The approach here is a stepwise mapping sequence, starting from the Mozambican land use and land cover map, where areas with different constraints are successively eliminated in order to obtain a pool of available lands for sustainable expansion of sugarcane. The limitations were classified as strong or moderate, with strong limitations defined as those that should be respected and moderate limitations defined, considered nonobligatory if care is taken to address local factors, for example, by choosing different production systems.

The mapping exercise consists of three major phases, as described in Figure 19.1. The first phase is to obtain the land use and land cover map of Mozambique, which serves as the basis for subsequent steps. The second phase involves removing the areas with strong restrictions for sugarcane expansion. The restrictions can be related to physical or legal aspects. Once strong restrictions are removed, we classify the area into different suitability levels – high, medium and low – according to methodology developed by

Figure 19.1 Simplified diagram of the mapping process.
Source: author.

the Brazilian Center for Sugarcane Technology (CTC). The next phase is the inclusion of moderate restrictions criteria.

Major findings

An updated land use and land cover map of Mozambique was not available at the beginning of this project. A prior mapping of land use and land cover from 1998 (JICA, 1998) acquired from the NMA/EMBRAPA was then used as base in the preparation of the current map of the Mozambican territory. The preparation of the land use and land cover maps for the territory of Mozambique was carried out in three stages: (i) reclassification of base map from 1998, (ii) update for 2001 and (iii) update for 2013.

Based on visual interpretation of images from Landsat-5 and Landsat-7 satellites, and on identification of landscape elements present in the base map of 1998, a new reclassification was performed, at spatial resolution of 250 m, according to the description of the classification system adopted by the International Geosphere-Biosphere Programme (IGBP).

In a second step, the reclassified map was updated in the year 2001 using a temporal sequence of the MCD12Q1 product (Land Cover Type) derived from MODIS sensor images acquired in 2000 and 2001. Later, the 2001 map was updated in the year 2013 using a temporal sequence of the MCD12Q1 product based on MODIS images acquired from 2002 to 2013.

Due to its relevance, sugarcane and annual crop areas were first mapped at higher spatial resolution (30 m) through visual interpretation of Landsat-7 images in the year 2001 and Landsat-8 images in the year 2013. Afterward, the classified sugarcane and annual crop areas were resampled to the same spatial resolution of the land use and land cover maps of 2001 and 2013 (250 m). The main sources of data and procedures are shown in Table 19.1.

The results shown in this section are two land use and land cover maps of Mozambique for the years 2001 and 2013. The territory was categorized into 11 equivalent thematic classes as shown in Table 19.2.

Table 19.1 Data used for the construction of the land use and coverage maps of Mozambique

Data source	Use
Land use and land cover map of 1998	Construction of base map 2001
Landsat-5 and Landsat-7	Reclassification of thematic classes for the 1998 base map
MODIS Land Cover Type product (MCD12Q1) from 2001 to 2013	Construction of pixel by pixel mosaic maps of 2001 and 2013
Landsat-7 (2001) and Landsat-8 (2013) images	Sugarcane and annual crop areas of 2001 and 2013 maps

Source: Data from the study based on JICA (1998), Tucker et al. (2004) and Friedl et al. (2010).

Table 19.2 Land use and land cover in Mozambique in the years 2001 and 2013 (1,000 ha and %)

Classes	2001		2013	
	1,000 ha	%	1,000 ha	%
Water	985	1.2	986	1.2
Forest	38	0.0	38	0.0
Shrublands	36,332	45.3	31,979	39.9
Savannas	27,264	34.0	27,267	34.0
Grasslands	6,693	8.3	9,558	11.9
Permanent wetlands	888	1.1	814	1.0
Croplands	690	0.9	859	1.1
Urban and built-up	134	0.2	134	0.2
Cropland/natural	6,856	8.6	8,125	10.1
Barren or sparsely	246	0.3	327	0.4
Sugarcane	33	0.0	72	0.1
Total	80,159	100	80,159	100

Source: Data from the study.

The land is not characterized by native forest vegetation, which represents a very small portion of total land cover in Mozambique. In 2001 and 2013, there is a large predominance of land coverage types without clear traces of anthropogenic presence, particularly in the shrubland class. Considering the geographic characteristics of the region, it is not possible to determine, if the savanna and grassland classes register or not, the changes in the landscape due to human disturbance.

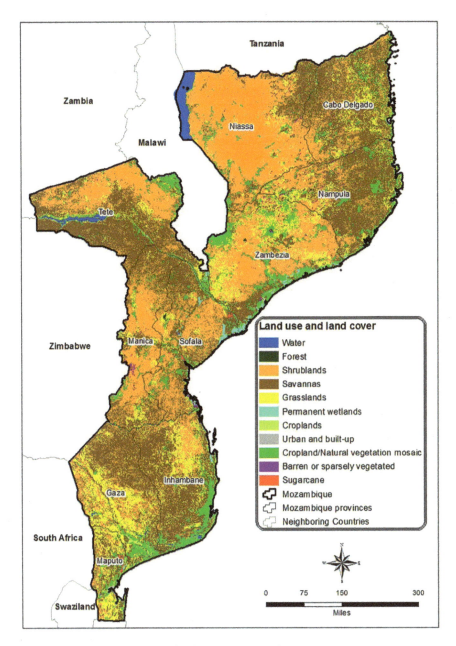

Figure 19.2 Map of land use and land cover in Mozambique in 2013.
Source: made by the author using data developed by Agrosatélite.

The grassland class showed the largest absolute growth in the period, mainly in areas previously identified as savannas or shrubland. Although the grassland class is clearly due to low occupation, such as low-density livestock production, it is not possible to determine what portion of the landscape changes are due to natural phenomena or human intervention. The second largest absolute growth was in the cropland/natural vegetation mosaic area. This class is composed of agriculture mosaics, forest and native vegetation, but no component represents more than 60% of the landscape. In Mozambique, these areas are characterized by small agricultural or pasture properties in the middle of small native vegetation areas. In general, such landscape can be attributed to subsistence agriculture.

The sugarcane class is very small, totaling 33,000 ha in 2001. The sugarcane area is easily identified in the higher-resolution images, since it consists of continuous and flat areas near the four mills that are currently operating. The mapping process identified much more pronounced changes in recent years since 2012, with the establishment of new areas since then.

The territorial distribution of land use classes for the year 2013 is shown in Figure 19.2.

Restrictions

The strong restrictions, composed of strong physical restrictions and strong legal restrictions, are presented in Figure 19.3. The chapter 18 explains the considered restrictions to this work.

Of the total Mozambican area (around 80 million hectares), around 10.5 million have some kind of strong physical restriction and almost 13 million hectares have some kind of strong legal restriction. Physical restrictions are widespread around the country. Legal restrictions are mostly reserves (Niassa National reserve and part of the Kruger Park). It is possible that some of the areas that are owned by local population have not been identified and registered in GIS maps. This analysis deals with this informational restriction by analyzing the presence of local population as moderate restriction.

A first set of moderate restrictions is presented in Figure 19.4. The climate can be considered as moderate restriction in areas that require irrigation. Those areas are mostly located in Gaza, Inhambane, Sofala and Tete. Non-mechanizable areas are located in the north of the country, concentrated in Niassa, Tete, Manica Zambezie and Nampula. Cabo Delgado has the larger concentration of high carbon areas, followed by Niassa and Sofala.

The last two moderate restrictions are presented in Figure 19.5. Small-scale agriculture is relatively widespread in the country. In general, such type of land cover is found in lowest altitudes, close to the Indian Ocean, but significant intensity is also found in other parts of the country such as interior of Zambezia, Cabo Delgado and Tete. Commercial agriculture (identified as cropland) is much smaller than small scale.

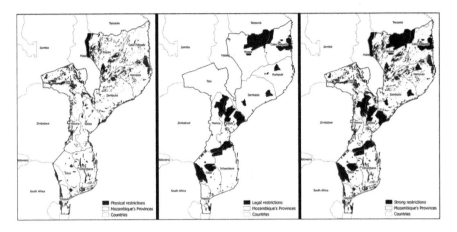

Figure 19.3 Strong restrictions (physical and legal).
Source: made by Agroícone using data developed by Agrosatélite.

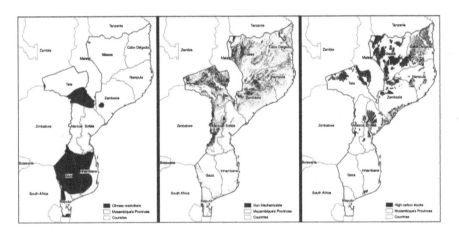

Figure 19.4 Moderate restrictions (climate, non-mechanizable and high carbon stocks).
Source: made by Agroicone. Harris, Grimland, Batjes and Brown are public data (articles) and Agrosatélite is a project partner.

Potential production

The production potential of Mozambique was done using the data from the parallel project (Batistella and Bolfe, 2010), obtained through soil maps of the Institute of Agricultural Research of Mozambique (IIAM) carried out in the 1:1,000,000 scale. The soils in those maps were classified in the international system, World Reference Base for Soil Resources (IUSS Working Group, 2014) as well as local classification considering drainage, presence of

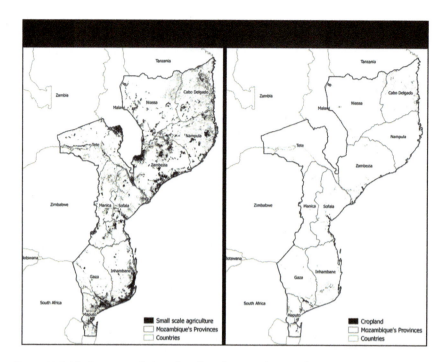

Figure 19.5 Moderate restrictions (small-scale agriculture and cropland).
Source: Agrosatélite. The data was developed by Agroicone. The data were developed by Agrosatélite. Agrosatélite is a project partner.

salts, rocks and other impediments. Then, a correlation between the soils data and the Brazilian Soil Classification System (Embrapa, 2013) was performed.

Areas without strong restrictions

The total area without "strong restrictions" (described above in chapter 18) is shown in Table 19.3. Of the total Mozambican area (around 80 million hectares), around 10.5 million have some kind of strong physical restriction and almost 13 million hectares have some kind of strong legal restriction. Of the 57 million hectares without strong restrictions, most (95%) are in the medium- and low-suitability classes. There are almost 5 million hectares in areas with high-potential suitability soils.

Table 19.3 also shows the distribution of areas on a regional perspective. A polarization of high-suitability areas were verified in the provinces of Zambezia and Cabo Delgado. Seventy-five percent of the medium-suitability areas are concentrated in four provinces: Niassa, Zambezia, Tete and Nampula.

When only strong restrictions are considered, the northern portion of Mozambique stands out. The provinces of Niassa, Cabo Delgado, Nampula and Zambézia and Tete province in the central-west region have significant areas

Table 19.3 Total area without strong restrictions in each Mozambican province (1,000 ha and %)

Province	Total (1,000 ha)	High (%)	Medium (%)	Low (%)	Total (%)
Cabo Delgado	5,773.9	1.9	5.0	3.1	10.1
Gaza	5,098.0	0.4	0.2	8.3	8.9
Inhambane	6,075.6	0.1	1.1	9.4	10.6
Manica	4,166.5	0.1	3.7	3.4	7.3
Maputo	1,771.3	0.1	0.5	2.5	3.1
Nampula	6,456.3	0.0	9.3	1.9	11.2
Niassa	7,371.0	0.3	10.9	1.7	12.8
Sofala	3,582.2	0.3	1.8	4.1	6.2
Tete	7,984.9	0.2	11.4	2.3	13.9
Zambézia	9,145.7	1.4	10.1	4.5	15.9
Total	57,425.3	4.9	54.0	41.0	100

Source: Data from the study.

with medium and high potentials of production and sugarcane (Figure 19.6). In Niassa and mainly Cabo Delgado, soils with high potential correspond mainly to Nitisols, clayey, deep, derived from basic igneous rocks and with no apparent problems of fertility. Fluvisols with mollic horizons (high base saturation and moderate to high content of organic matter horizon) also occurs, which, when deep and without problems related to drainage or the presence of salts, were also classified with high potential. In the medium potential, soils of several classes occur, especially Ferralsols, Lixisols, Luvisols, Acrisols and Fluvisols. Soils with low potential are more concentrated in the central and southern regions of the country, mainly in the provinces of Maputo, Gaza, Inhambane and Sofala. The chapter 18 describes soil reference groups cited in this study.

In total and considering strong restrictions, approximately 2.8 million hectares of high potential, 30.2 million of medium potential and 22.9 million of low potential were identified. If moderate restrictions, as showed in Figure 19.3, are included, then about 1.5 million hectares of high potential, 18.8 million of medium potential and 9.5 million of low potential were identified (Table 3). It indicates that, disconsidering other factors such as logistics and production model, the provinces of Tete, Nampula, Zambézia, Niassa and Cabo Delgado have the greatest potential areas for the expansion of bioenergy production of sugarcane. It is noteworthy that there are currently four plants in operation in Mozambique, which are part of areas considered as "restricted" (chapter 18). This project did not consider the expansion in similar areas considering that these areas must be preserved because they belong to the majority of floodplains, with high environmental importance, contrasting the objective of producing sustainable energy from renewable biofuels.

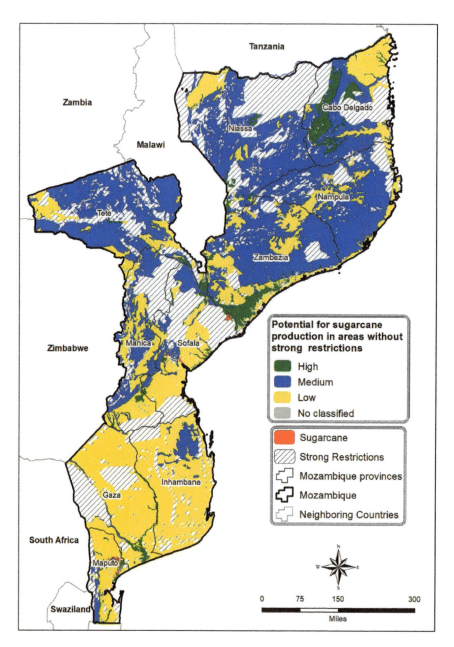

Figure 19.6 Potential for sugarcane production in Mozambique without strong restrictions. Source: Agroicone.

Areas without moderate restrictions

The second stage consists of the identification, qualification and analysis of the area without moderate restrictions, beyond the strong restrictions, already considered as shown in Figure 19.7. The results are shown in Table 19.4. Each line represents each moderate restriction separately, with exception of first and last lines. First line ignores moderate restrictions (no moderate restrictions active) and classifies all the areas that do not have strong restrictions. Last line considers that all restrictions (strong and moderate) are simultaneously active.

The first line classifies the total area without strong restrictions into high, medium and low suitability (2.8, 30.9 and 23.6 million hectares, respectively). The following lines identify the individual impact of each restriction mentioned in the chapter 18 in the land budget. Climate, for example, represents an additional restriction of 11.7 million hectares, reducing land availability from 57.4 to about 45.7 million hectares. The last line of the table considers that all restrictions are simultaneous in place.

The climatic restriction is the main individual restriction for the total suitable area, resulting in a restriction of 11.7 million hectares. Next, in a descending order of restrictions, are the areas with carbon stock equivalent to forests, areas with potential land use conflicts and high steepness areas. The class without any moderate restrictions had most areas excluded for obvious reasons.

By focusing on high-suitability areas, the restrictions have low impact when analyzed individually. High carbon stock areas are the main restriction, totaling 642,000 ha. Such low impact is partially explained by the fact that total suitable areas with high potential are significantly smaller than those with medium and low potentials. On the other hand, if all restrictions are combined, the areas with high potential are reduced by about half.

Conversely, it is possible to verify that there is a significant amount of area available for sugarcane production. Even if we consider the combination of all restrictions, the mapping indicates the availability of around 1.5 million

Table 19.4 Distribution of potential production according to active moderate restriction

Active moderate restriction	Total	High	Medium	Low
None	57,425	2,822	30,949	23,653
Climate	45,724	2,567	28,579	14,578
Non-mechanizable	52,834	2,765	27,455	22,613
Carbon	50,189	2,180	26,123	21,885
Small-scale agriculture	50,942	2,337	27,833	20,771
Cropland	56,863	2,715	30,783	23,364
All (simultaneously)	30,798	1,577	19,383	9,837

Source: Data from the study.

hectares with high potential and 19 million hectares with medium potential, totaling more than 21 million hectares with medium or high potential, always considering some kind of irrigation.

As can be seen in Table 19.5, medium potential areas without additional restrictions are relatively widespread throughout the country, predominating in the provinces of the north portion of the country, particularly in the provinces of Nampula, Niassa, Tete and Zambezia, which represent 80% of the areas with medium suitability.

On the other hand, areas with high suitability are concentrated in specific areas. The highest concentrations are in the provinces of Cabo Delgado and Zambezia, which concentrated 80% of areas with these characteristics. Table 19.6 summarizes the areas of high, medium and low potentials considering levels of restriction in Mozambique.

Table 19.5 Total area without strong or moderated restrictions in each Mozambican province (1,000 ha and %)

Province	Total (1,000 ha)	High (%)	Medium (%)	Low (%)	Total (%)
Cabo Delgado	3,940.1	2.3	6.8	3.7	12.8
Gaza	162.1	0.0	0.0	0.5	0.5
Inhambane	1,518.3	0.0	0.5	4.5	4.9
Manica	2,679.1	0.2	3.7	4.9	8.7
Maputo	880.8	0.1	0.5	2.3	2.9
Nampula	5,158.2	0.0	14.3	2.5	16.7
Niassa	4,122.5	0.2	12.1	1.1	13.4
Sofala	2,348.9	0.5	1.7	5.5	7.6
Tete	3,801.0	0.1	10.9	1.4	12.3
Zambézia	6,187.3	1.8	13.0	5.3	20.1
Total	30,798.2	5.2	63.3	31.5	100

Source: Data from the study.

Table 19.6 Area (ha) of high, medium and low potentials considering levels of restriction in Mozambique

Potential	Strong restrictions (1,000 hectares)	Strong and moderate restrictions (1,000 hectares)
High	2,822	1,577
Medium	30,949	19,383
Low	23,653	9,837
Total	57,424	30,797

Source: From this study.

Discussion

This study infers the land use and land cover dynamics in Mozambique between 2001 and 2013 in such a way that describes the land occupation trends. Then it establishes criteria to identify in a stepwise approach factors of strong to moderated restrictions to sustainable biofuels expansion. Following the recommendation of the Mozambican Biofuels Policy, sugarcane served as a reference crop for bioethanol production. Finally, the study draws scenarios of sustainable expansion that considers additional attention regarding socioeconomic and environmental aspects. The work is developed using geospatial techniques, combining GIS databases, literature review and taking into account recommendations of international institutes dedicated to biofuels sustainability issues.

The study concludes that, in last decade, Mozambique recorded small traces of change in land cover that can be directly attributed to anthropogenic action. Most of those are marked by the expansion of areas of small-scale agriculture and low-intensity pastureland. Changes due to commercial agriculture are not significant. Sugarcane areas expansion was timid and recent. Although new projects were initiated, the expansion of sugarcane area occurred close to existing mills and related to the recovery of the infrastructure built before the Civil War.

The mapping process resulted in an impressive amount of areas suitable for sustainable sugarcane expansion. From the 80 million hectares of country's total land, 57 million hectares is suitable for sugarcane expansion, of which at least 33 million hectares would be of medium or high potential. Even once all moderated restrictions are imposed, the total suitable area sums up to 20.9 million hectares of good and medium potential (in addition to 9.8 million hectares of low potential).

The values are consistent with the Strategic Plan for the Development of Agricultural Sector of Mozambique, that estimates 34 million hectares of arable land, and gives a more detailed land budget, in the sense that it allows for different inference, including mechanization potential (areas with low steepness), as well as other sustainability issues.

It is worth mentioning that Brazil, the largest sugarcane producer in the world and the seventh largest economy, has currently about 10 million hectares of sugarcane plantations and uses only about half of it for ethanol production. With this area, the country is able to supply about 18% of Brazilian Primary Energy Production and has already supplied more than 50% of total liquid fuel demand for light vehicles (EPE, 2015).

The study does not account for possible conflicts between agricultural and mining activities. Part of this potential problem is diminished because mining locations are located in areas with moderate restrictions (particularly in stepwise areas) and by the possibility of simultaneous underground and aboveground activities (as happened in gold mines in South Africa). Similarly, the study did not explore the risk of production in areas subjected to extreme events in a world under climate change.

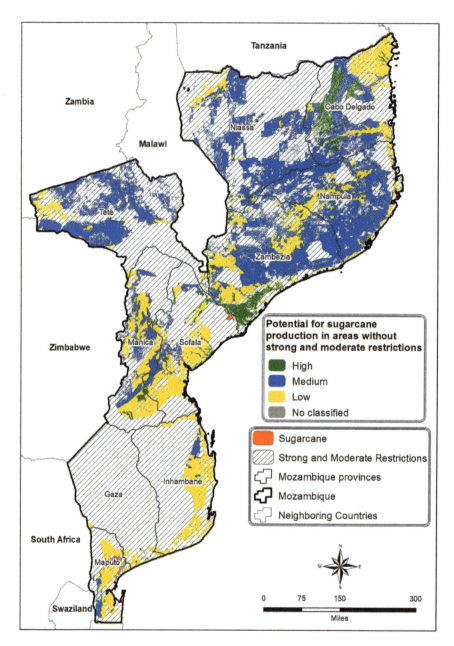

Figure 19.7 Potential for sugarcane production in Mozambique without strong and moderate restrictions.
Source: Agroícone.

Two observations must be made regarding the final land budget. First, it is important to mention the possible impact of recent programs of agricultural development, particularly the Prosavana project. The project is a partnership among Mozambique, Brazil and Japan and has a stated objective of developing 10 million hectares of agriculture (particularly commercial soybeans) at the Nacala Corridor (provinces of Cabo Delgado, Nampula, Zambézia, Niassa and Tete). Further, several institutions, particularly NGOs, are blaming the Mozambique government for ignoring the effects of large-scale agricultural projects on local population. Looking at our maps, these situations tends to be stressed in areas of medium and high potential, in mosaics of agriculture and mixed vegetation, that are concentrated on the same regions of the Prosavana project. Although production patterns can be designed in structures that favors low-skilled labor and participation of rural associations (particularly in sugarcane), it is necessary to check if local social infrastructure is strong enough to guarantee the rights of local communities.

Although we acknowledge that many questions still need to be addressed, this paper contributes towards those answers with updated maps, databases and layers of potentials and risks of biofuels expansion. It is now possible to spatially identify and quantify probable impacts of a bioenergy policy in Mozambique. It is our best hope that this work and consequent products will serve as a tool allowing for adequate national and regional planning.

References

Batistella, M.; Bolfe, E. L. (Org.). *Paralelos: Corredor de Nacala*. Campinas, SP: Embrapa Monitoramento por Satélite, 2010. 80p.

EMBRAPA - Empresa Brasileira de pesquisa Agropecuária. *Sistema brasileiro de classificação de solos*. 3rd ed. Brasília, 2013. 353p.

EPE - Empresa de Pesquisa Energética. Balanço energético nacional - Séries Completas. Available at https://ben.epe.gov.br/BENSeriesCompletas.aspx. Accessed in 2015.

Friedl, M. A.; Sulla-Menashe, D.; Tan, B.; Schneider, A.; Ramankutty, N.; Sibley, A.; Huang, X. MODIS Collection 5 global land cover: Algorithm refinements and characterization of new datasets. *Remote Sensing of Environment*, v. 114, pp. 168–182, 2010.

IEA – INTERNATIONAL ENERGY AGENCY - IEA. IEA Statistics. Disponível em 482 www.iea.org/statistics/statisticssearch/report/?year=2012&country=MOZAMBIQUE&pr483 oduct=Balances. Acesso em janeiro 2016.

IUSS Working Group WRB. 2015. World Reference Base for Soil Resources 2014, update 2015 International soil classification system for naming soils and creating legends for soil maps. World Soil Resources Reports No. 106. FAO, Rome.

Japan International Cooperation Agency – JICA. Uso e cobertura da terra de 488 Moçambique. Scale 1:1.000.000. 1998.

KPMG. *Mining Mozambique Country Guide*. Available in kpmg.com. Accessed in 2013.

Moçambique. Zoneamento agrário a nível nacional. Grupo de Trabalho Técnico. Moçambique, 2008.

Moçambique. Política de biocombustíveis. Boletim da República. I Serie, Número 20. 21 de 504 maio de 2009.

Moçambique. Cartaz de Dados sobre a População Moçambique em 2013. www.prb.org/pdf13/mozambique-population-datasheet-2013.pdf. Access in 2016.

Tucker, C. J.; Grant, D. M.; Dykstra, J. D. NASA's global orthorectified Landsat data set. *Photogrammetric Engineering & Remote Sensing*, v. 70, pp. 313–322, 2004.

UNDP – United Nations Development Program. Human Development Report 2016 – Human Development for everyone. UN, NY, 2016. 286p.

20 Case study
Potential of sugarcane production for Colombia

Felipe Haenel Gomes, Marcelo Melo Ramalho Moreira, Mariane Romeiro and Rafael Aldighieri Moraes

Colombian context

Colombia is a South American country, located between latitudes 17° N and 4° S, composed of 32 departments and a Federal District, which adds a total area of 1.14 million km². The population is estimated at 48.7 million people, with 75% living in urban areas. The largest city in Colombia is its capital, Bogota, and the other important cities are Medellin, Cali, Cartagena, Barranquilla, Ibagué, Manizales, Pasto, Pereira, Cúcuta and Bucaramanga. The country has six natural regions that differ in relief, climate and ecosystem: Amazon, Andean, Caribbean, Insular (islands of San Andrés and Providencia in the Caribbean, and the island of Malpelo in the Pacific Ocean), Orinoquia or Llanos Orientales (Figure 20.1).

The climate is tropical along both coasts and in the eastern plains, while the uplands can be considerably cooler, creating a spatial variation ranging from tropical to subtropical climate. In general, the distribution of relative air humidity in Colombian territory is similar to the maximum observed in the Amazon and Pacific regions. In the central regions and on the border with Venezuela, the humidity tends to be drier, with some islands of humidity higher than the contiguous regions.

The highest altitudes are in the Andean region (from 500 to 5,000 m), some isolated mountains in the Caribbean region, such as the Sierra Nevada de Santa Maria (from 1,000 to 5,000 m) and in the Amazon region, such as the Serrania de Chiribiquete (from 500 to 1,000 m). The rest of the country reaches altitudes between 0 and 500 m, with Orinoquia and Amazon regions being flattered and with a rich drainage network.

According to the National Administrative Department of Statistics (DANE, 2014), pasture areas are present in all departments, with a larger amount in the Orinoquia and east/southeast regions of the Andean region.

The rural population is predominant in the Andean region (emphasizing the departments of Antioquia, Cundinamarca and Boyacá), Pacifica (Nariño and Cauca) and the Caribbean (Córdoba) (DANE, 2005), which are aligned with the quantification of agricultural areas of DANE (2014).

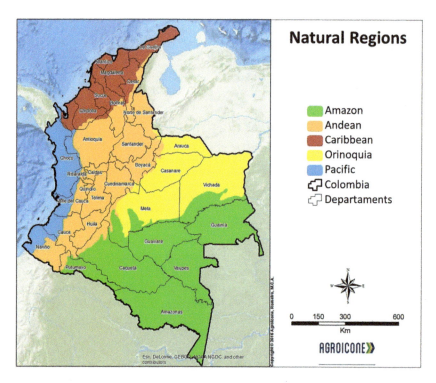

Figure 20.1 Map of the natural regions of Colombia.
Source: IGAC (2012)/Elaboration: Agroicone.

The Colombian sugar-energy sector is predominantly located (85%) in the Cauca River valley (Andean Region), which is formed by 47 municipalities from northern Cauca to the south of the department of Risaralda. The Valle del Cauca region planted 225,560 ha of sugarcane, distributed by departments, as shown in Figure 20.2 (DANE, 2014).

The production of "paneleira" sugarcane (destined to the production of rapadura – a kind of raw sugar) occupied 266,559 ha through 70,000 producers in 2011. This production is strong in the Andean region, especially in the departments of Antioquia, Boyacá, Cundinamarca and Santander (ICA, 2011; DANE, 2014).

About 25% of sugarcane area is owned by the mills and 75% is owned by more than 2,750 producers (Asocaña, 2012). Most of the producers supply 14 larger mills: Cabaña, Carmelita, Manuelita, María Luísa, Mayagüez, Pichichí, Risaralda, Sancarlos, Tumaco, Ríopaila-Castilla, Incauca, Providencia and Bioenergy.

In 2001, Colombian law 693 was approved, requiring that all gasoline used in urban centers should contain oxygenated components such as ethanol, which marked the entry of biofuels in Colombia. In addition to promoting

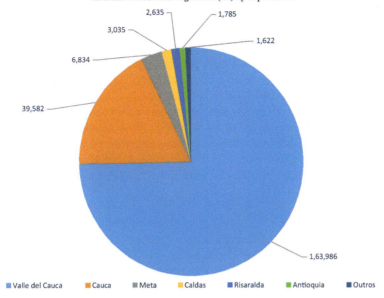

Figure 20.2 Sugarcane harvested area by department.
Source: DANE (2014)/Elaboration: Agroicone.

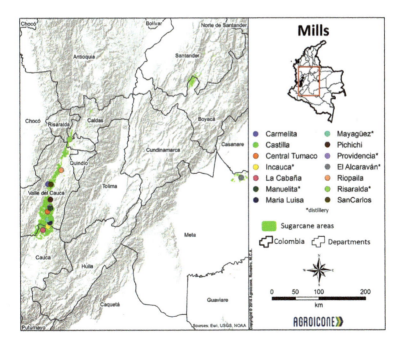

Figure 20.3 Map of sugar mills and sugarcane in Colombia.
Source: Agrosatélite (Sugarcane) and Power Plants (Asocaña)/Elaboration: Agroicone.

the use of biofuels, the law provided incentives for its production and commercialization. In 2004, the law 939 was approved, stimulating the production and commercialization of biofuels in diesel engines, including the transport sector that was not contemplated in the 2001 law. In addition, the law established that the domestically produced biofuel used in the blend with ACPM (Motor Fuel Oil) would be exempted from sales and global taxes, encouraging investments.

The demand for biofuels has increased considerably since 2005. Five mills attached distilleries to their sugar plants (Incauca, Manuelita, Providencia, Mayagüez and Risaralda), whereas the construction of the largest ethanol plant in Colombia (Bioenergy – the Alcaraván) started in 2011. The sugarcane mills and areas are shown in Figure 20.3.

Materials and methods

The methodology consists of a logical sequence of updating, elaboration, overlapping and interpretation of satellite images, maps and databases with different characteristics to identify geographies with good potential for sustainable sugarcane ethanol production. First, the land use and land cover map was updated for the year 2012/2014 using satellite images and deforestation data. Then, an edaphoclimatic map was obtained through a sequence of data crossing, including temperature, water deficit and soils. Starting from this map, the areas with strong restrictions were excluded, generating a map of the edaphoclimatic potential for the sugarcane expansion. Finally, areas with moderate constraints were excluded, obtaining the map of soil and climatic potential for the sustainable sugarcane expansion. The methodology used is presented in Figure 20.4.

The Colombian land use and land cover maps dated 2000/2002 and 2005/2009 were obtained from the National Institute for Hydrological, Meteorological and Environmental Studies (IDEAM, in Colombian acronym). A set of image classification procedures and geoprocessing techniques were applied to the acquired maps, allowing updating of preexisting land use information and the generation of maps for the years 2000/2002 and 2012/2014 for the Colombian territory. These procedures were performed in two stages: (i) update of the 2000/2002 map and (ii) update of the 2012/2014 map. The generated land use maps were crossed with the purpose of generating a transition matrix, making possible to analyze the changes in land use between 2002 and 2014.

The previous map of land use and land cover of 2000/2002 of IDEAM was used for the elaboration of the base map of land use for the Colombian territory. This data has undergone an aggregation process of thematic classes and class nomenclature adjustment in order to be consistent with the land use classification system adopted in the International Geosphere-Biosphere Program (IGBP).

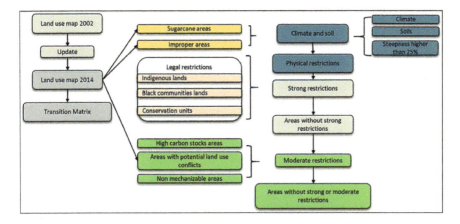

Figure 20.4 Scheme of the methodology used in the study.
Source: Agroicone.

The process of thematic classes aggregation was realized based on the visual images interpretation of the TM/Landsat5 and ETM+/Landsat7 sensors/satellites of 2002 and identification of the landscape elements present in the map classes of 2002. The classes that presented the elements of the landscape with similar characteristics were aggregated according to the description of the IGBP classification system.

Such a classification system was adopted because it is compatible with the system used by the Food and Agriculture Organization of the United Nations (FAO) in the Forest Resources Assessment reports, which are used as ancillary data of quantitative reference in this mapping. In addition, it is one of the systems adopted by the MODIS Land Cover Type MCD12Q1 product (Friedl et al., 2010), also used in the processing of land use maps for the territory of Colombia.

The 2012/2014 Colombian land use map was produced from the update of the land use map of 2005/2009 (IDEAM, 2012a, 2012b) according to the IGPB classification, using the same procedure adopted for updating of the 2000/2002 base map.

After the map update, a deforestation mask was used between the years 2009 and 2014 of the Global Forest Change 2014 (Hansen et al., 2013) to indicate areas where changes in forest cover occurred. In areas where deforestation occurred, land use was updated using MODIS MCD12Q1 data. Then, the sugarcane area was updated to 2014/2015 from images of the ETM+/Landsat 7 and OLI/Landsat 8 sensors.

The first stage of the study resulted in two land use and coverage maps of Colombia for the years 2000/2002 and 2012/2014, qualified in 11 equivalent thematic classes, which allowed to generate a transition matrix of use

Table 20.1 Land use and land cover in Colombia in the years 2000/2002 and 2012/2014 (1,000 ha and %)

Classes	2000/2002		2012/2014	
	1,000 ha	%	1,000 ha	%
Water	2,668	2.4	2,347	2.1
Forest	55,434	48.9	53,182	46.9
Shrublands	5,031	4.4	5,307	4.7
Savannas	16,803	14.8	17,504	15.4
Woody savannas	3,013	2.7	2,856	2.5
Grasslands	14,093	12.4	14,334	12.6
Permanent wetlands	1,281	1.1	1,072	0.9
Croplands	1,097	1	1,210	1.1
Urban and built-up	417	0.4	374	0.3
Cropland/natural	11,957	10.5	13,622	12.0
Barren or sparsely	1,113	1	1,092	1
Sugarcane	305	0.3	312	0.3
Mixed forest	118	0.1	117	0.1
Snow and ice	7	0	6	0
Total	113,335	100	113,335	100

Source: Data from the study.

between the years. The area of each class and the change occurred between 2000/2002 and 2012/2014 can be seen in Table 20.1.

All geoprocessing procedures were performed in ArcGIS 10.3 software. The database is in the WGS-84 datum, and the Albers projection (South America Albers Equal Area Conic) was used to calculate the areas.

Main findings

Land use and land use change

The Colombian territory is covered predominantly by forests, savannas, grasslands and cropland/natural areas. The forest areas are located in the Amazon and Pacific regions, while the savanna areas predominate in the *Llanos Orientales* and in the North of the Caribbean. Grasslands are distributed throughout the *Llanos Orientales*, *Andina* and *Caribe* regions. The cropland/natural areas, shrublands, the areas of agriculture and sugarcane are concentrated in the *Andina* region.

Forest areas had the largest area reduction (2.3 million ha), and their area was occupied by savanna, open shrub and mosaic.

Case study 249

The sugarcane area in 2012/2014 was estimated at approximately 312,000 ha, representing only 0.3% of the Colombian territory. The dynamics of soil use and occupation of sugarcane can be divided into two periods with quite different characteristics. In the first period (up to 2012/2014), the sugarcane area was concentrated only in the traditional region of Valle del Cauca, with a small expansion of sugarcane areas on mosaic areas of crops, shrubs and pastures, pasture and agriculture, while also losing area for grazing.

From 2012/2014, after the construction of the Bioenergy plant (the Alcaraván – Puerto Lopez, Meta department), sugarcane planting was advanced in pasture areas near the plant, altering the trend previously observed. There is an indication that, if the most recent pattern prevails, the expansion of sugarcane to ethanol may occur through expansion in pasture areas.

The updated land use and land cover map for the 2012/2014 period is shown in Figure 20.5.

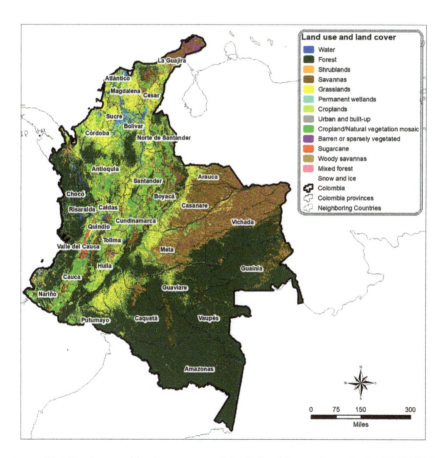

Figure 20.5 Land use and land cover map of the Colombian territory for the 2012/2014 period.
Source: Agrosatélite. The data were developed by Agrosatélite, a project partner.

Restrictions

Legally protected

In Colombia, the legally protected areas are formed by natural parks, indigenous lands and black communities. The country has an entity called National Natural Parks (PNN) that manages 56 natural parks, which are part of the Natural National Parks System (SPNN) and coordinates the National System of Protected Areas. This large natural area has 12.6 million ha, which represents approximately 10% of the country's territory (PNNC, 2016).

According to the General Census of the National Administrative Department of Statistics (DANE, 2005), the indigenous population occupies an area of 32 million ha, while 5.4 million ha belongs to the black communities.

Considering the overlap of indigenous lands or black communities in national parks, the total legally protected area is estimated at 46.2 million ha, representing 40.4% of Colombian territory. Protected areas are shown in Figure 20.6.

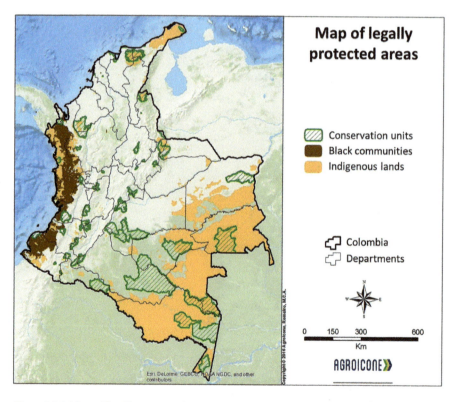

Figure 20.6 Map of legally protected areas.
Source: IGAC and SINAP/Elaboration: Agroicone.

Physical

The edaphic constraints were discussed in the chapter 18. The restrictions related to climate (temperature, rainfall and water deficit) are presented in Figure 20.7. We must remember that the low average temperatures and excess of rainfall were considered to be strong restrictions. Temperatures below 20°C are found in the regions of higher altitude, such as in the Andes Mountains, and render an area of 15 million ha inapt. In the Amazonian and Pacific regions, there is an excess rainfall, making 68 million ha unfit for sugarcane cultivation, and almost 3 million ha have precipitation below the ideal, and irrigation is necessary in these areas to have a high potential productivity. The water deficit makes it unfit for an area of 6 million ha, concentrated in the Caribbean region.

Other moderate restrictions

Figure 20.8 shows a graphic presentation a first set of moderate restrictions. The first figure on the left represents the non-mechanizable areas, where slopes area higher than 12%, which are concentrated in the Andean highlands. The high carbon areas – carbon stocks than 136 tonnes C/ha – are concentrates in areas surrounding Amazonian and Pacific natural zones. The areas with potential land use conflicts – areas currently occupied by agriculture – are concentrated in the Andean and Pacific zones.

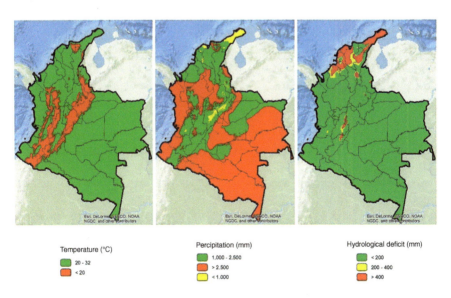

Figure 20.7 Climate related restrictions (temperature, precipitation and hydric deficit).
Source: Temperature and Precipitation. WorldClim/Elaboration: Agroicone.
   Hydric deficit – Source: Moraes (2016)/Elaboration: Agroicone.

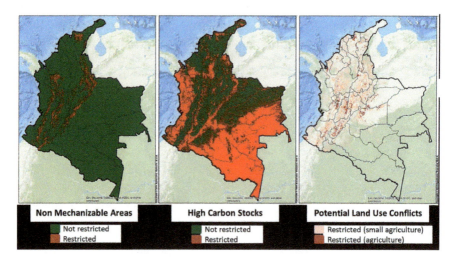

Figure 20.8 Non-mechanizable areas, high carbon stocks and areas occupied by agriculture.
Source: Agroícone. Based on Worldclim (2016), IDEAM and Harris (2009), Agrosatélite (2016). Agrosatélite is a project partner.

Potential production

For Colombia, the soil map made by IGAC (Geographic Institute Agustín Codazzi) in the 1:500,000 scale, acquired in a digital format, was used. The soils in those maps were classified on the basis of international systems – Soil Taxonomy (Soil Survey Staff, 1999) and World Reference Base for Soil Resources (IUSS Working Group, 2014), as well as local classification considering mainly color, texture and parent material. Then, a correlation between the soils data and the Brazilian Soil Classification System (Embrapa, 2013) was performed. It is noteworthy that the Colombian map, besides presenting a larger scale leading to greater data precision, presents the highest quality in the information of each soil spot, determining the percentage of associated soils, as in this level of detail the spots correspond to associations of more than one class of soil.

The edaphic suitability classification generated four groups of potential production: medium, low, restricted and unfit, as shown in Figure 20.9. As mentioned in chapter 18, The "inappropriate" level corresponds to areas with some kind of strong restriction for plant growth. In "restricted" areas, production is possible but requires significant interventions. Therefore, "inappropriate" and "restricted" level indicates areas that are strongly restricted.

The medium, low and restricted potentials refer to the production potential of that type of soil. The use of appropriate technologies (i.e., soil correction and fertilization) and good cultivation practices can allow good productivity in these potentials.

Case study 253

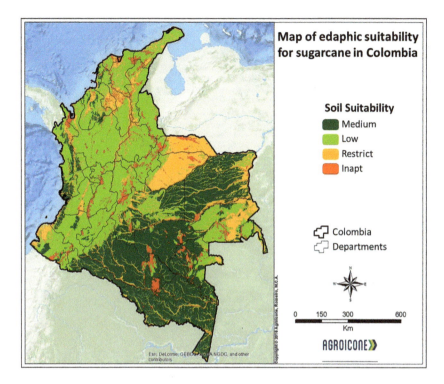

Figure 20.9 Map of edaphic suitability for sugarcane in Colombia.
Source: Pedológica. Elaboration: Agroicone. Pedológica is a project partner.

Soils with an average potential occupy an area of 37 million ha (32.5% of the territory) and are located predominantly in the Llanos Orientales and Amazon regions. On the other hand, the areas with low potential are located mainly in the Caribbean, Andean and Pacific regions, occupying an area of 48 million ha (42%). Restricted areas cover 22 million ha (20%) and inapt areas 6 million ha (5%).

In the case of Colombia, when only the strong restrictions are excluded (Figure 20.10), no areas with high potential were identified. This is probably because areas with greater potential are already occupied by sugarcane or are considered restricted due to problems of soil drainage. It is known that many of the areas currently occupied by sugarcane cultivation in Valle del Cauca, a traditional Colombian producer pole, are under the category of restricted, corresponding to areas that have been drained and systematized in order to use irrigation, leading to high productivity. There are no longer many similar areas available for expansion in this region. Another probable reason for not identifying the areas of high potential is the quality of the map, where, in addition to the main soil, it provided information about the other associated

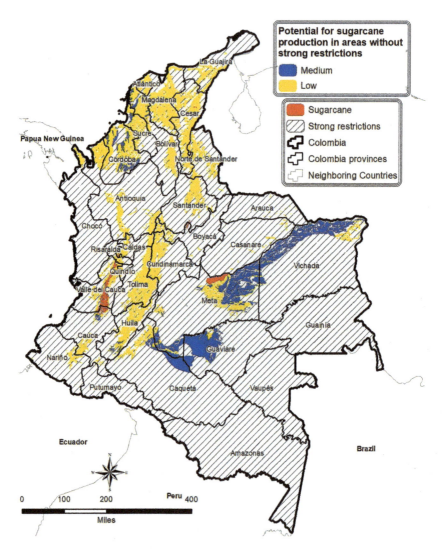

Figure 20.10 Potential for sugarcane production in Colombia without strong restrictions.

soils, and this work considered that the potential to be represented should occupy at least 50% of the soil spot in question. More detailed soil mapping can, however, indicate soils with high potential together with soils of lower potential not identified at the level of the actual map.

Then, considering strong restrictions, about a total of 4.8 million ha of medium potential and 12.4 million of low potential were identified and including moderate restrictions (Figure 20.11), approximately 2.5 million ha of medium potential and 4.0 million ha of low potential were identified (Table 20.2). It is observed that the areas with greater extensions of soils with average potential

Table 20.2 Area (ha) of high, medium and low potentials considering levels of restriction in Colombia

Potential	Strong restrictions (1,000 ha)	Strong and moderate restrictions (1,000 ha)
High	0	0
Medium	4,835	2,537
Low	12,443	4,026
Total	17,278	6,563

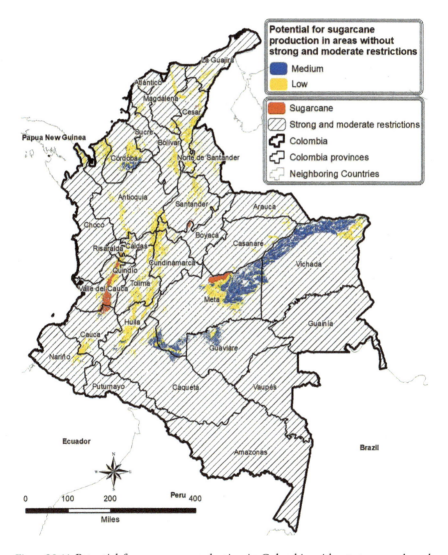

Figure 20.11 Potential for sugarcane production in Colombia without strong and moderate restrictions.
Source: Agroícone.

for implantation of the sugarcane culture are in the departments of Meta, Vichada, Guaviare and Caquetá. Considering the "strong and moderate" constraints, the departments of Meta and Vichada are the ones that present the largest extensions of soils with this potential, corresponding mainly to Ferralsols, as well as Acrisols and Lixisols, many of them with the fine texture. It is, therefore, the place where there is greater potential for expansion of the sugarcane crop for bioenergy purposes, and nowadays, there is already sugarcane planting in this region, and a Bioenergy plant that started grinding in 2016.

Discussion

It is noteworthy that the most suitable areas for expansion are concentrated in the departments of Meta and Vichada, mainly if moderate and strong restrictions are considered simultaneously. In this region, part of the Llanos Orientales, there is already a new distillery in operation, representing an increase in the area of sugarcane in the country. Although the area is suitable, soils do not present high potential, being limited by their chemical conditions, with low CEC and base saturation and often with the presence of exchangeable aluminum, but which can be easily corrected. They are, however, mostly fine textured, clayey and silty, with good water storage capacity. Care must be taken, however, for problems related to the management of these soils, due to the rainy season being extensive, with a small gap for harvesting operations, and that can cause serious soil compaction damage if the soil moisture is not observed. This fact was observed in a visit to the bioenergy plant, where they try to wait at least for 3 days without rain to harvest, even when they were only planting to obtain seedlings, as the visit occurred before the start of the plant operations. The biggest challenge in this case will be to provide a constant raw material for the industry, especially in the beginning and end of the dry season, when there is still a possibility of rainfall, when harvesting in high humidity soils can lead to very low yields in the next harvest.

It should be emphasized that in the Llanos Orientales, it is possible that the expansion of sugarcane culture occurs in the extensive areas of underutilized pasture. And projects such as the use of hydrolyzed bagasse combined or not with the intensification of agriculture with integrated systems can, in the end, even increase the production of meat considering the same area.

References

DANE – Departamento Administrativo Nacional de. Estadística (DANE). Censo de población y vivienda de Colombia. Año 2005. Análisis de la estructura y composición de las principales variables demográficas y socioeconómicas del Censo 2005. Informe final. Bogotá, D.C. 30 de setembro de 2008.

DANE – Departamento Administrativo Nacional de Estadística. Censo Nacional Agropecuario 2014. Inventarios agropecuarios en las unidades de producción agropecuaria (UPA). Report 9. Available: <https://www.dane.gov.co/index.php/estadisticas-por-tema/agropecuario/censo-nacional-agropecuario-2014#9>. Access: May 25, 2016.

Embrapa – Empresa Brasileira de pesquisa Agropecuária. Sistema brasileiro de classificação de solos. 3rd ed. Brasília, 2013. 353p.

Friedl, M. A.; Sulla-Menashe, D.; Tan, B.; Schneider, A.; Ramankutty, N.; Sibley, A.; Huang, X. MODIS Collection 5 global land cover: Algorithm refinements and characterization of new datasets. Remote Sensing of Environment, v. 114, pp. 168–182, 2010.

Hansen, M. C.; Potapov, P. V.; Moore, R.; Hancher, M.; Turubanova, S. A.; Tyukavina, A.; Thau, D.; Stehman, S. V.; Goetz, S. J.; Loveland, T. R.; Kommareddy, A.; Egorov, A.; Chini, L.; Justice, C. O.; Townshend, J. R. G. High-resolution global maps of 21st century forest cover change. Science, v. 342, pp. 850–853, 2013.

IDEAM – Instituto de Hidrología, Meteorología y Estudios Ambientales. Mapa de Coberturas de la Tierra. Metodología CORINE Land Cover adaptada para Colombia. Período 2000–2002. Escala 1:100.000. 2012a.

IDEAM – Instituto de Hidrología, Meteorología y Estudios Ambientales. Mapa de Coberturas de la Tierra. Metodología CORINE Land Cover adaptada para Colombia. Período 2005–2009. Escala 1:100.000. 2012b.

IUSS Working Group WRB. World Reference Base for Soil Resources 2014, update 2015 International soil classification system for naming soils and creating legends for soil maps. World Soil Resources Reports No. 106. FAO, Rome. 2015.

PNNC – Parques Nacionales Naturales de Colombia (PNNC). Disponível em: www.parquesnacionales.gov.co/portal/es/. Access: 06 de dez. 2016.

Soil Survey Staff. Soil taxonomy: A basic system of soil classification for making and interpreting soil surveys. 2nd edition. Natural Resources Conservation Service. U.S. Department of Agriculture Handbook 436. 1999.

21 Final comments

Marcelo Moreira

Cross-country comparisons are a tricky exercise once direct comparison is several times nonapplicable. However, the case studies allowed to identify some clear similarities and differences between the two countries.

Mozambique has extensive areas with possibilities for the expansion of sustainable production of sugarcane in areas with high and medium potentials, even considering moderate restrictions, thus avoiding agrarian conflicts, high costs with irrigation, areas with problems of mechanization and high carbon stock. This area extends in several provinces – Tete, Nampula, Zambezia, Niassa and Cabo Delgado, but also occurs in other provinces of the country.

There is a greater need for more detailed studies in Mozambique, not only of soil but also in relation to the survey of climatic data, as the country has a more limited database. In addition, despite presenting sugarcane production, it is not a country with this tradition, having been scene of conflicts and civil war that did not allow a continuous production.

Colombia has 17.3 million hectares for sugarcane expansion, of which 6.5 is considered if all strong and moderate restrictions are considered simultaneously. The regions close to Meta and Vichada were found to be potential areas for a new cluster formation with other mills, following the example of the first ethanol dedicated plant in the country. Even having less available area without high potential, and despite having very different conditions from its traditional producing region – the Cauca Valley – it is a country with a tradition of culture.

Soil and climate conditions contribute to the differences between the countries, and socioeconomic factors also differ. Lack of precipitation seems to be a great challenge in Mozambique, whereas excess of rainfall limits sugarcane production in a significant area of Colombia, driving sugarcane expansion out of the forested areas (Amazon). The lower density of human occupation in the savannahs (occupied by pastures, grasslands and prairies), where carbon stocks area lower than natural forests, makes it a potential area for biofuels expansion. Expansion of sugarcane area driven by ethanol will be a novelty in both countries.

Countries can learn from the recent Brazilian experience, with expansion on the Cerrados. The rapid area expansion in new areas, with changing technologies (mechanization), showed that first years of production might require care. Particularly unexpected yield patterns may appear in the first years. However, as the time goes by, yield in such regions turned to be even higher than in the traditional areas. The industry expansion into areas with extensive pastureland areas also brought positive spillovers such as higher yields for other crops and increase in local purchasing power. In agronomic terms, collaboration can speed up the search for adapted cultivars to the conditions of the Colombian savannah, more similar to the Brazilian *Cerrado*, and where the culture has expanded more in the last decade. It is also highly recommended that biofuels expansion is governed by formalized land use policies and supporting instruments.

Finally, this study aims to indicate possible expansion areas for the production of biofuels, but does not exempt the researchers from carrying out more detailed studies of edaphoclimatic conditions *in loco*, but has a fundamental role for a better choice of the sites to carry out these studies, which will give more precise information to be considered as an appropriate decision whether or not to implement a new enterprise.

Part IV
Impacts and feasibility of sugarcane production

22 Sugarcane production model and sustainability indicators

Manoel Regis L.V. Leal and
João Guilherme Dal Belo Leite

Introduction

Bioenergy is one of the oldest forms of energy used by humanity to fulfill some of its basic needs, such as heating, cooking, lighting, protection and other increasing varieties of use. Over the past few decades, it started to modernize and compete with existing nonrenewable alternatives, particularly fossil based. This growth in bioenergy production and use was closely followed by a fair amount of criticism, such as threats on food security, greenhouse gas (GHG) emissions higher than fossil alternatives, land grabbing, deforestation, degradation of water resources and biodiversity losses. Yet, critics often treat bioenergy as a single entity, instead of a multitude of options. Moreover, there are a large number of model-based studies that can be undermined by unreliable databases and arbitrary assumptions (Locke and Henley, 2014), lack of consistency in the results (CBES, 2009; EC, 2010) and absence of causality analyses.

Bioenergy, in general, and biofuels in particular, can be either bad or good depending on the local conditions, driving forces and objectives, and the production model. The choice of feedstock and the production system (with impacts over land use change, yields, land tenure, water use, production scale), combined with the processing technology option (plant scale, product recovery, use and disposal of residues and wastes) and the interactions with local communities (jobs, food and energy security, infrastructure improvements, education and health care), are the key issues for a sustainable biofuel production (Bogdanski, 2014).

If done properly, bioenergy can significantly contribute to the achievement of many of the United Nations Sustainable Development Goals (UN SDGs) (UN, 2015):

- SDG1 (No Poverty): bioenergy projects can create jobs, boost household income, improve food and energy security and promote rural development.
- SDG2 (Zero Hunger): the integrated production of bioenergy and food (Bogdanski, 2014; Bogdanski et al., 2010) can improve land use efficiency with positive spin-offs towards infrastructure (irrigation, roads, schools, hospitals) and access to electricity.

- SDG3 (Good Health and Well-Being): bioenergy projects can promote health services, which include rural clinics, power to preserve medicines and vaccines, and potable water to local communities.
- SDG4 (Quality Education): women and children can be relieved from the time-consuming task of collecting wood for cooking and heating since bioenergy can provide clean cooking fuels to replace traditional biomass (collected fuel wood). With more free time, children can attend school and adults engage in productive tasks (e.g. agriculture) and capacity building.
- SDG5 (Gender Equality): bioenergy can promote women and girls through better health, education and their engagement in income-generating activities.
- SDG6 (Clean Water and Sanitation): wider access to electricity, such as from biofuel production (coproduct), enables water treatment and distribution systems to local communities.
- SDG7 (Affordable and Clean Energy): this is central to bioenergy initiatives that can be produced locally, thus providing access to modern energy where it is mostly needed – rural areas of developing countries.
- SDG8 (Decent Work and Economic Growth): electricity and other forms of modern energy are the basic requirements for economic growth and the creation of adequate (quantity and quality) jobs, particularly in the rural areas. The bioenergy plant itself and the feedstock production are important sources of jobs and income.
- SDG9 (Industry, Innovation and Infrastructure): bioenergy projects can foster infrastructure development such as roads, potable water distribution systems and power grids. Modern and renewable energy access enables innovation with positive spin-offs on local businesses and industrial development.
- SDG10 (Reduce Inequalities): modern energy availability and access is a paramount condition to economic growth and social progress, thus essential to reduce inequalities among countries.
- SDG13 (Climate Action): bioenergy can help countries to meet their national GHG emission targets with worldwide positive consequences. The biofuel and feedstock choice and its land use change dynamics are crucial to a positive and high reduction in GHG emissions.
- SDG14 (Life below Water): bioenergy can produce good or bad impacts on water resources depending on its effects on deforestation, erosion, use of pesticides and fertilizers.
- SDG 15 (Life on Land): if managed properly, land use changes resulting from bioenergy projects can preserve important ecosystems. On the other side, if not, it can promote the deterioration of these ecosystems. Sustainability criteria can be useful to set projects on the right track.
- SDG17 (Partnerships for the Goals): bioenergy can have an important positive contribution by attracting foreign capital to poor countries, with further investment in new and modern technologies that entice further development in infrastructure and capacity building at different levels (local, regional and national).

To ensure bioenergy sustainability, there are several certification systems such as Bonsucro, Roundtable on Sustainable Biomaterials (RSB), International Sustainability and Carbon Certification (ISCC) and Global Bioenergy Partnership (GBEP), which have been successfully used under different conditions around the world.

The Food and Agriculture Organization of the United Nations (FAO) have developed two systems for the sustainable implementation of bioenergy programs: Integrated Food-Energy Systems (IFES) and Bioenergy and Food Security (BEFS). IFES (Bogdanski, 2014; Bogdanski et al., 2010) focuses on energy efficiency, lower land demand and the integration of energy and food production at the local level. BEFS (FAO, 2014) uses a similar approach, yet the optimization of food and biofuel systems is upscaled to the national level. There are plenty cases of both methodologies applied under real conditions in different countries and contexts (e.g. Maltsoglu and Beall, 2012; Maltsoglu and Khwaja, 2010).

So far, we have treated bioenergy in generic terms, but from now on we concentrate on sugarcane ethanol, which is the focus of the LACAf Project. In the following sections, we describe important aspects of the production models, based on the Brazilian and other countries experiences. Our aim is to support policymakers to better understand what drives sustainable biofuel production. Since our focus is on developing regions, i.e. Latin America, Caribbean and sub-Saharan Africa, rural development as well as climate change mitigation, fossil fuel reduction and environmental preservation were elected as main sustainability indicators.

A series of articles have been published as part of the LACAf Project scope. Our research tackles what we consider the most important factors for the selection of the biofuel production model that is best fitted to promote rural development, such as energy security (Leal et al., 2016; Leal and Pereira, 2014; Leite et al., 2016a), project scale (Kubota et al., 2017), smallholder participation (Leite et al., 2016b), agriculture mechanization (Cardoso et al., 2017), production model (Leal and Nogueira, 2014), sugar industry diversification (Leal and Leite, 2016; Leal et al., 2016) and food security (Leite et al., 2016b). In Chapter 28, we summarize the findings and conclusions in these articles to provide a consolidated view of the project results.

Ethanol production model concept

By *production model,* we mean the agricultural and industrial stages associated with sugarcane ethanol production, which also includes the socioeconomic interfaces between ethanol-producing enterprise, government and the local community (Table 22.1). Considerations about ethanol market (local, national or international) are very important in the selection of the production model. We could identify in the literature four basic project types (Gasparatos et al., 2015; von Maltitz and Stafford, 2011) based on processing capacity and agriculture production (smallholder or large commercial production) as shown in Table 22.1.

Table 22.1 Biofuel project types

Project type	Feedstock production	Property size (ha)	Biofuel use
1	Smallholder and outgrower	1–10	Local use at village or farm level
2	Large industrial farms	100–1,000	In the large commercial farms or enterprises like mines
3	Outgrowers or smallholders linked with commercial plantations or biofuel processing plant	1–10	National use or exports
4	Large-scale commercial plantations	100–1,000	National use or export

Source: von Maltitz et al. (2011).

The feasibility of the different biofuel project types in Table 22.1 is highly context specific. And depends on, for instance, jobs opportunities and labor availability, food security, energy security, access to technology, business risk and, above all, services provided to local communities, particularly in health and education. Moreover, there are two ways to kick off sugarcane ethanol production – greenfield and brownfield projects. In a greenfield project, agricultural and industrial activities are designed and implemented from scratch. Therefore, it assumes the conversion of land under traditional uses to sugarcane production and the construction of a brand new industrial plant. In brownfield projects, ethanol production starts by retrofitting existing sugar mills, often by annexing an ethanol distillery (fermentation and distillation units). This project type allows diversification strategies (alternatives to the sugar-based model) while reducing the demand for investment capital. Electricity surpluses may also be an alternative product, with several additional socioeconomic benefits (see Chapter 28). Brownfield project was the option to introduce sugarcane ethanol in many countries such as Brazil, Colombia, Guatemala, Argentina, Peru and Thailand, to mention a few, representing the majority of the sugarcane ethanol produced globally.

Bioenergy production systems

We discuss local level small-scale projects in Chapter 26. Technologies are discussed in Chapter 25 with special focus on economic, social and environmental impacts. At the national level, our approach suggests four different production systems:

- **High tech**: the largest scale possible, verticalization in sugarcane production and processing, maximum energy efficiency, full use of mechanization and automation.

- **Medium tech**: mixed sugarcane production (independent growers and mill production), scale compatible with local agricultural practices, mechanization and automation compatible with existing labor availability and qualification.
- **Social**: smallholder-independent cane production (cooperatives and associations) and outgrowers, integration of food and energy, local supply, maximum job generation and social responsibility initiatives (schools, health clinics, professional training, capacity building, etc.).
- **Optimum**: the characteristics of the three systems above are analyzed to provide the maximum benefits for the local communities, taking into consideration the government priorities and an extended negotiation among the stakeholders; nevertheless, the economic feasibility of the enterprise must be assured for long-term survival of the business. The burden of some social services provided to the communities must be divided among the enterprise (Social Responsibility Projects), government and other institutions. The IFES and BEFS FAO's methodology could be applied in the model optimization process to increase public acceptance, government assurance and planning, and attract foreign capital.

There are several factors that interfere in the sustainability of biofuels, and a complete analysis of them is out of the scope of this chapter. We have limited our analysis to what we consider the most relevant issues that are important to meet the sustainability certification requirements and to contribute to the UN SDGs.

With respect to the certification process, it is important to analyze some alternatives that can comply with the requirements of the main potential markets for export: the European Union (EU Renewable Energy Directive), the United States (US Renewable Fuel Standard) in general and California in particular (CARB Low Carbon Fuel Standard). To illustrate these requirements, we present the cases of the GBEP (2011) (Table 22.2) and BONSUCRO (2017). The former is a standard developed in a consensus system with the participation of 23 partner countries, under the request of the G8 group, and the latter is the most popular and adopted system for the certification of sugar/ethanol mills worldwide. Other sustainability certification systems such as RSB and the ISCC are also widely used.

BONSUCRO sustainability principles (Bonsucro, 2017) are as follows:

- Obey the law,
- Respect human rights and labor standards,
- Manage efficiency to improve sustainability,
- Manage biodiversity and ecosystem services,
- Continuously improve key areas of business,
- Adhere to EU Directives (optional).

Table 22.2 GBEP sustainability indicators

Indicators		
Environmental	Social	Economic
Lifecycle GHG emissions	Allocation and tenure of land for new energy production	Productivity
Soil quality	Price and supply of a national food basket	Net energy balance
Harvest levels of wood resources	Change in income	Gross value added
Emissions of non-GHG air pollutants, including air toxics	Jobs in the bioenergy sector	Change in the consumption of fossil fuels and traditional use of biomass
Water use and efficiency	Change in unpaid time spent by women and children collecting biomass	Training and re-qualification of the workforce
Water quality	Bioenergy used to expand access to modern energy services	Energy diversity
Biological diversity in the landscape	Change in mortality and burden of disease attributable to indoor smoke	Infrastructure and logistics for distribution of bioenergy
Land use and land use change related to bioenergy feedstock production	Incidence of occupational injury, illness and fatalities	Capacity and flexibility of the use of bioenergy

Source: GBEP (2011).

Sustainability criteria for ethanol production are very important, particularly if there is intention to export and attract foreign investments. In this respect, public acceptance, national and international, is much easier when sustainability can be clearly demonstrated.

Key issues in the selection of the production model

In the following paragraphs, we underline some key aspects in the selection of the production model based on lessons learned from LACAf Project (i.e. Latin America, the Caribbean and sub-Saharan Africa). The Brazilian case is briefly described below, because it is a fine case to show how the sugar-based production model evolved during almost one century to the current integrated model of sugar/ethanol/electricity production. Although it can be considered as a successful model for Brazil, it cannot be considered the best model for all countries. The local context and priorities will have to be considered in each case.

The development of the Brazilian sugarcane ethanol production model

The first government mandate in Brazil to add ethanol to the gasoline took place in 1931, motivated by the desire to reduce gasoline imports and to create an economic and easy use for the existing surpluses of sugar and sugarcane (Cortez, 2016). At the time, a blend of 5% ethanol was adopted in all imported gasoline (in 1938, it was extended to all gasoline consumed in the country). Several sugar mills annexed distilleries to the existing sugar plants, making possible a fast increase in ethanol production at low investment cost. In 1975, motivated by the sharp oil price increase in 1973, the government acted again to stimulate the increase in ethanol use to a higher level by setting a new target, i.e. 20% blend in 10 years. The sugarcane sector fully adhered to the new Ethanol Program because it was again facing problems with the volatile sugar market. The second oil shock in 1979 made the government stimulate the sector to speed up the ethanol production beyond the levels permitted by annexed distilleries. Then, autonomous distilleries that produced only ethanol started to be built until 1985 when oil prices dropped to pre-1973 levels. A stagnation phase was initiated and lasted until the beginning of the 21st century. During this period, sugar factories were annexed to the autonomous distilleries thus consolidating the Brazilian cane processing system with combined sugar and ethanol production. With the privatization of the electric power system and the following changes in the electric sector legal framework, the generation of power surpluses became the third product from sugar mills. Today, more than half of the mills in operation are selling electricity to the national grid.

Lessons can be learned from this short story of the Brazilian sugarcane sector, from sugar-based to a sucro-energetic industry: (1) diversification is an efficient way to reduce business risk by producing two or three different commodities to different markets (food and energy); (2) it can be done without reducing sugar production (i.e. efficiency gains); (3) scaling up agricultural and industrial stages such as cane preparation/juice extraction and energy production boosts efficiency and economic competitiveness and (4) plant flexibility as to the production rate of ethanol and sugar allows to balance output with market prices.

Table 22.3 shows the profile of the Brazilian sugar/ethanol sector in 2011.

Table 22.3 confirms the trend towards larger industrial plants (>1.2 million tons cane per year) and an overwhelming majority of mixed mills (i.e. sugar + ethanol + electricity). The most common sugar-based mills (sugar factory), production model worldwide, represent less than 4% of the milled cane.

Table 22.3 Brazilian sugar/ethanol sector profile in 2011

Mill type	No. of mills	Milled cane (Mt/y)	Milled cane (%)	Average size (t/y)
Mixed	257	378.4	67.4	1,472,000
Ethanol distillery	127	161.2	28.7	1,269,000
Sugar factory	18	21.4	3.8	1,188,000
Total	402	561.0	100.0	1,396,000

Source: CONAB (2013).

Regarding to cane supply, the Proálcool Program fostered cane production and processing done by two independent agents, but it did not work like that. Today, countrywide, on average, 1/3 of the cane is produced by the mills in their own land, another 1/3 is produced by the mills in rented land and the remaining 1/3 is supplied by independent growers, mostly as members of large associations. Although this is the countrywide average, there are many mills either producing 100% of the cane they crush or outsourcing 100% from independent cane growers. The production system is coupled with a cane payment system defined in common accord between millers and growers, which takes into account cane quality parameters, sugar and ethanol prices. The aim is to make a fair division of profits (or losses) between the two groups (millers and growers).

Final comments

In this chapter, we presented an overview of some relevant aspects for sugarcane ethanol production models. Realistic alternatives, however, require an in-depth analysis of the local conditions as there is no "one-size-fits-all" solution. We listed and briefly discussed the UN SDGs and explored some of the most important certification systems in use by the sugarcane sector in order to highlight the important indicators to the selection of the production model.

The basic concepts of the different production models were presented to serve as benchmarks to kick off sugarcane ethanol production. Technology, scale, sugarcane supply and smallholder inclusion are important aspects in this discussion.

In Chapter 28, we will provide more detailed information about alternatives evaluated under LACAf using the data available for Brazilian conditions (Cardoso et al., 2017; Kubota et al., 2017; Leal et al., 2016; Leite et al., 2016a, 2016b), but adjusted to policymakers in other developing regions. It is our understanding that Latin America and Caribbean have already selected their preferred production model, which consists of annexing distilleries to existing sugar mills operating with final molasses, and increasing progressively ethanol production by using B molasses or cane juice. Autonomous distilleries are only installed after the exhaustion of this model capacity to increase production or if a larger step in ethanol production is desirable. This is the recent case of Colombia and reflects the situation when the ethanol production needs to be increased significantly without affecting the sugar production.

Last, but not least, public acceptance is a critical success factor. In this respect, sustainability certification and the application of the FAO's IFES and/or BEFS methodologies can be used to produce ample evidence of the project impacts.

References

Bogdanski A, 2014. *Food and Fuel: Learning Lessons from Assessing Integrated Fuel-Energy Systems*. Sustainable Aviation Fuels Forum, 20–22 October 2014, Madrid, Spain.

Bogdanski A, Dubois O, Jamieson C, Krell R, 2010. *Making Integrated Food-Energy Systems Work for People and Climate: An Overview*. Environment and Natural Resources Management Working Paper No. 45. Food and Agriculture Organization of the United Nations (FAO), Rome.2010. 136p.

Bonsucro, 2017. *Bonsucro Production Standard* (available at www.bonsucro.com).

Cardoso TF, Watanabe MDB, Souza A, Chagas MF, Cavalett O, Morais ER, Leal MRLV, Braunbeck OA, Cortez, LAB, Bonomi A, 2017. Economic, environmental, and social impacts of different sugarcane production systems. *Biofuels Bioproducts & Biorefining*. doi: 10.1002.

CBES, 2009. Land-Use Change and Bioenergy: report from the 2009 workshop. *Center for Bioenergy Sustainability, Oak Ridge National Laboratory (ORNL)*, ORNL/CBES-001, US/bbb.1829Department of Energy (available at www.ornl.gov/sci/berd/cbes.shtml)

CONAB, 2013. *Perfil do Setor do Açúcar e do Álcool no Brasil*. Vol 5 – Safra 2011/2012. Companhia Nacional de Abastecimento – CONAB, Brasília, 2013, p. 88.

Cortez LAB, 2016. *Universidades e empresas: 40 anos de ciência e tecnologia para o etanol brasileiro (in Portuguese)*. Luís Augusto B. Cortez (organizer). Editora Edgar Blücher Ltda. São Paulo.

EC, 2010. Report from the Commission on indirect land-use change related to biofuels and bioliquids. *European Commission COM*: 811 final, Brussels, 22/12/2010.

FAO, 2014. *FAO's BEFS Approach Implementation Guide, Food and Agriculture Organization of the United Nations*, Rome, 25p.

Gasparatos A, von Maltitz GP, Johnson FX, Lee L, Mathai M, Oliveira JAP, Willis KJ, 2015. Biofuels in sub-Sahara Africa: Drivers, impacts and priority policy areas. *Renewable and Sustainable Energy Reviews* 45: 879–901.

GBEP, 2011. *The Global Bioenergy Partnership Sustainability Indicators for Bioenergy*. FAO, Rome, 2011, 213p.

Kubota AM, Leite JGDB, Watanabe M, Cavalett O, Leal MRLV, Cortez LAB. 2017. The role of small-scale biofuel production in Brazil: Lessons for developing countries. *Agriculture* 7: 61. doi:10.3390/agriculture7070061.

Leal MRLV and Leite JGDB, 2016. *Ethanol as a major co-product*. Proceedings of the International Society of Sugar Cane Technologists, vol. 29.

Leal MRLV, Leite JGDB, Chagas MF, da Maia R, Cortez LAB. 2016. Feasibility assessment of converting sugar mills to bioenergy production in Africa. *Agriculture* 6: 45. doi:10.3390/agriculture6030045.

Leal MRLV and Nogueira LAH, 2014. The sustainability of sugarcane ethanol: The impacts of the production model. *Chemical Engineering Transactions* 37: 835–840.

Leal MRLV and Pereira TP, 2014. Biofuels and energy poverty. *Chemical Engineering Transactions* 37: 835–840.

Leite JGDB, Leal MRLV, Nogueira LAH, Cortez LAB, Dale BE, da Maia RC, Adjorlolo C. 2016a. Sugarcane: a way out of energy poverty. *Biofuels Bioproducts & Bioprocessing.* doi: 10.1002/bbb.1648.

Leite JGDB, Leal MRLV, Langa FM. 2016b. *Sugarcane outgrower schemes in Mozambique: findings from the field.* Proceedings of the International Society of Sugarcane Technologists, vol. 29, 434–440.

Locke A and Henley G, 2014. Biofuels and local food security: what does the evidence says? Overseas Development Institute (ODI) Briefing 86, March 2014, 4p.

Maltsoglu I and Beall E, 2012. Steps toward Sustainable Bioenergy in Sierra Leone. *Renewable Energy Training Program, Module 8/Bioenergy,* December 2012. Food and Agriculture Organization of the United Nations (FAO).

Maltsoglu I and Khwaja Y, 2010. *Bioenergy and Food Security: The BEFS Analysis for Tanzania.* Environment and Natural Resources Management Working Paper No. 35. Food and Agriculture Organization of the United Nations (FAO). Rome. 248p.

UN, 2015. *Transforming our world: The 2030 Agenda for Sustainable Development.* United Nations. 41p (www.available at sustainabledevelopment.un.org)

Von Maltitz G and Stafford W, 2011. *Assessing opportunities and constraints for biofuel development in sub-Saharan Africa.* Working Paper 58, CIFOR, Bogor, Indonesia. 66p (available at www.cifor.cgiar.org).

23 Sugarcane and energy poverty alleviation

João Guilherme Dal Belo Leite and Manoel Regis L.V. Leal

Introduction

Access to sufficient and safe energy is a major barrier to global sustainable development (WB and IEA, 2017). In many regions of the world, rural and urban families struggle to meet their basic human needs (IEA, 2006). For instance, in sub-Saharan Africa (SSA), 58% of the population (588 million) do not have access to electricity, and 78% of the population, or 783 million, still collect wood, crop residues and dung for cooking, heating and charcoal production. Globally there are 25 countries where more than 90% of the population cook with traditional biomass and 20 of those are in SSA (WB and IEA, 2017). According to the World Health Organization, 4.3 million deaths per year result from inefficient cooking practices, i.e., indoor air pollution (WHO, 2014). Moreover, the procurement of biomass for cooking is a labor-intensive activity, requiring on average 1.4 hours per day, that can distance children from schools, women from income-generating activities and, in some regions, become a major driver for deforestation (WB and IEA, 2017).

For many developing nations, improving access to safe modern energy requires strategies to overcome economic and distribution limitations, as well as local preferences (IEA, 2002; Karekezi et al., 2004). In rural areas of SSA, nearly 90% of households rely on traditional biomass due to a detrimental combination of poverty and biomass abundance. The popular view of wood, charcoal and dung as "cheap" energy can compromise the competitiveness of modern alternatives.

However, there are promising initiatives to promote modern renewable energy in developing countries. The sugarcane industry is one example in which existing agro-industrial settings may be used and upgraded to produce electricity and ethanol in countries such as Mozambique (Karekezi et al., 2009).

In this chapter, our objective is to explore the potential for modern energy production from sugarcane in SSA. We built different scenarios from an average sugarcane mill to quantify socioeconomic and environmental impacts associated with wider access to electricity and ethanol for transportation and cooking.

Key concepts and their relevance

Energy security is a broad and complex concept that can range from national to household dimensions. In the former, it embraces the long-term and short-term policies that ensure a sustainable supply of energy, as well as strategies to balance supply-demand shocks.

Energy poverty at the household level, as defined in the Sustainable Development Goal 7 (SDG7 Affordable and Clean Energy), means insufficient access to safe and modern energy sources to meet basic human needs such as cooking and heating. In energy-poor regions, people meet their energy needs with traditional sources, particularly firewood and charcoal (DFID, 2002).

As mentioned above, energy poverty may vary from local to national level. It means that there are countries which are net exporters of energy, such as oil and coal, but remain energy poor at the household level (e.g., Nigeria, Angola and Mozambique). On the opposite side, there are countries such as those in EU and Japan that are major importers of energy, but their population has universal access to modern energy. Therefore, affordability instead of availability is a key to reduce energy poverty and improve access. Policymaking is often necessary to make energy affordable for the poor (REN21, 2017). People in populated areas, such as peri-urban zones and rural villages, may spent hours in the procurement of wood and crop residues, as well as paying high prices for energy (Saghir, 2005).

On the other hand, modern energy can support agricultural development (e.g., irrigation) and local small businesses, and provide conditions for essential health and education services (Modi et al., 2005). The combination of social services and economic activity creates an enabling environment for development, allowing more households to afford modern energy. Besides, energy-poor households are not always income poor. In both rural Asia and Africa, there are relatively well-off households keen to access modern energy sources (Hankins, 2000).

Traditional biomass

The historical prevalence of traditional sources of energy over modern ones has profound consequences to development, particularly in SSA (Chapter 24). Economic growth is significantly limited in energy-poor nations as economic prosperity requires secure access to safe and reliable sources of energy (Stern, 2011). This approach is supported by the concept of "productive uses of energy", which results from access to modern energy in nonresidential activities and its consequences towards agricultural productivity gains, economic growth and employment (WB, 2002).

The correlation between energy consumption and economic growth is an indicator of economic development. International databases show that people living in countries that are highly dependent on traditional biomass are more likely to live in poverty (WB, 2014).

The "energy ladder" theory attempts to explain the transition from traditional to modern sources of energy. This theory foresees a shift between energy carriers driven by household income or economic prosperity. As households become better off, they gradually replace traditional sources of energy (e.g., firewood, charcoal, crop residues and dung) by modern ones (e.g., electricity and LPG) (Smeets et al., 2012). The approach proposed by the energy ladder suggests a rather static transition, which implies a complete replacement of energy sources as households climb up the ladder. Nevertheless, over the last decade, a number of studies favored a more dynamic and diverse process of fuel substitution. With prosperity, families tend to adopt a diverse mix of fuels, traditional and modern, influenced by socioeconomic and agroecological conditions (Hiemstra-van der Horst and Hovorka, 2008).

Sugarcane: cultivating an alternative

According to some estimates, there are up to 6 million ha suitable for sugarcane production in southern Africa (Watson, 2011), which is four times the current cultivated area in the entire continent (FAOSTAT, 2014).

Sugar production is traditionally important to SSA, yet the region remains a net importer. Ethanol from sugarcane is a more recent phenomenon in SSA. Production started in the 1970s (Johnson and Matsika, 2006), as an alternative to imported fossil fuels.

Despite the opportunities, ethanol and other energy sources from sugarcane remain far from explored. Malawi and Mauritius are among the few successful cases where sugarcane-based energy (i.e., ethanol and electricity) significantly contributes to the national energy matrix through a set of policies to promote renewable energies (Johnson and Silveira, 2014). Overall, SSA is littered with unsuccessful cases and halted by funding constraints, crop failure, excessive red tape and an unclear/unreliable policy (Locke and Henley, 2012).

There are, however, signs of improvement partially driven by the ever-increasing energy gap, which batters SSA countries. Governments have implemented and reformed their political frameworks in order to accommodate and spur modern, renewable energy production. For instance, from 2011 to 2014, Mozambique introduced a renewable energy feed-in tariff, and South Africa reformed its REFIT policy into an auction-based renewable energy procurement program (BR, 2014; Ebarhard, 2014). Other developments can be found in Swaziland, Zambia and Zimbabwe pushed by hydro, solar and wind projects (REN21, 2017).

Our approach

To assess the potential for bioenergy production in SSA, we define a baseline scenario, based on an existing sugar mill using information collected in Mozambique and South Africa from 2014 to 2015. During this period, we visited sugarcane mills in northern South Africa and southern Mozambique.

We also interviewed local researchers, agricultural organizations and sugarcane smallholder farmers from three associations in Maputo province, Mozambique. A semi-structured questionnaire guided the interviews to access agricultural and industrial characteristics of sugarcane production and processing, as well as opportunities for modern energy (i.e., ethanol and electricity).

The baseline scenario represents a sugar-based production model, which produces – through cogeneration using bagasse as fuel – just enough electricity and process steam to meet the mill's needs. We also built alternative scenarios from the baseline. Power scenario introduces an efficient high-pressure steam boiler and turbine generator, enabling the mill to increase electricity output without changing cane input. Ethanol scenario incorporates ethanol production from molasses, a coproduct from the sugar production. In this scenario, we assumed that 90% of the ethanol production would be oriented to transportation and 10% to household consumption (i.e., cooking fuel). This is rather arbitrary assumption, as the fuel/cooking ratio depends on biofuel policies and energy access (e.g., LPG distribution and affordability), which is biased by the larger potential market for transportation ethanol. The production of ethanol also requires further investments in fermentation and distillation equipment to process molasses. Ethanol plus scenario advances on ethanol production by converting 50% of the total sugarcane juice into fuel, expanding energy output and significantly reducing sugar production by 50% (Table 23.1).

Data collection (i.e., fieldwork) plus literature review also served to parameterize the baseline and the alternative scenarios. Apart from the fieldwork in South Africa and Mozambique, we consulted Brazilian sugarcane mills and firms to complement and validate some of the technical data, especially on electricity cogeneration.

Agriculture and processing coefficients are as follows:

- Sugarcane area: 15,000 ha (Cepagri, 2013);
- Sugarcane yield: 80 t ha^{-1} (Cepagri, 2013; and interviews);
- Operating days: 187 representing a 220 days harvesting season with 85% milling time (interviews);

Table 23.1 Description of baseline and alternative sugarcane-based scenarios

Scenarios	Description
Baseline	All sugarcane is converted into sugar with no electricity surpluses; molasses is a feed coproduct (e.g., livestock supplement)
Power	The same production of sugar as in baseline, but with surpluses of electricity due to an industrial upgrade (high-pressure boiler and turbine generators)
Ethanol	Built over power scenario, with the production of ethanol from molasses expected for transportation (90%) and cooking (10%)
Ethanol plus	Built over ethanol scenario, with the uptake of 50% of the sugarcane juice

- Sugarcane milling capacity: 5,350 t day^{-1};
- Boiler steam pressure and temperature: 21 bar/300°C for baseline scenario and 67 bar/520°C for all other scenarios (Ramjeawon, 2008; and interviews).

Energy efficiency and yield coefficients are as follows:

- Process steam consumption: 500 kg t^{-1} cane (BNDES/CGEE, 2008);
- Electricity surplus: 0 for the baseline and 57 kWh t^{-1} cane for all other scenarios (Ramjeawon, 2008)[1];
- Ethanol from molasses: 0 for baseline and 12 l t^{-1} cane for all other scenarios (our own estimate);
- Ethanol from juice: 0 for baseline and power scenario and 85 l t^{-1} cane for the ethanol and ethanol plus scenarios (interviews);
- Sugar yield: 120 kg t^{-1} cane (Cepagri, 2013).

Ethanol and sugar productivities represent current management under regional technologies. Electricity cogeneration assumes a modern high steam pressure boiler (67 bar) supplying steam to electric turbine generator to produce power surpluses. Similar industrial arrangement has been used in Mauritius (Ramjeawon, 2008), but its utilization remains limited in most of SSA countries.

The impacts of the different scenarios were assessed based on the number of households (equivalent to five people) supplied with electricity and modern cooking fuel (i.e., ethanol), as well as the contribution from ethanol to replace fossil fuels (i.e., gasoline). Moreover, we estimated the avoided deforestation due to modern energy use for cooking as a replacement for traditional biomass. We used "R" to build and run all scenarios.

Household modern energy minimum demand and equivalence ratio are as follows:

Demand

- Electricity:[2] 500 kWh year^{-1} (IEA, 2014);
- Fuelwood: 2,260 year^{-1} (UN, 2007);
- Cooking ethanol: 360 l year^{-1} (UN, 2007).

Equivalence ratio

- Ethanol/fuelwood:[3] 1 l ethanol = 6.4 kg fuelwood (Gaia, 2015; UN, 2007);
- Forest area/fuelwood: 1 ha = 20 t fuelwood (FAO, 1949).

Results and discussion

Our simulations show that there are opportunities to expand modern energy production in SSA regions without compromising sugar production (Table 23.2). For instance, electricity production may increase by up to 57 GWh per year. Ethanol production from sugarcane molasses reaches 10,800 m^3 (transportation fuel) and 1,200 m^3 (cooking fuel). Although this is a modest production, in these scenarios (i.e., power and ethanol) the production of sugar remains the same as in baseline. Ethanol plus converts 50% of the sugarcane juice into ethanol, thus significantly expanding transportation and cooking fuel to 38,000 m^3 and 4,200 m^3, respectively. However, the production of ethanol costs a reduction in sugar production by 50% (Table 23.2). It is important to remember that the simulated outcomes consider the production of sugarcane of 15,000 ha under average by SSA technology and by a single sugarcane mill (Table 23.2), which is equivalent to a standard cane project in a country such as Mozambique.

At the local level, electricity provides larger benefits. From power to ethanol plus scenarios, up to 222,000 households may come out of darkness. Moreover, as much as 12,000 households could quit traditional biomass with the use of cooking ethanol (Table 23.3).

Apart from the household benefits of wider bioenergy access, significant environmental gains can emerge when local communities replace firewood with modern cooking fuels. In the most ambitious scenario (ethanol plus), up to 1,300 ha year^{-1} of deforestation could be avoided due to the use of cooking ethanol (Table 23.4). From the household perspective, avoided deforestation means less labor demanded from women and children to collect firewood and a healthier living environment, free of noxious indoor air pollution.

Table 23.2 Production of sugar, electricity and ethanol from a standard SSA sugar mill (baseline) and alternative scenarios

Scenario	Sugar production (tonnes)	Electricity surplus (GWh)	Transportation ethanol (m^3)	Cooking ethanol (m^3)
Baseline	120,000	0	0	0
Power	120,000	57	0	0
Ethanol	120,000	57	10,800	1,200
Ethanol plus	60,000	57	38,267	4,252

Table 23.3 Number of households supplied with electricity and cooking ethanol under different sugarcane-oriented scenarios

Scenario	Electricity	Cooking ethanol
Baseline	0	0
Power	222,000	0
Ethanol	222,000	3,335
Ethanol plus	222,000	11,811

Table 23.4 Potentially avoided deforestation (ha year^{-1}) by substituting cooking firewood for cooking ethanol

Scenario	Avoided deforestation (ha year^{-1})
Baseline	0
Power	0
Ethanol	384.2
Ethanol plus	1,360.6

In some regions, however, traditional biomass may not cause deforestation. Environmental preservation can be maintained in relatively humid and sparsely populated rural areas, as well as in peri-urban zones where public/private initiatives to preserve and restore local forests and woodlands are in place, e.g., Cape Town and Pointe-Noire (Du Toit et al., 2012). On the other hand, the combination of semiarid conditions, densely populated cities or villages and poor policymaking is a major threat to natural wooded vegetation, e.g., Kinshasa and Abuja (FAO, 2012).

At the economic side, business opportunities can emerge from bioenergy production. For instance, local small firms can benefit from the electricity generated by the mill. Wider access to electricity can also provide fundamental services, such as sanitation and potable water access, as well as irrigation to smallholder farmers.

At the national level, ethanol as a transportation fuel may reduce foreign exchange demand required to finance imported refined oil products, foster energy security and reduce GHG emissions, thereby mitigating climate change. In the ethanol scenario, the production of ethanol provides sufficient fuel to achieve 2% ethanol blending in gasoline (E2) in a country such as Mozambique, where 100% of the gasoline is imported (IEA, 2013). Remember that in this scenario there is no impact on the mills production of sugar (Table 23.2). Furthermore, E7 could be achieved in Mozambique under ethanol plus.

Challenges and drawbacks

Despite the opportunities, expanding bioenergy production from sugarcane projects in SSA is not an easy task. Power generation requires attention to issues such as reliability and stability as electricity surpluses are available only during the harvest season (6–8 months year^{-1}). Hydropower may be an alternative in regions where there is an abundance of power during the sugarcane off-season. Sugarcane mills may also complement off-season shortages through alternative fuels, such as sugarcane straw (leaves and tops). This may be a viable option in regions where sugarcane-mechanized harvest is well advanced. In spite of the technical challenges to efficiently collect and use the straw, its application for power generation has been growing in Brazil and in South Africa (Azad et al., 2014; BNDES/CGEE, 2008; Leal et al., 2013). Infrastructure is another challenge. Expanding and/or erecting new grid lines requires investment capital and electricity prices at an affordable range for local consumers.

Investment capital is also a major limiting factor for ethanol production. According to some estimates, there is a need of 150 million USD to erect a new sugar + ethanol mill, able to process 1 million tonnes of sugarcane per year (BNDES, 2014). At the national level, investment costs should be balanced against expenditures required to import refined oil products.

Although renewables such as sugarcane ethanol are increasingly cost competitive with fossil fuels (Goldemberg et al., 2004), a well-designed political framework is necessary to ensure the long-term demand required to build and strengthen agricultural feedstock production as well as the industrial sector. Except for Malawi, which has been blending sugarcane ethanol into gasoline since the 1980s, mandatory blending policies remain underdeveloped in most of African countries.

Finally, introducing ethanol for cooking in rural Africa faces both economic and cultural challenges. Cooking ethanol requires modern stoves that burn in a safe and controlled way. In rural areas, both fuel and stove present a cost barrier. Empirical experience, mainly from NGOs and development practitioner organizations, indicates wider penetration of modern fuels and stoves among more affluent households, often in (peri)urban regions (Johnson and Takama, 2012). Cultural aspects may also affect the overall impact of modern cooking energy sources. Rural and urban households might find it difficult to prepare some traditional dishes on ethanol stoves or simply prefer traditional biomass as part their cooking ritual (Khandker et al., 2012).

Final considerations

Our modest simulation for a standard sugarcane mill in SSA highlights some promising results. An average size mill processing 1 million tonnes of sugarcane a year could provide electricity to 222,000 households (during the milling season), while supplying other 12,000 with modern cooking fuels (i.e., cooking ethanol). On the environmental side, up to 1,300 ha per year of deforestation may be avoided. The production of ethanol for transportation may enable countries to implement ambitious blending policies (e.g., E7 in Mozambique) and alleviate expenditures with the gasoline importations.

Despite significant socioeconomic and environmental benefits, the development of modern energy in rural Africa – through bioenergy production – remains challenging. High investment demands coupled with an unstable socioeconomic and political environment are important challenges. Moreover, lack of supportive policies and production costs that compromise competitiveness with fossil energy (though not explored in this study) may undermine business viability, which may be further reduced by high opportunity costs for sugar and molasses (inter)national markets. Low household income limits families' ability to access electricity, modern cooking fuels and stoves. Moreover, proper electricity distribution and access require additional investment at the local and the regional level.

Therefore, reaping the opportunities offered by sugarcane bioenergy projects requires effective policies to create an enabling environment for investment in modern energy, particularly in rural areas. Countries such as Mauritius, South Africa and Malawi are leading the way and may serve as a benchmark for others to follow.

Notes

1 After accounting for sugarcane irrigation energy consumption (3 kWh t^{-1}).
2 Consumption of two mobile phones, a fan and two compact fluorescent light bulbs during five hours a day, an efficient refrigerator and a small television or computer (household = five people).
3 Firewood consumption based on a three-stone stove.

References

Azad AK, Islam S and Amin L (2014) Chapter 16: Straw availability, quality, recovery, and energy use of sugarcane. In: Hakeem KR, Jawaid M, and Rashid U (eds), *Biomass and bioenergy: processing and properties*, Cham, Heidelberg, New York, Dordrecht, London: Springer.

BNDES (2014) *Estudo de viabilidade de produção de biocombustíveis na UEMOA (União Econômica e Monetária do Oeste Africano)*. Rio de Janeiro.

BNDES/CGEE (2008) *Sugarcane-based bioethanol: energy for sustainable development*. Nogueira LAH (ed.), Rio de Janeiro.

BR (2014) Boletim da República. Publicação oficial da República de Moçambique. Decreto n. 58/2014 de 17 de Outubro, Regulamento que Estabelece o Regime Tarifário para as Energia Novas e Renováveis (REFIT). *Imprensa Nacional de Moçambique, E.P. I Série - Número 84.*

Cepagri (2013) *Balanço anual Açúcar-2013, Moçambique*. Maputo: Centro de Produção de Agricultura (Cepagri).

DFID (2002) *Energy for the poor: underpinning the Millennium Development Goals*. London: Department of International Development.

Du Toit B, Swart J and de Waal TJ (2012) Of wood from invasive exotic trees to cover urban woodfuel consumption: Cape Province. In: Marien J, Gaughier M, Abhervé-Quinquis A (ed.), *Urban and peri-urban forestry in Africa: the outlook for woodfuel*, FAO Forest Department, Rome.

Ebarhard A (2014) Feed-in tariffs or auctions? Procuring renewable energy supply in South Africa. *Energize RE Renewable Energy Supplement*: 36–38.

FAO (1949) *Unasylva – A review of forestry and forest products*. Rome: FAO Forestry Department.

FAO (2012) *Urban and peri-urban forestry in Africa: the outlook for woodfuel. Urban and peri-urban forestry working paper no. 4.* Rome.

FAOSTAT (2014) *Production – Crops*. Available at http://faostat3.fao.org/download/Q/QC/E Accessed in January 04th, 2015.

Gaia (2015) *Project Gaia: energy revolution*. Available at: www.projectgaia.com/ Accessed in March 2015.

Goldemberg J, Coelho ST, Nastari PM, et al. (2004) Ethanol learning curve – the Brazilian experience. *Biomass and Bioenergy* 26(3): 301–304. Available from: www.sciencedirect.com/science/article/pii/S0961953403001259.

Hankins M (2000) *A case study on private provision of photovoltaic systems in Kenya. Energy services for the world's poor*. Washington, DC: World Bank.

Hiemstra-van der Horst G and Hovorka AJ (2008) Reassessing the 'energy ladder': Household energy use in Maun, Botswana. *Energy Policy* 36(9): 3333–3344. Available from: www.sciencedirect.com/science/article/pii/S0301421508002280.

IEA (2002) Chapter 13: Energy and poverty. In: *World Energy Outlook*, Paris: OECD/IEA.

IEA (2006) *World Energy Outlook-2006*. Paris.

IEA (2013) Energy Statistics – Mozambique. Available at: www.iea.org/statistics/statisticssearch/report/?country=MOZAMBIQUE&product=oil&year=2012. Accessed in November 2015.

IEA (2014) *Africa Energy Outlook*. Paris.

Johnson FX and Matsika E (2006) Bio-energy trade and regional development: the case of bio-ethanol in southern Africa. *Energy for Sustainable Development* 10(1): 42–53. Available from: www.scopus.com/inward/record.url?eid=2-s2.0-34147139017&partnerID=40&md5=cc955a83a803c266fa3459ff3a6a0c3e.

Johnson FX and Silveira S (2014) Pioneer countries in the transition to alternative transport fuels: Comparison of ethanol programmes and policies in Brazil, Malawi and Sweden. *Environmental Innovation and Societal Transitions* 11: 1–24. Available from: www.sciencedirect.com/science/article/pii/S2210422413000579.

Johnson FX and Takama T (2012) Economics of modern and traditional bioenergy in African households: consumer choices for cook stoves. In: Janssen R and Rutz D (eds), *Bioenergy for sustainable development in Africa*, Dordrecht, Heidelberg, London, New York: Springer.

Karekezi S, Kithyoma W and Kamoche M (2009) Evaluating biomass energy cogeneration opportunities and barries in Africa: the case of bagasse cogeneration in the sugar industry. In: UNPD and UNEP (eds), Bio-carbon opportunities in Eastern & Southern Africa - harnessing carbon finance to promote sustainable forestry, agroforestry and bioenergy, United Nations Development Programme, New York.

Karekezi S, Lata K and Coelho ST (2004) *Traditional biomass energy – improving its use and moving to modern energy use*. Bonn: REN21, p. 60.

Khandker SR, Barnes DF and Samad HA (2012) Are the energy poor also income poor? Evidence from India. *Energy Policy* 47(0): 1–12. Available from: www.sciencedirect.com/science/article/pii/S0301421512001450.

Leal MRLV, Galdos MV, Scarpare FV, et al. (2013) Sugarcane straw availability, quality, recovery and energy use: A literature review. *Biomass and Bioenergy* 53: 11–19. Available from: www.sciencedirect.com/science/article/pii/S0961953413001396.

Locke A and Henley G (2012) *Scoping report on biofuels projects in five developing countries*. London: Overseas Development institute (ODI).

Modi V, McDade S, Lallement D, et al. (2005) *Energy and the millennium development goals*. New York: Energy Management Assistance Programme, United Nations Development Programme, UN Millennium Project and World Bank.

Ramjeawon T (2008) Life cycle assessment of electricity generation from bagasse in Mauritius. *Journal of Cleaner Production* 16(16): 1727–1734. Available from: www.sciencedirect.com/science/article/pii/S0959652607002314.

REN21 (2017) *Renewables Global Status Report (GSR2017)*. Paris: Renewable Energy Policy Network for the 21st Century (REN21).

Saghir J (2005) *Energy and poverty: Myths, links, and policy issues. Energy sector notes. No. 4, energy and mining sector board*. Washington, DC: World Bank.

Smeets EMW, Johnson FX and Ballard-Tremeer G (2012) Keynote introduction: traditional biomass and improved use of biomass for energy in Africa. In: Janssen R and Rutz D (eds), *Bioenergy for sustainable development in Africa*, Dordrecht, Heidelberg, London, New York: Springer.

Stern DI (2011) The role of energy in economic growth. *Annals of the New York Academy of Sciences*, Blackwell Publishing Inc. 1219(1): 26–51. Available from: doi:10.1111/j.1749-6632.2010.05921.x.

UN (2007) *Feasibility study for the use of ethanol as a household cooking fuel in Malawi. The United Nation Development Program UNDP/Malawi: growing sustainable business for poverty reduction program in Malawi*. United Nations.

Watson HK (2011) Potential to expand sustainable bioenergy from sugarcane in southern Africa. *Energy Policy* 39(10): 5746–5750. Available from: www.sciencedirect.com/science/article/pii/S0301421510005653.

WB (2002) *Rural electrification and development in the Philippines: measuring the social and economic benefits. Energy Sector Management Assistance Programme (ESMAP), 255/02*. Washington, DC: World Bank.

WB (2014) Poverty headcount ration at $1.25 a day (PPP) (% of population) from 1980–1984 to 2010–2014. Available at http://data.worldbank.org/indicator/SI.POV.DDAY. Accessed in December 2014.

WB and IEA (2017) *Sustainable Energy for All 2017-Progress toward Sustainable Energy*. Washington, DC: The World Bank.

WHO (2014) *Burden of disease from household air pollution for 2012*. Geneva: World Health Organization.

24 Why modern bioenergy and not traditional production and use can benefit developing countries?

*Luís A. B. Cortez, Manoel Regis L. V. Leal,
Luiz A. Horta Nogueira, Arielle Muniz Kubota
and Ricardo Baldassin Junior*

Introduction

Bioenergy is the single largest renewable[1] energy source today, providing nearly 9% of the world's primary energy production. However, as presented in Figure 24.1, traditional bioenergy (TB) still plays an important role being responsible for 47% of total renewable energy use in 2015 (REN21, 2017). Nevertheless, substantial 20% reduction of TB contribution is projected for 2040 (IEA, 2015).

TB is often used more intensively in less developed countries supplying energy for heating and cooking, which results in severe environment impacts (IEA, 2014). Some characteristics of the use of TB worldwide are described below:

- In 2013, one in three people worldwide has relied on traditional use of solid biomass for cooking. In Africa, this dependence achieves two in three African, and eight in ten people living in the sub-Saharan region. In India, a population similar to that of the European Union and the US lives without clean cooking facilities (IEA, 2015);
- In 2012, 7 million premature deaths worldwide were attributed to indoor and outdoor air pollution, where 4.26 million were attributed to household air pollution caused by burning traditional biomass, affecting 0.5 million children and 1.8 million women, mainly in Asia and Africa (WHO, 2015);

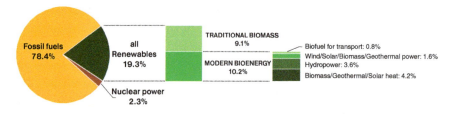

Figure 24.1 Estimated renewable energy share of global energy consumption (2015).
Source: Adapted from REN21 (2017).

Why modern bioenergy and not traditional production 285

- The depletion of forest cover in large scale is not attributed to the use of biomass for cooking; however, the wood exploration may cause local/regional deforestation, land degradation, air pollution and several biodiversity impacts (IEA, 2007);
- Fuelwood collection is a time-consuming and exhausting task, mainly done by women and children. In the sub-Saharan African region, women and children travel up to 11 km and load around 20–38 kg of wood daily. Serious long-term physical damage and limiting of opportunities (education and income-generating activities) are the major losses (IEA, 2007).

About 1.2 billion people still lack access to modern energy services worldwide (0.5 billion in sub-Saharan Africa), and 2.7 billion people lack sufficient energy to cook with anything but charcoal or biomass (Kammen 2006 and

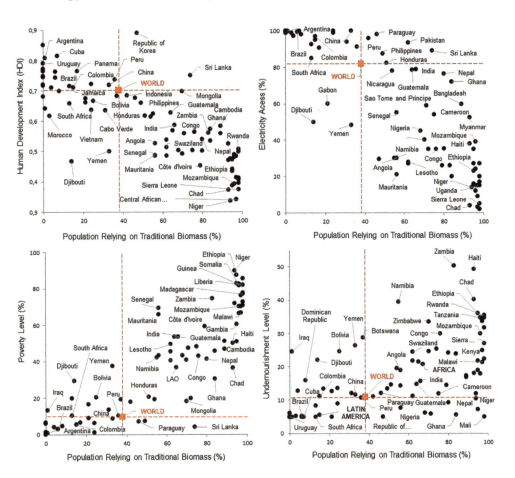

Figure 24.2 Developing countries: use of TB vs. poverty, undernourishment, electricity access and quality of life.
Source: Data from IEA (2014), UNDP (2014).

Ostheimer, 2014). Although Africa has an estimated 6 million ha of potential land for bioenergy (Watson, 2011), it essentially produces TB, having only few modern bioenergy (MB) experiences, such as in Malawi where bioethanol is produced to reduce 20% of gasoline imports (Jumbe, 2014).

Furthermore, it is exactly in Africa (particularly in the sub-Saharan region) where most of countries depend on TB and where the poverty, hunger and low quality of life are more severe compared to the rest of the world.

Life quality indicators in all African countries, with the exception of South Africa, are very distant from the world average (Figure 24.2), even though there has been a slow improvement over time (Rosling, 2017).

Although the share of TB in the total primary energy supply is for Africa and South Asia is 65% and 35%, respectively, this share is slowly declining (IEA, 2015; Karekezi et al., 2004), which is a sign of a promising future.[2] However, the use of TB is still recommended as a proper way to reconcile food and biofuels, and the negative effects of TB use are often minimized or simply neglected (Dubois, 2014).

Bioenergy production models: TB, MB and variations

Characterizing bioenergy production models

Which bioenergy production model can be recommended to less developed countries in order to maximize socioeconomic, environmental benefits, food and energy securities and development goals?

This is a commonly asked question to bioenergy specialists; the answer is frequently not known, and it strongly depends on the local conditions and driving forces. The scientific community seems to consider two basic models and several sub-models for bioenergy production and use (Bogdanski et al., 2010; Goldemberg and Coelho, 2004; Raswant et al., 2008).

The first is the so-called TB and is commonly found in developing countries, particularly in Africa, India and some regions of China, Southeast Asia and Latin America (Gurung and Oh, 2013). The concept of TB can go from purely extractive to family agriculture found in less developed countries and often related to cultural aspects. In the upstream, TB is characterized as gathering or poorly managed agriculture. In the downstream, it is characterized by inefficient and polluting combustion.

TB is typically used to generate heat used for cooking and heating, and it is usually associated with high losses (UN-Energy, 2005). Moreover, associated healthy problems in the collecting operations (long daily walks, cutting wood and heavy duty) and in the biomass conversion (biomass population, smoke during combustion causing eyes and lung problems including blindness and asthma) are also a severe social drawback of TB use (Gurung and Oh, 2013). Moreover, TB production often occurs in non-legalized land in areas where land tenure is weak. TB production is often informal, without taxes, marginalizing government intervention (Johnson et al., 2014).

On the other hand, TB use involves less technical difficulties being more appropriated to less skilled labor. Moreover, it is commonly believed that TB can more easily be compatible with complementary food production and agroforestry systems, therefore enhancing food security and biodiversity. However, these have not been demonstrated as even in areas where TB is the main energy source still suffers from food and energy insecurity.

As opposed, MB is typically found in developed countries, such as the US and Europe but also in intermediate economies such as Brazil, Colombia, Argentina and Mexico. In the upstream, MB uses conventional to modern technologies (systematic production for agricultural practices; use of essential inputs, such as seeds of genetically improved plants, fertilizers and machinery), often practiced in large-scale agriculture and based on a highly efficient energy crop such as sugarcane, eucalyptus, corn and sugar beet. Typical products of MB are biofuels, electricity, chemicals, buildings and industry in general (Larson and Kartha, 2000).

In the downstream, MB typically uses highly efficient conversion technologies for heat, liquid fuel and electricity production. Nevertheless, MB does not necessarily mean an efficient and sustainable system (Goldemberg and Coelho, 2004). Therefore, different options (e.g. energy crops, technological conversion processes) and practices (e.g. agricultural and land use management, coproducts/residues treatment and/or disposal, other) may result in higher/lower environmental/social impacts, waste of resources and time, which misses the opportunity to generate income, create wealth and eradicate poverty.

Based on these concepts, bioenergy systems models could be classified into four types, according to agricultural scale (production level), industrial scale (conversion level) and technological/management levels (technical support level) as represented in Figure 24.3.

In the TB quadrant, the system is characterized by low technology in both the agricultural (upstream) and processing (downstream) stages as well as no mechanization, low-skilled labor and small-scale production.

Nevertheless, it is possible to improve TB systems by increasing technology and scale as characterized by the systems at the upper left quadrant. The differences occur mainly in the upstream where the system uses a centralized industrial facility to convert crops cultivated from a group of individual outgrowers into modern energy (liquid and gas fuels, electricity, heat). This model considers large-scale industrial facilities, usually based on advanced technologies and processes, followed by industrial management supports. In this case, the industry based on large-scale production is crucial to the end use energy source competitive and feasibility (Mayer et al., 2016). Nevertheless, the downstream is similar to the TB system, in which biomass production is based on small scale with low (or no) mechanization, followed by low-agriculture management support. The agricultural step is usually done by smallholders or family farming, and the biomass commercialization is centralized in cooperatives or associations (e.g. biodiesel production in Brazil).

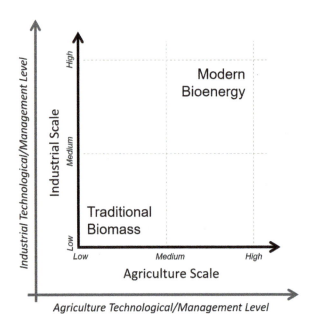

Figure 24.3 Types of bioenergy systems models according to agriculture, industrial scales and technological/management levels.

In the other two quadrants, the systems can achieve significant improvements by having more efficient conversion processes and/or agriculture techniques. The so-called Improved Biomass Technologies (IBTs), for instance, aim more efficient equipment for TB combustion and are an example of system placed at the right lower quadrant. Compared to TB systems, the major improvements on IBTs occur downstream, for example, using more efficient technologies for direct combustion of biomass such as improved cook stoves and kilns (Karekezi et al., 2004). The IBTs have been studied and developed for use in African and Indian regions in order to replace the TB use (Urmee and Gyamfi, 2014).

Nevertheless, to be feasible, competitive and sustainable, biomass production should aim to be practiced in medium/large scale, followed by a high-agriculture management support and techniques (e.g. the production of pellets from planted forest residues), which are the characteristics of the right upper quadrant.

Myths and Facts about TB and MB

Can TB be adapted to perform as MB?

In principle, the improvement could be achieved by having more efficient processes or by using "hybrid systems". According to the IEA (2014), "the deployment of advanced biomass cookstoves, clean fuels and additional off-grid biomass electricity supply in developing countries are key measures to improve the current situation and achieve universal access to clean energy facilities by 2030".

Also, the IBTs refer to improved and efficient technologies for direct combustion of biomass, e.g. improved cook stoves, improved kilns, etc.

Another experience is being suggested by SE4ALL High Impact Opportunities (Ostheimer, 2014) based on clean cooking solutions, increased agricultural productivity, energy from municipal solid waste, sustainable aviation biofuels, cellulosic ethanol and off-grid electrification.

On the other hand, the Inter-American Development Bank (IDB) has developed a financing system to favor sustainable bioenergy projects (Carvalho, 2014). The IDB criteria are based on a "scorecard for sustainable biofuels" ranking them as far as cultivation (based on land use), conversion yield (GJ/ha year), crop life cycle and possible dual cropping.

Bogdanski et al. (2010) present FAO's Integrated Food-Energy Systems (IFES) also in an attempt to propose reconciling food and biofuels production. IFES refer to farming systems designed to integrate, intensify and thus increase the simultaneous production of food and energy in two ways:

Type 1 IFES are characterized through the production of feedstock for food and for energy on the same land, through multiple-cropping patterns or agroforestry systems.
Type 2 IFES seek to maximize synergies between food crops, livestock, fish production and sources of renewable energy. This is achieved by the adoption of agro-industrial technology (such as gasification or anaerobic digestion) that allows maximum utilization of all by-products, and encourages recycling and economic utilization of residues.

According to FAO, bioenergy must be additional to food. However, TB "good practices" actually transfer to the small producers, families and most of the time the women, the responsibility to produce food and energy.

Can TB provide a living source of money when done in small production of few hectares, typically managed by a family?

No, the benefits of TB extraction are too low. One important point is that energy needs intrinsically to be produced at low cost, low effort and therefore the revenue obtained in few hectares will not be sufficient to maintain a family or a community. This is why, typically, economies of scale tend to favor large-scale projects. Therefore, small-scale energy systems are usually intended for self-sufficiency and not for commercialization (FAO, 2010).

Is MB always good and TB always bad?

No, MB can also be bad if poorly planned and/or managed (Karekezi et al., 2004). However, for TB to be "good" it needs to be highly subsidized and integrated to an extent that it is difficult to distinguish from MB (e.g. cooperatives systems). Therefore, a cooperative is a way to combine positive aspects of small production and management with large-scale MB management.

TB is often understood as a cultural activity, and therefore many organizations do not position against it and often protects TB because they don't want to deal with issues like sovereignty, self-determination, independence. The problem is that these so-called cultural activities such as wood gathering by women and wood-fired cooking in closed environments often bring very negative consequences for women/children, for health and other socioeconomic aspects (FAO, 2010), bringing the family and the less developed rural societies to a long-term imprisonment.

Why MB is more likely the way to go even in less developed countries?

The main reason why less developed countries need MB rather than TB is that they need to adopt a more developed agriculture, which leads to systematic/progressive improvements that are only possible with economy of scale. Another fundamental reason to go to MB is that modern production systems most frequently optimize labor and free women and children from the semi-slave type of activities to other more noble and financial worthy tasks (IEA, 2007; Karekezi et al., 2004). MB also tends to reduce production costs and, if done correctly, tends to reduce negative environmental impacts/restore biodiversity (Larson and Kartha, 2000). According to FAO,

> biofuels accounted for the fastest-growing market for agricultural products around the world and was a billion-dollar business. Increasing oil prices in recent years had had devastating effects on many poor countries, some of which spent six times as much on fuel as they did on health. In that regard, the modern form of bioenergy could create great opportunity.
>
> (Raswant et al., 2008)

Can the correct use of fossil energy in developing countries displace TB?

Many times, the correct use of fossil energy can greatly benefit developing countries. Brazil introduced LPG in both urban and rural areas during the last five decades. Until the 1950s, Brazilian energy matrix depended heavily on fuel wood and charcoal. However with the fast urbanization process, LPG was introduced changing habits of Brazilian population in nearly all regions, including the Amazon. Today practically all Brazilian territory is covered by LPG distribution system, which involves oil companies (Coelho and Goldemberg, 2013; Jannuzzi and Sanga, 2004; Lucon et al., 2004). This example not only demonstrates that using fossil energy, in its early stages of development, can have positive impacts leading to better living conditions, particularly to the low-income population.

Diesel can also be used to slow down deforestation. Diesel has been introduced in the Amazon area in South America for power generation. Today, nearly 10% of all diesel consumed in Brazil is in the Amazon region, helping to protect the forest and giving better living conditions for the population, despite the great logistic difficulties associated with diesel transportation in the region.

Later, when a country starts to improve socioeconomic conditions, some complementary bio-based fuels can be used, such as bioethanol gel in ethanol cooking stoves, and, in a next step, MB can be introduced completing the transition to a more affluent society.

Potential impacts of MB in developing countries: Brazilian case

Production costs reduction

Cost reduction can be considered a critical factor for commodities, particularly for bioenergy, because it competes with low-cost fossil energy.

Figure 24.4 presents three examples of agricultural commodities produced in large scale and at high productivity and low cost in Brazil: sugarcane, soybean and corn. Both sugarcane and soybean are used in Brazil as energy crops.

One point worthy to be mentioned is that raw material is a major cost component in bioenergy. Sugarcane represents nearly 70% of ethanol cost and soybean around 90% of biodiesel cost. Therefore, low-cost feedstock production is a key requirement for biofuel production (Chum et al., 2015; Long et al., 2015).

Salary and improvements in job quality

Another very important factor regarding MB relates to expected social impacts. Figure 24.5 shows the creation of direct formal jobs and correspondent salaries in Brazilian agricultural sector. Note the decline of jobs in the sugarcane sector due to introduction of mechanization notably after 2006.

However, the formal job reduction was followed by improvements on job quality (mainly safer jobs on harvesting), and average salaries (almost 80% higher), only inferior to the cotton sector. The positive benefits in formal jobs creation and salaries improvements were also observed in the soybean sector, another important large-scale energy crop in Brazil (Coelho et al., 2006).

Deforestation reduction

Another issue to be considered is the important contribution of MB use in Brazil in the reduction of deforestation (Figure 24.6). Expansion of Brazilian agricultural production in recent years has been accompanied by a significant

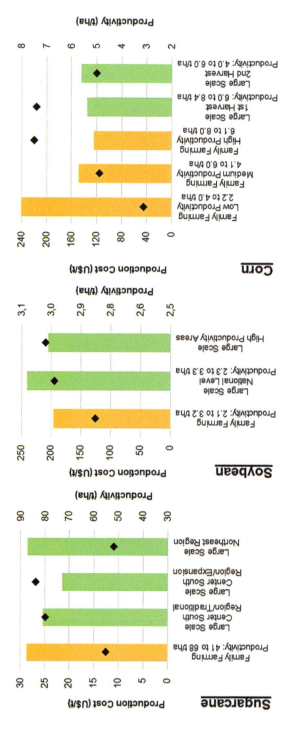

Figure 24.4 Production costs (dots) and productivity (columns) of sugarcane, soybean and corn in Brazil according to agricultural scale (2014). Source: Data from CONAB (2015), PECEGE (2014).

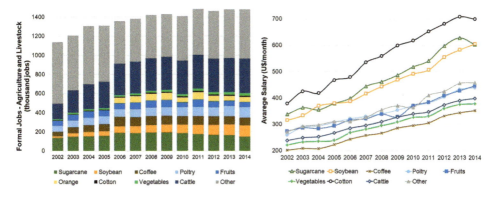

Figure 24.5 Evolution of number of formal jobs and average salary in Brazilian agriculture and livestock (2002–2014).
Source: Data from MTE (2015) and BC (2015).

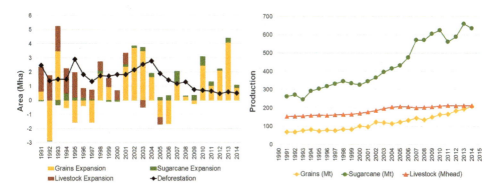

Figure 24.6 Evolution of expansion areas and deforestation, and agricultural and livestock production in Brazil (1991–2014).
Source: Data from CONAB (2015), IBGE (2015) and FAO (2014).

deforestation (Nepstad et al., 2008). Agricultural expansion in Brazil is occurring almost exclusively over grazing and degraded/abandoned areas. Considering the period (1991–2014), the sum of deforestation can be considered equal to the increase of pasture + grains + sugarcane. Moreover, system intensification, characterized by developed and modernized agriculture as it occurs in Brazil, is a way of increasing the production without the need to increase area and, thus, reducing deforestation (Laurance et al., 2014).

Comparing social indicators TB, MB and variations

Table 24.1 presents a summary of the comparison between different bioenergy production models (TB, cooperatives, IBT and MB). The main social indicators of traditional biomass on the quality of life in rural areas (time

Table 24.1 Summary of the advantages/benefits and disadvantages of energy models

Indicator	TB	Cooperatives	IBTs	MB
1. Formal jobs creation	Zero	Positive impact. However, the number of jobs is limited. In the agricultural phase, considering family farming and smallholders, the formal job creation may be zero.	Positive impact. However, the number of jobs is limited. In the industrial phase, considering family farming and smallholders, the formal job creation may be zero.	Positive impact, if guided by efficient public policies and law enforcement.
2. Salary improvements	Zero	Positive impact. However, the income improvements may be affected by low productivity and high production costs.	IBTs not necessarily promote improvements in salary (e.g. production of pellets for supplying personal/local demands).	Positive impact, if guided by efficient public policies and law enforcement. The mechanization can improve the quality of jobs and offer higher salaries.
3. Reduction in agriculture production costs	Not applicable	Positive impact. However, limited considering the low-scale production (low productivity).	Positive impact, if consider the use of agricultural residues (e.g. forest residues).	Positive impacts. However good practices, adequate crops and advanced technologies are crucial.
4. Deforestation alleviation	Negative impact	Positive impacts. However, high efforts are required (law enforcement) to ensure good practices.	Positive impacts, if consider good practices in agriculture.	Positive impact, if guided by efficient public policies and law enforcement.
5. Life improvements in rural areas	Negative impacts	Positive impacts, if guided by "friendly business models" (market price regulation, purchase agreements, agriculture technical support, etc.) and by efficient public policies and laws.	Medium impacts. IBTs can reduce the efforts and malware of the use of TB in cooking; however, the energy source used (e.g. pellets) does not consider a modern energy.	High impacts, with more access of technology.
6. Bioenergy competitiveness	Zero	Medium.	Low, if consider the use of agricultural/industrial wastes and residues.	High. However, depends on good practices, adequate crops and advanced conversion technologies and processes.
7. Capacity to supply energy in large scale (regional/national level)	Zero	Medium. Large-scale supply will require a high number of cooperatives and a well-organized infrastructure.	Low. The use of agricultural/industrial wastes and residues in most cases is limited (availability and economic feasibility).	High, including exports.

Source: Cortez et al. (2014); FAO (2010); FAO (2015); Goldemberg and Coelho (2014); Gurung and Oh (2013); IEA (2006); IEA (2014); IEA (2015); Karekezi et al. (2004); Mayer

dispensed by women/children collecting firewood, health pollution effect, etc.) are compared with the beneficial local use of modern energy in cooking (ethanol, pellet, LPG, electricity, etc.).

It is also presented a list of initiatives around the world with "solutions" to the problem of traditional biomass.

The Role of Science and Technology and Innovation in developing modern sustainable bioenergy

Having presented the main arguments in favor of MB compared to TB for developing countries, it is also essential to develop a long-term strategy regarding adequate plant varieties, adequate infrastructure and qualified labor and scientists.

Investments in science and technology development have greatly helped Brazil to improve its agriculture and its energy matrix. With solid R&D investments in agriculture (e.g. Embrapa for soybeans and sugarcane breeding programs: IAC, CTC, Ridesa) (Cortez et al., 2014), notably after 1970, Brazil could implement MB programs and reduce hunger (FAO, 2015), consequently increasing food (IBGE, 2014) and energy securities. The São Paulo Research Foundation (FAPESP) is investing significant amount of resources to improve MB approaching it to the industrialized agriculture concept (BIOEN-FAPESP, 2017).

It is of course questionable how much can be done by S&T to improve TB in developing countries. It seems that investments in proper infrastructure (roads, harbors and also social infrastructure) and production of qualified labor at all levels (workers to operate trucks, agricultural machinery, mills, etc.) and of course universities to produce engineers and scientists are absolutely necessary in order to build a truly sustainable production system.

Conclusions

Reconciling food security and energy security goals may represent a very difficult task with TB. Instead, contrarily to what many policy advisors recommend, the best solution, even for less developed countries, seems to be adopting MB model which is ideally an integrated food production combined with the introduction of adequate fossil fuel to improve the quality of life.

TB is frequently used in less developed countries because of their apparently low cost and lack of available other energy options. However, TB true cost is not evident and does not appear as direct short-term cost to the users.

In conclusion, there is a correlation between hunger and TB use throughout the world. Moreover, the recommendation of TB use to improve food security is most likely to result in aggravating the problem. One important issue for less developed countries is to promote development and legalize all civil society transaction, including land tenure issues and formal commercial trade.

On the other hand, the recommendation of MB to less developed countries, utilizing available knowledge, can yield very positive social and economic impacts, helping to introduce these economies in the 21st century.

Notes

1 Although bioenergy is generally considered as "renewable", depending on the way it is produced or extracted from nature, the "renewable" label does not necessarily apply. Goldemberg and Coelho (2004) pointed out that possibly as much as 25% of TB cannot be considered as renewable.
2 World energy consumption is steadily increase in all regions but faster in non-OECD countries which already responds for more than 50% of global energy use. Renewable energy and natural gas are the energy sources presenting the fastest growth rate, although fossil fuels still represent 78% of global energy use (DOE/EIA, 2016).

References

BC, Banco Central do Brasil. "Indicadores econômicos consolidados", 2015. Available at: www.bcb.gov.br/pec/Indeco/Port/indeco.asp.

BIOEN-FAPESP, Fapesp Bioenergy Program-São Paulo Research Foundation, 2017. Available at: www.bioenfapesp.org/.

Bogdanski. A, Dubois, O.; Jamieson, C.; Krell, R. "Making Integrated Food-Energy Systems Work for People and Climate, An Overview", FAO-Food and Agriculture Organization of the United Nations, Rome, 116, 2010.

Carvalho, A. V. "Financing Biofuels in Latin America and the Caribbean". Inter-American Development Bank – IDB, Workshop on Biofuels and Food Security Interactions-International Food Policy Research Institute Washington DC, November 19–20, 2014. Available at: www.iadb.org/biofuelsscorecard.

Chum, H. L.; Nigro, F.E.B.; McCormick, R.; Beckham, G.T.; Seabra, J.E.A.; Saddler, J.; Tao, L.; Warner, E.; Overend, R.P. "Conversion Technologies for Biofuels and Their Use" In: Souza G.M.; Victoria, R.L.; Joly, C.A.; Verdade, L.M. (Eds.), *Bioenergy & Sustainability: Bridging the Gaps*, SCOPE and FAPESP, Paris, 2015, pp. 374–467.

Coelho, S.T.; Goldemberg, J. Energy access: lessons learned in Brazil and perspectives for replication in other developing countries. *Energy Policy*, 61(2013), 1088–1096.

Coelho, S.T.; Goldemberg, J.; Lucon, O.; Guardabassi, P. Brazilian sugarcane ethanol: Lessons learned, *Energy for Sustainable development*, 10(2), 26–39, 2006.

CONAB, National Supply Company. "Custo de produção", 2015. Available at: www.conab.gov.br/conteudos.php?a=1546&t=. Accessed in 2017.

Cortez, L.A.B.; Souza, G.M.; Cruz, C.H.B.; Maciel, R. "Chapter 2: An assessment of Brazilian Government initiatives and policies for the promotion of biofuels through research, commercialization and private investment support". In: Silva, S. S. and Chandel A. K. (Eds.), *Biofuels in Brazil: Fundamental Aspects, Recent Developments, and Future Perspectives*, Springer, Switzerland, 2014.

DOE/EIA, US Energy Information Administration. "International Energy Outlook 2016", 290p, 2016. www.eia.gov/outlooks/ieo/pdf/0484(2016).pdf.

Dubois, O. "What FAO Thinks and Does about Biofuels and Food Security", Presentation at the Workshop on Biofuels and Food Security Meeting – IFPRI, Washington DC, 19 November, 2014.

FAO, Food and Agriculture Organization of the United Nations. Making integrated Food-Energy work for people and climate: An overview. Rome, 2010.
FAO, Food and Agriculture Organization of the United Nations. "FAOSTAT", 2014. Available at: http://faostat3.fao.org/home/E.
FAO, Food and Agriculture Organization of the United Nations. "The State of Food Insecurity in the World", Rome, 75p, 2015.
Goldemberg, J.; Coelho, S. "Renewable energy—traditional biomass vs. modern biomass", *Energy Policy* 32, 711–714, 2004.
Gurung, A.; Oh, S. E. Conversion of tradition biomass into modern bioenergy systems: A review in context to improve the energy situation in Nepal. *Renewable Energy*, 50, 206–213, 2013.
IBGE, Brazilian Institute of Geography and Statistics. "Pesquisa Nacional por Amostra de Domicílios (PNAD): Segurança Alimentar 2013", Rio de Janeiro, 2014.
IBGE, Brazilian Institute of Geography and Statistics. "SIDRA – Banco de Tabelas Estatísticas", 2015. Available at: https://sidra.ibge.gov.br/home/pms/brasil.
IEA, International Energy Agency. "World Energy Outlook 2006", OECD/IEA, Paris, 596p, 2007.
IEA, International Energy Agency. "World Energy Outlook 2014", OECD/IEA, Paris, 726p, 2014.
IEA, International Energy Agency. "World Energy Outlook 2015", OECD/IEA, Paris, 700p, 2015.
Jannuzzi, G.M.; Sanga, G.A., LPG subsidies in Brazil: an estimate, *Energy for Sustainable Development*, 8 (3), 127–129, 2004.
Jumbe, C. "Economic Security and Development" PP presentation at the Workshop on Biofuels and Food Security Interactions – International Food Policy Research Institute Washington DC, November 19–20, 2014.
Kammen, D. M. "Bioenergy and Agriculture: Promises and Challenges for Food, Agriculture, and the Environment: Bioenergy in Developing Countries: Experiences and Prospects", *Focus* 14, Brief 10 of 12, 2006. Available at: http://cdm15738.contentdm.oclc.org/utils/getfile/collection/p15738coll2/id/128353/filename/128564.pdf.
Karekezi, S.; Lata, K.; Coelho, S.T. "Traditional Biomass Energy: Improving its Use and Moving to Modern Energy Use" Secretariat of the International Conference for Renewable Energies, Bonn 2004.
Laurance, W.F.; Sayer, J.; Cassman, K.G. Agricultural expansion and its impacts on tropical nature, *Trends in Ecology & Evolution*, 29(2), 107–116, 2014.
Larson, E. D.; Kartha, S. Expanding roles for modernized biomass energy. *Energy for Sustainable Development*, 4(3), 15–25, 2000.
Long, S.; Karp, A.; Buckeridge, M.S.; Davis, S.C.; Jaiswal, D.; Moore, P.H.; Moose, S.P.; Murphy, D.J.; Onwona-Agyeman, S.; and Vonshak, A. "Feedstocks for Biofuels and Bioenergy". In: Souza G.M.; Victoria, R.L.; Joly, C.A.; Verdade, L.M. (Eds.), *Bioenergy & Sustainability: Bridging the Gaps*, SCOPE and FAPESP, Paris, 2015, pp. 302–346.
Lucon, O.; Coelho, S.T.; Goldemberg, J. LPG in Brazil: lessons and challenges, *Energy and Sustainable Development*, 8(3), 82–90, 2004.
Mayer, F. D.; Brondani, M.; Hoffmann, R.; Feris, L. A.; Marcilio, N. R.; Baldo, V. Small-scale production of hydrous ethanol fuel: Economic and environmental assessment, *Biomass and Bioenergy*, 93, 168–179, 2016.

MTE, The Ministry of Labour and Employment, "CAGED – Cadastro Geral de Empregados e Desempregados", 2015. Available at: http://trabalho.gov.br/portal-pdet/.

Nepstad, D.C.; Stickler, C.M.; Soares-Filho, B.; Merry, F. Interactions among Amazon land use, forests and climate: prospects for a near-term forest tipping point, *Philosophical Transactions of the Royal Society B*, 363, 1737–1746, 2008.

Johnson, O.; Sola, P.; Odongo, F.; Kituyi, E.; Njenga, M.; Iiyama, M. Sustainable energy from trees Adopting an integrated approach to biomass energy, Policy Brief No. 20, 2014.

Ostheimer, G. J. Bioenergy for Sustainable Development, Workshop on Biofuels and Food Security Interactions – International Food Policy Research Institute Washington, DC, November 19–20, 2014.

PECEGE, Programa de Educação Continuada em Economia e Gestão de Empresas. "Custos de produção de cana-de-açúcar, açúcar, etanol e bioeletricidade no Brasil: Fechamento da safra 2013/2014". PECEGE-USP, Piracicaba, 54p, 2014. Available at: www.pecege.org.br/.

Raswant, V.; Hart, N.; Romano, M. "Biofuel expansion: challenges, risks and opportunities for rural poor people, how the poor can benefit from this emerging opportunity" Round Table, 31st Session of IFAD's Governing Council, Feb. 24th, 2008.

REN21, Renewable Energy Policy Network for the 21st Century. "Renewables 2017: Global Status Report", Paris, 301 p, 2017

Rosling, H. "New insights on poverty". Available at: www.ted.com/talks/hans_rosling_reveals_new_insights_on_poverty. Accessed in 2017.

UNDP, United Nations Development Programme. "Human Development Report 2014-Sustaining Human Progress: Reducing Vulnerabilities and Building Resilience", New York, 225p, 2014.

United Nations-Energy (UN-Energy), The Energy Challenge for Achieving the Millennium Development Goals, New York, United Nations Development Programme, 2005.

Urmee, T.; Gyamfi, S. "A review of improved cookstove technologies and programs", *Renewable and Sustainable Energy Reviews*, 33, 625–635, 2014.

Watson H. K., Potential to expand sustainable bioenergy from sugarcane in southern Africa, *Energy Policy*, 39, 5746–5750, 2011.

WHO, World Health Organization. "World health statistics 2015", *Geneva*, 161p, 2015.

25 Sugarcane bioenergy production systems

Technical, socioeconomic and environmental assessment

Terezinha de Fátima Cardoso, Marcos Djun Barbosa Watanabe, Alexandre Souza, Mateus Ferreira Chagas, Otávio Cavalett, Edvaldo Rodrigo de Morais, Luiz A. Horta Nogueira, Manoel Regis L. V. Leal, Oscar Antonio Braunbeck, Luís A. B. Cortez and Antonio Bonomi

Introduction

The main drivers of biofuel adoption policies are energy security, reduction of greenhouse gas (GHG) emissions and, for underdeveloped countries, rural development and economic opportunities (Demirbas, 2009; Ackrill and Kay, 2012).

In this perspective, Brazil takes on a relevant role in the biofuels technological and economic development (Silva, 2008), mainly through the ethanol industry using sugarcane as feedstock.

Brazilian sugarcane ethanol is considered a reference in renewable energy from agricultural products that has been commercially successful. Representing the main agricultural product and relevant feedstock for agroindustry, the sugarcane has a very important role in the Brazilian economy, considering sugar, ethanol and bioelectricity production. The sugarcane production cost typically represents around 70% of the final cost of sugar or ethanol production (PECEGE, 2017).

Brazilian sugarcane sector has experienced several changes over the years. Historically, the technology of sugarcane production has been based on manpower and associated with the preharvesting burning of straw to reduce the risk of poisonous animals, improve field conditions for rural workers and decrease the production cost. Over the last decade, however, a variety of economic, social and environmental issues have pushed the sugarcane sector to mechanical-based agricultural operations, especially in Center-South region of Brazil, mainly those of harvesting and planting (UNICA, 2013).

The mechanical harvesting participation in São Paulo state – responsible for more than 50% of sugarcane Brazilian production – increased from about

33% in 2008 to nearly 95% in 2016 (CONAB, 2017). Although mechanical harvesting appears to consolidate its path in the sugarcane sector, many questions can still be raised regarding its sustainability.

Sugarcane mechanization is related to lower production costs when compared with the manual system (Garcia and Silva, 2010), mainly due to the increase in surplus electricity production using straw. Moreover, mechanical harvesting is associated with environmental benefits, such as reduction of GHG emissions and particulate material due to the elimination of sugarcane burning (Capaz et al., 2013). Although mechanization in rural areas leads to lower job creation, this impact would be minimized by additional job opportunities in sectors such as machinery and inputs to the agricultural production (Martinez et al., 2013). Moreover, mechanization would promote better working conditions and higher income when compared with the manual sugarcane production system (Moraes et al., 2015).

In order to assess the broader impacts of different sugarcane technologies, this chapter analyzes the economic, social and environmental aspects of manual- and mechanical-based sugarcane production systems in Brazil, as well as their effects on the ethanol production system when a vertically integrated production model is considered. This chapter aims at identifying strengths and weaknesses of these technologies and enlightens decision-making processes in other countries with substantial potential for sugarcane production expansion for bioenergy, such as Mozambique, Colombia and Guatemala, among others.

Materials and methods

The Virtual Sugarcane Biorefinery (VSB) was used to perform the simulations which give support to the technology assessments. The VSB, developed by the Brazilian Bioethanol Science and Technology Laboratory/Brazilian Center for Research in Energy and Materials (CTBE/CNPEM), is an integrated computer simulation platform that evaluates technologies in use or underdevelopment, for the entire sugarcane production chain, considering the three axes of sustainability: economic, environmental and social (Bonomi et al., 2016).

Scenarios description

Agricultural phase

Seven sugarcane production scenarios were defined in this study. Annual yield of 80 tons of sugarcane stalks per hectare (TC/ha) was assumed for all scenarios, considering the average of five harvests per sugarcane cycle, and the average transport distance from the field to the sugarcane industry was assumed to be 25 km (Bonomi et al., 2016; Cardoso et al., 2017). In the

Figure 25.1 Evaluated scenarios – description based on main agricultural operations.

green cane systems, i.e., the management without preharvesting burning, the amount of straw (green leaves, dry leaves and tops) corresponds to about 140 kg of dry matter per ton of stalk (Hassuani et al., 2005).

The main differences among the scenarios are highlighted in Figure 25.1. These conditions were based on different assumptions for sugarcane planting operations, sugarcane preharvesting burning, harvesting and straw recovery technologies.

In this study, 12 ton per hectare of seedlings for semi-mechanized planting and 20 ton per hectare for mechanized planting were considered. The straw recovery systems assessed were the integral harvesting and the baling system. The straw is windrowed when moisture is below 15% and then collected and compacted in bales, which, in turn, are subsequently loaded and transported to the mill separately from the stalks (Cardoso et al., 2015; Bonomi et al., 2016). It was assumed that 50% of the total straw available on the field is transported to the sugarcane mill. A 10% loss of sugarcane stalks due to harvesting process inefficiencies was assumed in all scenarios, except Scenario Mechanized Integral whose straw recovery technology is based on the reduction of harvester's primary extractor speed, which, in turn, reduces stalk losses to around 6% (Hassuani et al., 2005; Cardoso et al., 2017).

Table 25.1 highlights the main agricultural parameters associated with the fertilization operations and harvesting efficiencies. Compared to the burned sugarcane (i.e., sugarcane harvested after pre-harvesting burning), green cane requires a slightly higher fertilizer input due to the remaining aboveground straw which decreases the capacity of fertilizer absorption by the soil (Rossetto et al., 2008). In the scenarios with straw recovery, nutrients removed along with straw were assumed to be replaced by synthetic fertilizers (Cardoso et al., 2013). Regarding harvesting efficiencies, burned sugarcanes are related to better yields both in the manual and mechanized scenarios because the absence of straw facilitates the harvesting operations.

Industrial phase

In order to assess the broader impacts of different harvesting technologies, sugarcane production scenarios are assumed to be vertically integrated to the industrial conversion processes. The sugarcane production scenario affects the investment on industrial equipment. For instance, straw recovered using green integral sugarcane harvesting technology must be separated from stalks using a dry cleaning station. Moreover, ethanol and electricity yields may vary according to the biomass inputs associated with the different agricultural production scenarios.

Table 25.1 Main agricultural parameters considered in the scenarios

	Scenarios						
	Manual			Mechanized			
Parameters	Burned	Green	Bales	Burned	Green	Bales	Integral
Mineral fertilizers application (ratoon, kg ha^{-1})							
N	100	120	153	100	120	153	154
P_2O_5	0	0	5	0	0	5	5
K_2O	120	150	184	120	150	184	190
Harvesting efficiency (ton day^{-1})							
Manual (per worker)	8.6	4.0	4.0				
Mechanized (per harvester)				697	581	581	604

Source: Based on Rossetto et al. (2008) and Cardoso et al. (2017).

Table 25.2 Main operational characteristics of the industrial phase

Processing capacity (10^6 ton cane year^{-1})	2
Mill drives	Electric drives
Dehydration	Molecular sieves
Boiler pressure (bar)/efficiency (%, LHV basis)	65/87.7
Reduction of 2.5 bar steam consumption due to heat integration (%)	20
Days of operation (effectives)	200

Source: Bonomi et al. (2016).

In the VSB, the industrial conversion scenarios were simulated using AspenPlus® in order to establish complete mass and energy balances of sugarcane processing operations. Despite variations in industrial equipment and adjustments related to different ethanol and electricity yields, all industrial scenarios considered an autonomous distillery, according to Table 25.2.

Techno-economic analysis

A cash flow analysis was performed to assess the economic feasibility of the different sugarcane harvesting technologies. Considering the assumption by which every scenario is a vertically integrated production model, sugarcane production and straw recovery costs were calculated (using CanaSoft model included in the VSB platform) considering the technological specificities of each agricultural scenario. These biomass production costs were further used to calculate the operating expenses with biomass in the industrial phase. Other operating costs – such as labor, utilities, chemical inputs and maintenance – for the industrial phase were calculated according to the database available in the VSB (Bonomi et al., 2016). The revenues from anhydrous ethanol and electricity were calculated according to market prices of US$ 0.58 per liter (CEPEA, 2014) and US$ 57.40 per MWh (MME, 2014), respectively. The exchange rate considered in this study was 2.30 BRL (Brazilian Real) per US$. These values used in the techno-economic assessment considered July 2014 as the reference date.

For the ethanol production cost, operating and capital expenses were taken into consideration to compute the total production costs. The operating expenses are calculated by adding variable costs (such as sugarcane, chemical inputs and utilities) and fixed costs (mainly maintenance and labor) of a distillery, on a yearly basis. Most of the VSB studies assume a 25-year project lifetime and no financial leverage, i.e., the firm is totally financed by equity. Therefore, the yearly capital cost of a biorefinery was estimated by considering the annual payment that would be necessary to remunerate the total investment as if it was a loan (12% per year interest rate over a 25-year period).

The production costs are obtained by adding operating and capital expenses. All operating and capital expenses were allocated according to the ethanol and electricity participations on the total revenues. In the case of ethanol production, the cost per liter would be the total allocated annual cost divided by the number of liters of ethanol produced over the year.

Environmental assessment

Environmental analysis was performed using the environmental life cycle assessment (LCA) methodology. LCA is a method for determining the environmental impact of a product (good or service) during its entire life cycle or, as in the case of this study, from production of raw materials, transport of inputs and outputs to the industrial processing. The software package SimaPro® (PRé Consultants B.V.) and selected categories from ReCiPe Midpoint (H) V1.05 life cycle impact assessment have been used as tools for the environmental impact assessment. The evaluated environmental impact categories were as follows: terrestrial acidification potential (AP) measured in kg of kg SO_2 eq., particulate matter formation (PMF) measured in kg of PM_{10} eq., climate change (CC) measured in kg of CO_2 eq., ozone depletion potential (ODP) measured in kg of CFC-11 eq. and fossil depletion (FD) measured in kg of oil eq. Identification of significant issues, conclusions and recommendations are made in the interpretation step of the LCA methodology. The approach applied is compliant with the ISO 14040–14044 standards and follows the current state of the art of LCA methodology documents (ISO 14040, 2006; ISO 14044, 2006).

According to the LCA methodology, allocation is required for multi-output processes. In this study, economic allocation based on the market value (ethanol and electricity) of the process output was applied in each scenario, as specified in the ISO 14040–14044 documents (ISO 14040, 2006; ISO 14044, 2006).

Social assessment

The social assessment in this study was performed using the social life cycle assessment (S-LCA) methodology. S-LCA aims at assessing social and socioeconomic aspects of products, including their potential positive and negative impacts along their life cycle (UNEP/SETAC, 2009). According to Macombe and Loeillet (2013), S-LCA has also been able to estimate important social effects on the mostly affected actors (e.g., workers) by considering changes in the organizations' behavior.

One of the features of the S-LCA is the estimation of social effects of changes considering present and future scenarios (Macombe, 2013). This method allows for anticipating social consequences of a given change, for example, the adoption of a new technology. The social effects assessed in the sugarcane production systems were the total number of jobs created, number of occupational accidents and average wage of workers. This assessment relies on detailed sugarcane production models (electronic spreadsheets) for calculating the total working hours and sugarcane production costs based on the characteristics of each scenario. These outputs were then used to estimate the number of jobs and the average wage of workers.

The data on occupational accidents in the sugarcane sector was estimated in a two-step procedure. First, a linear correlation between the incidence of accidents (number of accidents per worker) in the sugarcane production sector (MPS, 2015) and the level of mechanization (Nunes Jr, 2012) was established. This correlation reveals that the higher the mechanization level, the lower the probability of occupational accidents. Assuming this correlation as reasonable, the second step was to estimate the number of accidents in each agricultural scenario. In the case of industrial stage, the number of occupational accidents, from MPS (2015), was maintained constant for all scenarios because the same industrial plant configuration is considered.

Risk assessment

Uncertainties related to agricultural parameters associated with both mechanized and manual operations in sugarcane production scenarios were considered. The Latin Hypercube method embedded in @Risk 6.2® software was employed to assess the impact of uncertainties on both sugarcane stalks and straw production costs. As shown in Table 25.3, seven parameters were related to triangular distributions based on literature and consultancy to experts. The main uncertainties considered in this study are those related to harvesting operations which, in turn, affect the sugarcane production costs: sugarcane yield, harvester speed (which is directly related to the harvester yield in CanaSoft model), manual cutting yield, diesel price, harvest operator salary and capital cost related to the investment on machinery.

A total of 5,000 simulations were performed to estimate the uncertainties related to the sugarcane production and recovery costs – sugarcane stalks and straw – in scenarios using both manual and mechanical operations. It is important to point out that the results in the analysis of the vertically integrated production models (agricultural and industrial stages) will embody only the uncertainty related to the biomass production costs.

Table 25.3 Ranges considered for parameters in the risk assessment of agricultural scenarios

Parameter	Unit	Min	Avg.	Max	References
Salary of harvester operator	R$/hour	4.17	7.14	16.07	IEA (2015)
Manual cutting yield (burned sugarcane)	tons/day	6.5	8.56	12	Cardoso et al. (2017)
Manual cutting yield (green sugarcane)	tons/day	1	3.5	5	Cardoso et al. (2017)
Harvester speed	m/s	0.9	1.25	1.5	Cardoso et al. (2017)
Sugarcane yield	TC/ha/year	70	80	100	CONAB (2013)
Diesel price	R$/L	1.704	1.985	2.518	ANP (2012)
Discount rate (cash flow analysis)	% per year	10%	12%	14%	Cardoso et al. (2017)

Results

Production costs of sugarcane biomass

The sugarcane production cost breakdown highlights the main sugarcane production operations such as planting, fertilization, harvesting and transportation. It is possible to observe that manual harvest (Scenario Manual Burned) leads to higher sugarcane production costs when compared with mechanical harvesting (Scenario Mechanized Green).

Regarding straw production costs presented in Table 25.4, it is possible to observe that both scenarios, Manual Bales and Mechanized Bales, lead to very similar straw recovery costs – roughly US$ 36 per metric ton (dry basis). Scenario Mechanized Integral, on the other hand, presented the lowest straw recovery cost (roughly US$ 26/t_{db}) mainly due to lower stalk losses in the harvest operation and also because additional costs are proportionally divided between straw and extra stalks, according to their mass (wet basis).

Table 25.4 shows that transport costs per hectare increase in the integral mechanized scenario as much as the load density decreases due to the straw added to the load. However, it is important to emphasize that the cost per ton of stalk decreases because Scenario Mechanized Integral presents the highest amount of biomass transported per hectare, according to the data shown in Table 25.5.

Table 25.4 Main components of sugarcane stalks and straw production costs according to CanaSoft model

Production costs (US$ per ha)	Manual Burned	Manual Green	Manual Bales	Mechanized Burned	Mechanized Green	Mechanized Bales	Mechanized Integral
Agricultural operations	811.28	1,105.67	1,197.15	702.88	784.96	876.45	832.89
Inputs	412.44	474.81	549.33	434.27	496.51	571.59	549.56
Labor[a]	481.80	763.57	775.87	141.26	140.50	152.80	157.32
Transport[b]	360.05	361.28	387.39	247.84	249.07	275.19	342.30
Total	2,139.21	2,497.21	2,689.32	1,937.93	2,083.50	2,276.19	2,283.47
Stalks (US$ per tons)	27.57	32.18	32.18	25.50	27.41	27.41	26.95
Straw (US$ per tons$_{db}$)			36.54			36.66	26.07

db: dry basis.
[a] Included transport.
[b] Included inputs (Exchange rate 1 US$ = 2.30 BRL – July 2014).

Table 25.5 Sugarcane stalk harvest and straw recovery

	Scenarios						
	Manual			Mechanized			
	Burned	Green	Bales	Burned	Green	Bales	Integral
Sugarcane stalks (TC ha^{-1})	80	80	80	80	80	80	83.7
Straw recovery (tons$_{wb}$ ha^{-1})	–	–	6.0	–	–	6.0	7.9

wb: wet basis.

It is possible to observe that different agricultural technologies lead to different costs for sugarcane production as well as straw recovery costs. Scenario Mechanized Burned presented the lowest sugarcane production cost (US$ 25.50 per metric ton) mainly because of harvester efficiency which is higher with burned cane fields when compared to the green cane harvesting scenarios. The second lowest production cost is associated with Scenario Mechanized Integral (US$ 26.95 per metric ton) because straw recovery under the integral harvesting system decreases sugarcane stalk losses (Table 25.5).

Considering that the higher the stalk yield of a given scenario, the smaller the area required to produce sugarcane – considering a constant industrial processing capacity – production costs will decrease. Moreover, smaller areas imply on additional cost reduction because of shorter transportation distances.

Figure 25.2 shows the results according to the risk assessment including uncertainties on sugarcane yield (TC/ha), harvester speed (m/s), manual cutting yield (TC/worker/day), diesel prices (R$/L), harvest operator wages (R$/hour) and the discount rate (% per year) as previously described in Table 25.3. The highest uncertainties on sugarcane production costs are clearly associated with scenarios Manual Green and Manual Bales. These scenarios are highly reliant on manual operations whose uncertainties on parameters are relatively high, especially the manual sugarcane harvesting yield which varies from 6.5 to 12 tons per worker per day. Considering that manual operations importantly contribute to the overall green sugarcane production costs, such uncertainties were expected to be higher in scenarios Manual Green and Manual Bales.

On the other hand, scenarios with more intensive employment of mechanical operations (Mechanized Burned, Mechanized Green, Mechanized Bales and Mechanized Integral) are related to relatively lower levels of uncertainties because the parameters associated with mechanical operations are either related to a lower range of uncertainties or cause comparatively lower impact on the total production costs.

Regarding the straw recovery costs, uncertainties were higher in scenario Mechanized Bales. This result is related to the approach used to calculate

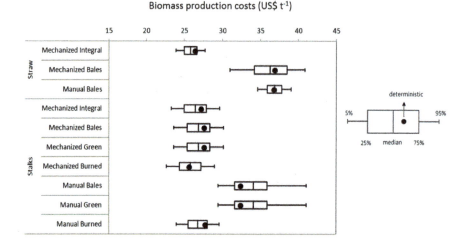

Figure 25.2 Sugarcane biomass production costs US$ per ton considering risk assessment. Stalks production is represented in white bars, while straw recovery is represented in gray bars. 1: Manual Burned; 2: Manual Green; 3: Manual Bales; 4; Mechanized Burned; 5: Mechanized Green; 6: Mechanized Bales; 7: Mechanized Integral.

Source: Cardoso et al. (2017).

straw recovery costs. They are obtained by the difference between the scenarios with straw recovery and without straw recovery. In both scenarios, stalk production cost is the same. The difference between these scenarios will be the cost of straw, which, in turn, is allocated entirely to the amount of straw transported from the field to the industry. For this reason, the greater the difference between stalk and straw production costs, the greater the uncertainty associated with the straw recovery cost.

Considering that scenario Mechanized Bales has the highest difference between those costs, the uncertainties associated with straw recovery cost will be higher. In other scenarios, such as scenario Manual Bales, for example, the difference between straw and stalk costs is lower; consequently, the opposite situation is observed.

Technical results

The technical results related to the industrial phase (see Figure 25.3) were obtained from process simulations that highlight the electricity surplus and anhydrous ethanol production for the industrial plants associated with the different agricultural scenarios. It is clear that the agricultural stage affects mostly the electricity surplus, mainly because the straw recovered from the field to the sugarcane industry will be burned to generate bioelectricity.

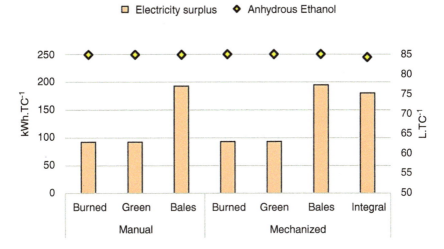

Figure 25.3 Industrial yields of considered scenarios.

The scenarios with straw recovery – Manual Bales, Mechanized Bales and Mechanized Integral – are related to the highest electricity production levels. The ethanol yields, on the other hand, resulted in roughly 85 liters per metric ton of sugarcane stalk in all scenarios. The electricity surplus is higher due to the technology adopted for the industrial phase, which has a 20% reduction of process steam consumption because of energy integration in the mill (Bonomi et al., 2016).

Economic assessment

The economic assessment was performed to understand the impact of different agricultural production technologies on the industrial phase. In order to perform the cash flow analysis – whose results are presented in Table 25.6 – it was necessary to estimate the total investment required for the industrial plants. The capital expenditures (CAPEX) associated with scenarios Manual Bales, Mechanized Bales and Mechanized Integral were slightly higher because they accounted for the additional investment in straw reception in the industry and also in the power and heating unit (Combined Heat and Power (CHP)) generating additional surplus electricity.

When uncertainties of biomass production costs are considered, the highest internal rate of return (IRR) is observed in Scenario Mechanized Bales (see Table 25.6 and Figure 25.4). Although scenario Mechanized Burned achieved a higher deterministic IRR, scenario Mechanized Bales is most likely to achieve higher IRRs when all parameters' ranges are considered in the risk assessment. According to the results in Table 25.6 and Figure 25.4, it is possible to observe that scenarios Manual Green and Manual Bales presented

Table 25.6 Main results of the economic analysis of the vertically integrated scenarios

		Scenarios						
		Manual			Mechanized			
Economic results		Burned	Green	Bales	Burned	Green	Bales	Integral
CAPEX[a]	US$ mi	188.90	188.90	200.61	188.89	188.89	200.82	203.84
IRR	% per year	13.25	10.51	11.73	14.43	13.37	14.33	14.12
NPV[b]	US$ million	17.8	−20.3	−4.0	35.5	19.6	36.2	33.2
Ethanol cost[c]	US$ L^{-1}	0.50	0.55	0.53	0.47	0.49	0.48	0.48

Exchange rate 1 US$ = 2.30 BRL – July 2014.
[a]Total investment in the industrial plant.
[b]Considering a 12% minimum acceptable rate of return per year.
[c]Ethanol production cost considering both the operating and capital costs.

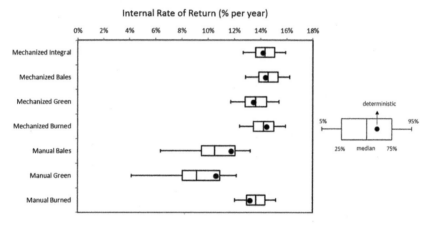

Figure 25.4 IRRs of vertically integrated systems considering the uncertainties on biomass production costs. 1: Manual Burned; 2: Manual Green; 3: Manual Bales; 4; Mechanized Burned; 5: Mechanized Green; 6: Mechanized Bales; 7: Mechanized Integral.

Source: Cardoso et al. (2017).

the lowest IRRs. Assuming a minimum acceptable rate of return of 12% per year, these scenarios would most likely to be unsustainable from an economic point of view due to its negative net present value (NPV).

In Figures 25.3 and 25.5, the results obtained from the deterministic approach are very close to the median in all scenarios, except in scenarios Manual Green and Manual Bales. In these scenarios, deterministic calculations of biomass production costs were underestimated and, as a consequence, the deterministic IRRs were overestimated when compared to their medians. It occurs mostly because of the deterministic value associated with the manual harvesting yield which was closer to the maximum value considered the estimated range. Bearing in mind that the manual harvesting yield has a high impact in the total biomass production costs, this assumption significantly affects the calculations based on the deterministic approach.

Environmental assessment

In general, scenarios with straw burning have higher impacts in the CC category, due to uncontrolled GHGs emissions in field burning. Even using higher amounts of fertilizers, green cane scenarios presented environmental advantages in CC category. Integrating the industrial impacts, the lower impacts were observed in scenarios with straw recovery (Manual Bales, Mechanized Bales and Mechanized Integral), due to higher electricity production in these scenarios and consequentially lower impacts allocation for ethanol (Cardoso et al., 2017).

Comparative environmental impact scores per unit of ethanol produced in each of the seven evaluated scenarios calculated using the LCA methodology are presented in Figure 25.5. These results are in accordance with other publications (Capaz et al., 2013; Martinez et al., 2013) that indicate mechanical harvesting as being associated with environmental benefits, such as reduction of GHG emissions and particulate material due to the elimination of sugarcane burning.

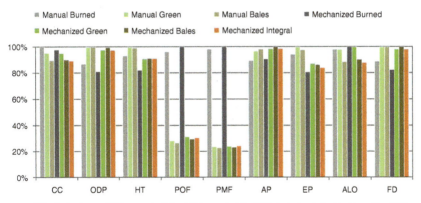

(CC: climate change; ODP: ozone depletion; HT: human toxicity; POF: photochemical oxidant formation; PMF: particulate matter formation; AP: terrestrial acidification; ALO: agricultural land occupation; FD: fossil depletion)

Figure 25.5 Environmental impacts per unit on mass of ethanol produced in the evaluated scenarios. CC: climate change; ODP: ozone depletion potential; PMF: particulate matter formation; AP: terrestrial acidification potential; FD: fossil depletion.

Source: Cardoso et al. (2017).

Compared to the scenarios where bagasse and straw are controlled burnt in industrial boilers, emissions from preharvesting burning of sugarcane lead to very high impact on the PMF category for scenarios Manual Burned and Mechanized Burned, as shown in Figure 25.5. ODP, AP and FD categories presented similar results for scenarios comparison. For ODP and FD categories, lower impacts are observed in scenarios with burned sugarcane practices, due to lower amounts of fertilizer used in burned sugarcane cultivation. In the AP category, scenarios with preharvesting burning of sugarcane present lower impacts since lower amounts of straw remain in the field.

Taking into consideration the evaluated environmental impact categories, it is not possible to identify the best scenario. However, bearing in mind the restrictions for sugarcane burning and practical difficulties of manual harvesting of green cane, a scenario with mechanical harvesting of green cane and integral straw recovery system presents, in general, the best comparative balance of environmental impacts.

Social assessment

The highest level of job creation was associated with manual sugarcane harvesting operations. The scenarios of manual green cane harvesting (Manual Green and Manual Bales) present low efficiency as an intrinsic characteristic of such agricultural operation. Consequently, more jobs are created. Compared with these two scenarios, scenario Manual Burned has a lower level of job creation mainly because manual harvesting is more efficient in a burned sugarcane field, as shown in Table 25.1. The other scenarios with Mechanized technology, associated with mechanical harvesting, are related to a much lower level of job creation mainly due to their higher reliance on mechanical operations (Table 25.7).

Likewise, scenarios of manual harvesting (Manual Burned, Manual Green and Manual Bales) are related to a higher level of occupational accidents due

Table 25.7 Workers in agricultural phase in the evaluated scenarios, considering a production of 2 million tons of sugarcane stalks per year

	Scenarios						
	Manual			Mechanized			
	Burned	Green	Bales	Burned	Green	Bales	Integral
Manual operations	1,179	2,161	2,161	77	49	49	46
Mechanized operations	196	204	234	152	161	189	147
Transport	188	188	200	90	90	102	121
Total	1,562	2,554	2,595	319	300	339	315

to two main reasons: first, more workers are hired in the agricultural phase, increasing the sample space. The second reason is that the lower the level of mechanization – which is the case in manual harvesting scenarios – the higher the probability of occupational accidents per worker. These two effects are combined to explain the higher level of occupational accidents in the scenarios of manual harvesting. The opposite effect is observed in the scenarios of mechanized harvesting: low levels of job creation and reduction of the total number of accidents in higher levels (Figure 25.6). The results are presented per million liters of ethanol.

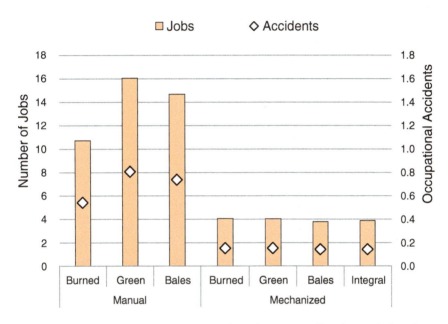

Figure 25.6 Number of jobs and occupational accidents, per million liters of ethanol.

Table 25.8 Average wage of workers in the different scenarios

Average Wage (US$ per month)	Manual			Mechanized			
	Burned	Green	Bales	Burned	Green	Bales	Integral
Agricultural phase	322.73	326.40	327.08	339.90	350.78	359.99	354.25
Industrial phase	434.48	434.48	434.48	434.48	434.48	434.48	434.48
Average	347.54	342.46	342.83	395.02	400.37	401.91	399.56

Exchange rate 1 US$ = 2.30 BRL – July 2014.

The results related to the social assessment are also in accordance with other publications (Martinez et al., 2013; Moraes et al., 2015), which indicate that although mechanized sugarcane systems create less jobs, better working conditions and workers with higher income are also observed, especially in the agricultural phase (Table 25.8).

Conclusions

The reported results highlight that the manual harvesting scenarios were related to a higher risk on biomass and ethanol production costs due to the uncertainties associated with manual operations, especially those employed in green sugarcane harvesting.

Evaluating the vertically integrated production systems (agricultural + industrial), manual technologies were related to the highest job creation levels; however, lower IRRs and higher ethanol production costs were also observed. In general, mechanized scenarios were associated with lower ethanol production costs and higher IRRs due to low-biomass production cost, high ethanol yield and high electricity surplus.

Bearing in mind the restrictions for sugarcane burning and practical difficulties of manual harvesting of green cane, scenario Mechanized Integral with mechanical harvesting of green cane and integral straw recovery system presents, in general, the best comparative balance of environmental impacts.

It is very difficult to predict the complete social implications of a given technology. The methods applied in this study, however, highlighted both strengths and weaknesses of different harvesting technological configurations considering the effects on the working conditions. This study showed that manual cutting technology is associated with positive effects on employment rates. On the other hand, harvesting mechanization scenarios were related to better working conditions since less occupational accidents and higher average wages are observed. Therefore, it will depend on the social context the decision on which alternative is more socially appropriate.

The results presented in this chapter can provide the decision maker with an overview on the economic, environmental and social impacts of sugarcane production technologies considering a broader perspective of vertically integrated production models. Emphasizing that the main purpose of this study was to provide quantitative subsidies for specific decision-making processes, further interpretation on the meaning of results presented in this chapter may vary according to the local economic situation, environmental conditions and social context of sugarcane industry.

References

Ackrill, R.; Kay, A. Sweetness and power – public policies and the 'biofuels frenzy'. *EuroChoices*, vol. 11(3), pp. 23–28, 2012.

ANP. Agência Nacional do Petróleo, Gás Natural e Biocombustíveis. Boletim Anual de Preços de Petróleo, Gás Natural e Combustíveis nos Mercados Nacional e Internacional-2012. Disponível em: www.anp.gov.br. Accessed March/2015.

Bonomi, A.; Cavalett, O.; da Cunha, M. P.; Lima, M.A.P. (Eds.). *Virtual Biorefinery: An Optimization Strategy for Renewable Carbon Valorization*. Series: Green Energy and Technology, Springer International Publishing, 1st ed. 2016, XL, p. 285. doi:10.1007/978-3-319-26045-7.

Capaz, R. S.; Carvalho, V. S. B.; Nogueira, L. A. H. Impact of mechanization and previous burning reduction on GHG emissions of sugarcane harvesting operations in Brazil. Applied Energy, vol. 102, pp. 220–228, 2013.

Cardoso, T. F.; Cavalett, O.; Chagas, M. F.; Moraes, E. R.; Carvalho, J. L. N.; Franco, H.C.J.; Galdos, M. V.; Scarpari, F.; Braunbeck, O. A.; Cortez, L. A. B.; Bonomi, A. Technical and economic assessment of trash recovery in the sugarcane bioenergy production system. *Scientia Agricola*, vol. 70, pp. 353–360, 2013.

Cardoso, T.F.; Chagas, M.F.; Rivera, E.C.; Cavalett, O.; Morais, E.R.; Geraldo, V.C.; Braunbeck, O.; DA Cunha, M.P.; Cortez, L.A.B.; Bonomi, A A vertical integration simplified model for straw recovery as feedstock in sugarcane biorefineries. *Biomass & Bioenergy*, vol. 81, pp. 216–223, 2015.

Cardoso, Terezinha F.; Watanabe, Marcos D.B.; Souza, Alexandre; Chagas, Mateus F.; Cavalett, Otávio; Morais, Edvaldo R.; Nogueira, Luiz A.H.; Leal, M. Regis L.V.; Braunbeck, Oscar A.; Cortez, Luis A.B.; Bonomi, Antonio. Economic, environmental, and social impacts of different sugarcane production systems. *Biofuels Bioproducts & Biorefining-Biofpr*, vol. 1, pp. 1, 2017.

CEPEA. *The CEPEA registers*. Center for Advanced Studies on Applied Economics (CEPEA) (2014). www.cepea.esalq.usp.br. Accessed 10 Jul 2014.

CONAB. Acompanhamento da safra brasileira cana, v. 4- Safra 2017/18, n. 2- Segundo levantamento, Brasília, pp. 1–73, agosto 2017.

CONAB: Perfil do Setor de Açúcar e Álcool no Brasil – Safra 2011/2012. Responsáveis técnicos: Ângelo Bressan Filho e Roberto Alves de Andrade. Brasília, vol. 5, pp. 1–88, 2013.

Demirbas, A. Political, economic and environmental impacts of biofuels: a review. *Applied Energy*, vol. 86, Supplement 1, pp. 108–117, 2009.

Garcia, R. F.; Silva, L. S. Avaliação do corte manual e mecanizado de cana-de-açúcar em Campos dos Goytacazes, RJ. *Engenharia na Agricultura*, vol. 18, no. 3, pp. 234–240, 2010.

Government of Brazil, MME Energy auctions. Ministry of Mines and Energy (MME). (2014) www.mme.gov.br/programas/leiloes_de_energia/. Accessed Jul 2014.

Hassuani, S.J., Leal, M.R.L.V., Macedo, I.C. (Eds.). *Biomass power generation. Sugar cane bagasse and trash*. Published by UNDP-UN and Centro de Tecnologia Canavieira-CTC, Piracicaba, Brazil, 2005.

IEA. Instituto de Economia Agrícola. Banco de dados. Available in: www.iea.sp.gov.br/out/bancodedados.html. Accessed march/2015.

ISO. International Organization for Standardization. Environmental management – Life Cycle assessment – Principles and framework – ISO 14040. ISO, Genebra, 2006.

ISO. International Organization for Standardization. Environmental management – Life cycle assessment – Requirements and guidelines. ISO 14044. ISO, Genebra, 2006.

Macombe, C. (org.), *Social LCAs: Socio-Economic Effects in Value Chains*, Cirad, Montpellier. 2013.

Macombe, C. and Loeillet, D., Social life cycle assessment, for who and why? In: Macombe, C. (org.), *Social LCAs: Socio-Economic Effects in Value Chains*, Cirad, Montpellier. 2013.

Martinez, S. H.; Eijck, J. V.; Cunha, M. P.; Walter, A. C. S.; Guilhoto, J. J. M.; Faaij, A. Analysis of socio-economic impacts of sustainable sugarcane-ethanol production by means of inter-regional input-output analysis: demonstrated for Northeast Brazil. *Renewable & Sustainable Energy Reviews*, vol. 28, pp. 290–316, 2013.

Moraes, M. A. F. D.; Oliveira, F. C. R.; Diaz-Chavez, R. A. Socio-economic impacts of Brazilian sugar cane industry. *Environmental Development*, vol. 16, pp. 31–43, 2015.

MPS, 2015. Anuário Estatístico da Previdência Social – AEAT. Ministério da Previdência Social. http://www.previdencia.gov.br/estatisticas/, (Accessed 01/20/2015).

Nunes Jr., 2012. [Performance Indicators of Sugarcane Agroindustry: Seasons 2012/2013 and 2013/2014] (in Portuguese). Indicadores de desempenho da agroindústria canavieira. Grupo IDEA, Ribeirão Preto.

PECEGE. *Custos de produção de cana-de-açúcar, açúcar, etanol e bioeletricidade no Brasil: fechamento da safra 2016/2017*. Piracicaba: Universidade de São Paulo, Escola Superior de Agricultura "Luiz de Queiroz", programa de Educação continuada em Economia e Gestão de Empresas/Departamento de Economia, Administração e Sociologia. Relatório apresentado à Confederação da Agricultura e pecuária do Brasil (CNA) como parte integrante do projeto Campo Futuro. 18 p., 2017.

Rossetto, R.; Dias, F. L. F.; Vitti, A. C.; Cantarella, H.; Landell, M. G. A. Manejo conservacionista e reciclagem de nutrientes na cana-de-açúcar tendo em vista a colheita mecânica. *Informações Agronômicas*, Piracicaba, vol. 1, pp. 8–13, 2008.

sSilva, R.D. *Setor sucroalcooleiro no estado de São Paulo: mensurando impactos socioeconômicos*. Observatório do Setor Sucroalcooleiro. Ribeirão Preto, pp. 1–16, 2008.

UNEP/SETAC. Life Cycle Initiative. Guidelines for Social Life Cycle Assessment of Products, United Nations Environment Programme. 2009.

União da Indústria de Cana-de-Açúcar (UNICA), 2013. Apresentação da estimativa de safra 2013/2014. www.unica.com.br/documentos/apresentacoes/pag=2. Accessed: feb/2015.

26 Sustainability of scale in sugarcane bioenergy production

Mateus Ferreira Chagas, Otávio Cavalett, Charles Dayan Faria de Jesus, Marcos Djun Barbosa Watanabe, Terezinha de Fátima Cardoso, João Guilherme Dal Belo Leite, Manoel Regis L. V. Leal, Luís A. B. Cortez and Antonio Bonomi

Introduction

Many countries and companies are heavily investing in biofuels for transport, motivated by concerns and opportunities related to global climate targets, energy security, and rural development (Howarth et al., 2009). In Latin America, particularly, biofuel market is established with greater coverage in Brazil, and in some extent or incipient in countries such as Colombia, Argentina, Venezuela, Costa Rica and Guatemala (Janssen and Rutz, 2011).

In this context, the ethanol (and electricity) production from sugarcane in Brazil is an important reference, both in terms of being commercially competitive and achieving significant ongoing reductions in greenhouse gas (GHG) emissions relative to fossil fuels use (Cavalett et al., 2013). The Brazilian sugarcane ethanol could be emphasized as a model for implementing biofuel production systems in other regions of the world with high potential for renewable biofuel development, such as other Latin American and African countries (Lynd et al., 2015). However, specific social, political, economic and environmental factors, as well as existing practices and technology in agriculture and industry must be always taken into account.

Among these factors, the scale of biofuel processing plants has a great importance, especially because economies of scale and technical efficiencies normally encourage ethanol cost reductions. Hamelinck and Faaij (2002, 2006) modeled decreasing costs of biofuels, including, in particular, ethanol from Brazilian sugarcane, due to upscaling and technology development. Similar results were observed by Hettinga et al. (2009) not only for Brazilian sugarcane ethanol but also for U.S. corn ethanol, showing that production costs decreased as industrial processing capacity increased. Cost reductions and environmental benefits were also related to learning curves attributed to size (Goldemberg et al., 2004; Van den Wall Bake et al., 2009; Canter et al., 2015) and learning effects (Chen et al., 2015; Daugaard et al., 2015).

Large-scale plants, on the other hand, require larger biomass volumes, which normally are scattered at the agricultural field. Some studies indicate that increasing the industrial capacity causes both higher biomass production cost and environmental impacts due to increased transportation distances (Wang et al., 2014; Bonomi et al., 2016).

In this sense, small-scale biofuel initiatives to produce sugarcane ethanol are claimed to be a sustainable opportunity for ethanol supply, particularly for regions with price restriction or no access to modern biofuels, such as communities located far from the large ethanol production centers in Brazil and family-farm communities in sub-Saharan Africa, respectively. For instance, decentralized microdistilleries can reduce transportation costs and have simpler and less expensive process technologies (lower investment) with the potential to increase rural employment (Ruan et al., 2008; Bruins and Sanders, 2012; Maroun and Rovere, 2014; Lynd et al., 2015; Mayer et al., 2015a, 2015b). Taking this into consideration, small-scale biofuel production (hereafter called microdistilleries) can be also considered as a promising alternative, but they normally rely on more rudimental technologies, which may result on lower agricultural and industrial yields and, consequently, loss of economic and environmental competitiveness (Van den Wall Bake et al., 2009; Mayer et al., 2015a).

Regarding the effect of scale on the environmental impacts of ethanol, several studies have presented sustainability impacts of ethanol production systems considering different scales and technologies (Goldemberg et al., 2004; Van den Wall Bake et al., 2009; Mayer et al., 2015a, 2015b). Although, Mayer et al. (2015a) indicated that small-scale ethanol production systems result in a climate change impacts about 20 times higher than ethanol produced in large-scale systems (due to the lower agricultural machinery efficiency and absence of electricity cogeneration).

In this sense, a comprehensive and structured assessment of economic and environmental sustainability as a function of scale of production is considered particularly important for the planning expansion of biofuel production in several countries of the world. Therefore, the objective of this study is to assess the sustainability impacts of the sugarcane biorefineries as a function of the scale and technological level of the industrial plants and sugarcane production systems.

Materials and methods

In all the assessments performed in this study, the Virtual Sugarcane Biorefinery (VSB) was used. VSB is a simulation platform developed by CTBE (Brazilian Bioethanol Science and Technology Laboratory) to evaluate technical, economic, environmental and social impacts of different biorefinery production routes, considering the integrated sugarcane and other biomass production chains.

Sugarcane production technologies

The CanaSoft (agricultural model of the VSB platform) was used to calculate production costs and environmental inventories associated with the sugarcane

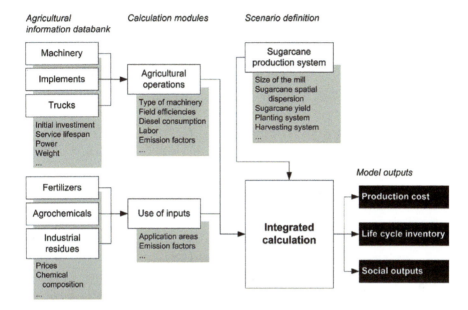

Figure 26.1 CanaSoft model structure.
Source: Cavalett et al. (2016).

production phase (Jonker et al., 2015; Cavalett et al., 2016). Figure 26.1 shows different modules of the CanaSoft model. A more detailed description of the model is found in Cavalett et al. (2016).

A sugarcane production scenario considering high mechanization levels in planting and harvesting is considered for large-scale plants. Different processing capacities were considered, including its effects on the sugarcane transportation distances. A second sugarcane production scenario was used to represent sugarcane production in small agricultural production scale to supply ethanol microdistilleries, considering its specificities as detailed further in this study.

Large-scale distilleries

The industrial conversion system is representative of a modern autonomous ethanol distillery in Brazil exporting surplus electricity to the grid. Main characteristics of this plant are based on Chagas et al. (2016) and Bonomi et al. (2016) and are summarized in Table 26.1. In this scenario, sugarcane yield was assumed as 80 t ha^{-1} y^{-1}, and no straw recovery is considered.

Mass and energy balances for the industrial conversion systems are obtained from computer simulations using Aspen Plus® software inside the VSB platform (Cavalett et al., 2012; Dias et al., 2012, 2013; Morais et al., 2016). Other

Table 26.1 Main agricultural and industrial parameters for large-scale scenarios

Scale of the distillery (10^6 t year^{-1})		1.0	2.0	4.0	6.0
Agricultural	Transportation distance (km)	17	25	35	43
	Planting	Mechanized			
	Harvesting	Mechanized			
	Sugarcane	Green			
	Fertilizers – planting (kg ha^{-1})	(N: 30; P_2O_5: 180; K_2O: 120)			
	Fertilizers – ratoon (kg ha^{-1})	(N: 120; P_2O_5: 0; K_2O: 150)			
Industrial	Mills drives	Electric drives			
	Dehydration process	Molecular sieves			
	Boilers	65 bar			
	Steam consumption (reduction)	20%			

technical parameters for these simulations were based on Dias et al. (2016). For computing the transport distance from the sugarcane fields to the industrial plant as function of plant capacity, correlations described by Cavalett et al. (2016) were employed.

Microdistilleries

The sugarcane production system in the microdistillery scenarios consider semi-mechanized planting and manual harvesting of unburned sugarcane. In the industrial conversion system, main differences are related to process control and efficiencies in several unit operations, especially in the sugarcane crushing (single-stage sugarcane milling tandem), fermentation and distillation (lower yields due to rudimentary equipment and lack of control systems). Data for microdistilleries project and operation were obtained in literature and from primary data collected from a microdistillery project developed in the South of Brazil (Oliveira, 2009; COOPERBIO, 2011; Fleck et al., 2011; Santos, 2011). Despite the fact that microdistilleries' energy input is, in many related cases, supplied by firewood while sugarcane bagasse may be used as cattle feed, in this study it was assumed that sugarcane bagasse is used in the very small-scale boilers to produce steam and electricity to the process. Another assumption to facilitate comparison is that all the extracted sugarcane juice is only used to produce ethanol with fuel specification (hydrous 93% ethanol). It was considered a microdistillery with a crushing capacity of 3,000 metric tons of sugarcane per year, with an industrial investment of US$ 77,400. Since microdistilleries technologies are yet not consolidated, a sensitivity analysis, considering agricultural and industrial yields, was performed, as presented in Table 26.2.

Table 26.2 Sensitivity analysis parameters for microdistillery yields

Parameter	Typical value	Sensitivity analysis
Sugarcane yield (t ha^{-1} year^{-1})	65	50–95
Industrial yield (L ethanol t^{-1})	60	55–80

Sustainability assessment methods

Economic analysis

The economic sustainability is evaluated using the total production cost of ethanol fuel. The main parameters of economic viability assessment are described by Watanabe et al. (2016). The fixed capital investment required for different scales was determined based on literature review and consultation with experts, equipment providers and engineering companies. The reference prices used in this study are US$ 0.48 per liter of ethanol and US$ 56.62 per MWh of electricity (exchange rate: US$ 1.00 = R$ 3.22). The minimum acceptable rate of return (MARR) was considered as 12% per year. In this study, it is assumed that sugarcane production system is vertically integrated to the industry, i.e., sugarcane production costs were used instead of historical sugarcane market prices. The sugarcane production costs include differences among agricultural management practices (such as harvesting and transportation) for each scenario.

Environmental analysis

The evaluation of environmental impacts is performed by using the life cycle assessment (LCA) methodology. The method has been largely applied to quantify environmental aspects of biofuel production systems and compare environmental benefits of biofuels and fossil alternatives (Seabra et al., 2011; Alvarenga et al., 2013; Cavalett et al., 2013; Tsiropoulos et al., 2014; Watanabe et al., 2016).

Sugarcane and ethanol life cycle datasets were compiled according to data from agricultural and industrial simulations, coupled with *ecoinvent* v 2.2 database (Frischknecht et al., 2005), modified by Chagas et al. (2012). The selected impact categories from ReCiPe Midpoint H v 1.05 (Goedkoop et al., 2013) methods are climate change, particulate matter formation, terrestrial acidification, freshwater eutrophication, and fossil depletion. Economic allocation was used whenever electricity is a coproduct of the industrial plant.

Results and discussion

Large-scale distilleries

The ethanol and electricity yields are presented in Table 26.3. These results are obtained from mass and energy balances generated in the computer modeling of the scenarios using the VSB platform.

Table 26.3 Technical results – ethanol and electricity yields per ton of sugarcane and ethanol and electricity annual production

		Scale of the distillery (10^6 t year^{-1})			
		1.0	2.0	4.0	6.0
Ethanol	Yield (L t^{-1})	86.1			
	Annual production (10^6 L year^{-1})	86.1	172.2	344.4	516.6
Electricity	Yield (kWh t^{-1})	115.2			
	Annual production (GWh year^{-1})	115.2	230.3	460.7	691.0

Economic assessment

Figure 26.2 shows the agricultural investment and total sugarcane production cost according to the scale of the industrial plants. Scenarios with higher crushing capacities present higher agricultural investments per ton of sugarcane, as well as sugarcane production costs due to larger cultivated areas, which implies in longer sugarcane transport distances.

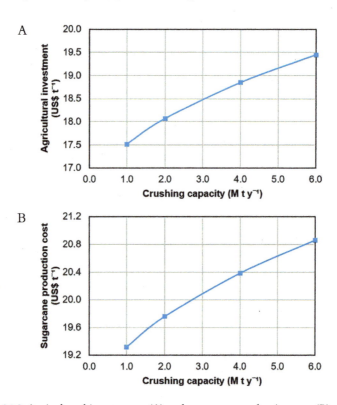

Figure 26.2 Agricultural investments (A) and sugarcane production cost (B).

Results at industrial plant indicate that lower investments per ton of processed sugarcane are required in scenarios with higher crushing capacities (Figure 26.3A), as a result of economies of scale. This effect is less pronounced at crushing capacities above 4.0 Mt y^{-1} because, at certain point, increase in industrial capacity will require more duplication and lesser larger equipment, reducing the expected effects of economies of scale.

Large-scale plants benefit industrial investments; however, they require more biomass, which normally is sparsely distributed at the agricultural field implying on higher feedstock transportation costs. The total production cost of ethanol is presented in Figure 26.3B and couples with agricultural and industrial effects regarding the scale of the plants (Figures 26.2B and 26.3A, respectively), since a vertically integrated business model is considered. Although feedstock production costs are higher for larger plants, ethanol production costs are highly influenced by the scale of the industrial plants and technical efficiencies to convert biomass into products, as presented in Figure 26.3B. Reduction in ethanol production costs is not observed for the largest plant (6.0 Mt y^{-1}) since the effect of scale in the industrial investment is balanced by the increase in biomass production costs.

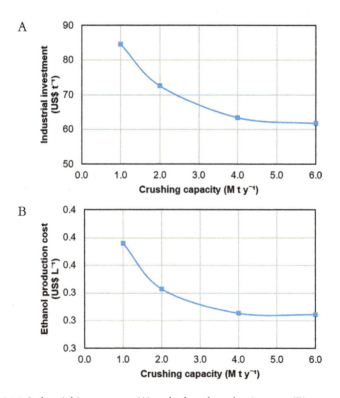

Figure 26.3 Industrial investments (A) and ethanol production cost (B).

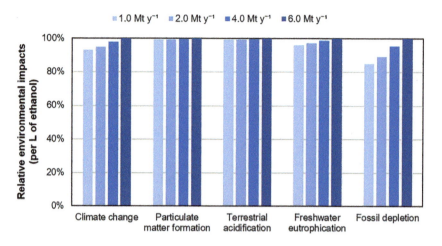

Figure 26.4 Comparative environmental impacts per liter of ethanol as a function of scale.

Comparing the results with the ethanol market price for producers (US$ 0.42 L^{-1}), ethanol production in distilleries with crushing capacity smaller than 1.0 Mt per year is not economically competitive. Results also suggest that optimum scale to produce ethanol at lower costs is between 4.0 and 6.0 Mt per year. In fact, the majority of the plants built most recently in Brazil are within this crushing capacity range. It is important to emphasize that this conclusion is valid for the assumptions and conditions considered in this study, which does not impair the existence of larger plants with even lower total ethanol production cost. In some particular cases, the spatial dispersion of sugarcane can be very low, so the effect of transportation on sugarcane production cost becomes less important.

Environmental assessment

The environmental impact comparison of scales in Figure 26.4 highlights that plants with higher milling capacities (regardless of the technology level) lead to higher environmental results mainly due to impacts associated with higher diesel consumption for transporting sugarcane in the agricultural phase.

Microdistilleries

Considering the same methodology used to calculate sugarcane production costs applied to the large-scale scenarios, the sugarcane production cost in small-scale units is US$ 24.0 t^{-1}. This value is 15% higher than the highest value observed for large-scale scenarios (US$ 20.9 t^{-1} in optimized 6.0 Mt y^{-1} scenario). This is mainly due to higher labor cost (due to lower labor productivity for manual harvesting of unburned sugarcane), lower transportation efficiency (besides shorter distance) and lower efficiency of other agricultural operations that are expected for small-scale agricultural systems in comparison to large-scale systems.

Since biomass production costs are highly related to sugarcane productivity, a sensitivity analysis is shown in Figure 26.5. This analysis included is also an important component of sugarcane production cost, that is, the opportunity cost of land, which is normally neglected by small-scale farming accountability. Figure 26.5 illustrates the sensitivity assessment for sugarcane production cost that varies from US$ 33.2 to 21.6 per ton of sugarcane for productivities of 50 and 95 t ha^{-1} y^{-1}, respectively. On the other hand, the exclusion of the cost of land leads to lower costs, from US$ 27.0 to 18.4 for productivities of 50 and 95 t ha^{-1} y^{-1}, respectively. This result indicates that sugarcane production in small-scale farms would only be competitive with large-scale farms in the case of high sugarcane productivities and without considering the cost of land.

A sensitivity analysis was also performed to assess the effects of sugarcane production cost and industrial yields on the ethanol production cost (Figure 26.6). Ethanol production costs were then compared to the ethanol market price for producers (US$ 0.42 L^{-1}) and at fuel stations (around US$ 0.68 L^{-1}) as a parameter of competitiveness.

When the market price for producers is used as a reference, microdistilleries would be cost-competitive only at sugarcane production cost below US$ 22.6 t^{-1} for industrial yield of 80 L t^{-1}. If a more realistic industrial yield of 70 L t^{-1} is assumed, microdistilleries would be only competitive with sugarcane costs below US$ 18.6 t^{-1}. However, these industrial yields are still optimistic considering existing microdistilleries experiences, and would hardly be reached under real-world operating conditions. Considering typical industrial yield for microdistilleries as 65 L t^{-1}, ethanol would not be economically viable at producer selling price.

Considering ethanol prices at fuel pump, microdistilleries would be cost-competitive when combining intermediate sugarcane costs (around

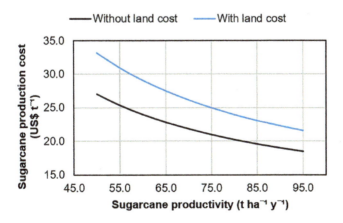

Figure 26.5 Sensitivity assessment for sugarcane production cost in microdistilleries.

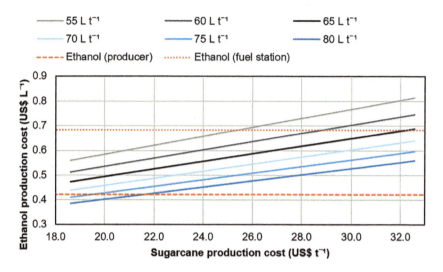

Figure 26.6 Sensitivity analysis for ethanol production cost in microdistilleries.

US$ 26.0 t^{-1}) and reasonable industrial yields (nearly 55 L t^{-1}). For sugarcane costs up to US$ 32.5 t^{-1}, competitiveness can be achieved only with industrial yields above 65 L t^{-1}. These sensitivity results show that, when consumer prices for ethanol are considered, microdistilleries may be cost-competitive to produce ethanol fuel. It is important to highlight, however, the limitations of this assumption, as legislation, taxes, transportation costs and fuel commercialization margins, among other components of fuel prices, should be considered.

However, as indicated in Kubota et al. (2017), other higher value non-energy sugarcane products, such as cachaça (sugarcane-based spirit) and brown sugar, integrated with ethanol production could make small-scale scenarios economically more attractive, especially if markets with higher willingness to pay – such as the external markets – are considered. These authors also add that product diversity in small-scale projects could increase biofuel access as well as food and energy security and sustainability.

Concerning the environmental results for microdistilleries, a similar sensitivity analysis was performed considering the climate change impacts in Figure 26.7. The results indicate that only with industrial yield around 80 L t^{-1} and sugarcane productivity above 90 t ha^{-1} y^{-1}, ethanol production in microdistillery scenarios would present lower impacts than in large-scale scenarios (around 20 g CO_2 eq. MJ^{-1}). For the worst environmental performances (low sugarcane productivities and low ethanol yield), ethanol produced in microdistilleries would be very close to the upper limit of GHG emissions to classify the ethanol as an advanced biofuel, according to the U.S. EPA. It is important to highlight that the direct and indirect effects of land use change were not included in this assessment.

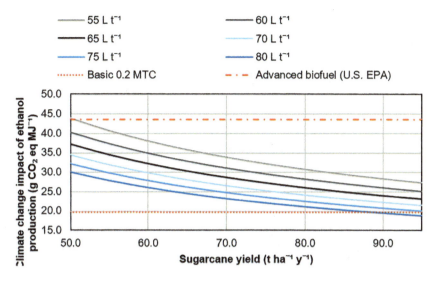

Figure 26.7 Sensitivity analysis for climate change impact of ethanol production in microdistilleries.

Conclusions

Although sugarcane production costs are higher with increasing industrial scale, the reduction of investment per ton of processed sugarcane favors larger plants. Results indicate that optimum scale to produce ethanol at lower costs is between 4.0 and 6.0 Mt per year. Sugarcane distilleries with crushing capacity smaller than 1.0 Mt per year presented ethanol production costs higher than producer selling prices, demonstrating the economic inefficiency of these plants. For microdistilleries, low agricultural and industrial efficiency lead to unfavorable economics. By comparing the results of ethanol production costs, small-scale plants can only be considered viable under very specific contexts, which will depend on a particular applicability and favorable market conditions. Regarding environmental impacts, larger crushing capacities lead to slightly higher environmental impacts due to larger distances for sugarcane transportation. In the case of microdistilleries, similar to economic results, lower yields lead to unfavorable environmental performance when compared to large-scale plants.

References

Alvarenga RAF, Dewulf J, De Meester S, Wathelet A, Villers J, Thommeret R, Hruska Z (2013) Life cycle assessment of bioethanol-based PVC. Part 1: Attributional approach. *Biofuels, Bioproducts and Biorefining*, 7 (4), 386–395. DOI: 10.1002/bbb.1405.

Bonomi A, Cavalett O, da Cunha MP, Lima MAP (Eds.). Virtual Biorefinery: An Optimization Strategy for Renewable Carbon Valorization. Series: Green Energy and Technology, Springer International Publishing, 1st Ed, 2016, 285 p. DOI: 10.1007/978-3-319-26045-7.

Bruins ME, Sanders JPM (2012) Small scale processing of biomass for biorefinery. *Biofuels, Bioproducts & Biorefining*, 6, 135–145. DOI: 10.1002/bbb1319.

Canter CE, Dunn JB, Han J, Wang Z, Wang M (2015) Policy implications of allocation methods in the life cycle analysis of integrated corn and corn stover ethanol production. *BioEnergy Research*, 9 (1), 77–87. DOI: 10.1007/s12155-015-9664-4.

Cavalett O, Chagas F, Magalhães PSG, Carvalho JLN, Cardoso TF, Franco HCJ, Braunbeck OA, Bonomi A. The Agricultural Production Model. In: Bonomi A, Cavalett O, da Cunha, MP, Lima MAP (Org.). *Virtual Biorefinery: An Optimization Strategy for Renewable Carbon Valorization*. Series: Green Energy and Technology. Springer International Publishing, 1st Ed, 2016, pp. 13–51. DOI: 10.1007/978-3-319-26045-7_3.

Cavalett O, Chagas MF, Seabra JEA, Bonomi A (2013) Comparative LCA of ethanol versus gasoline in Brazil using different LCIA methods. *The International Journal of Life Cycle Assessment*, 18 (3), 647–658. DOI: 1007/s11367-012-0465-0.

Cavalett O, Junqueira TL, Dias MOS, Jesus CDF, Mantelatto PE, Cunha MP, Franco HCJ, Cardoso TF, Maciel Filho R, Rossell CEV, Bonomi A. (2012) Environmental and economic assessment of sugarcane first generation biorefineries in Brazil. *Clean Technologies and Environmental Policy*, 14 (3), 399–410. DOI: 10.1007/s10098-011-0424-7.

Chagas MF, Bordonal RO, Cavalett O, Carvalho JLN, Bonomi A, La Scala Junior N (2016) Environmental and economic impacts of different sugarcane production systems in the ethanol biorefinery. *Biofuels, Bioproducts & Biorefining*, 10 (1), 89–106. DOI: 10.1002/bbb.1623.

Chagas MF, Cavalett O, Seabra JEA, Silva CRU, Bonomi A (2012) Adaptação de inventários de ciclo de vida da cadeia produtiva do etanol. In: *III Congresso Brasileiro em Gestão do Ciclo de Vida de Produtos e Serviços*. Maringá, Brazil, DentalPress Publisinhg, 2012, vol. 1. pp. 1–6.

Chen X, Nuñez HM, Bing X (2015) Explaining the reductions in Brazilian sugarcane ethanol production costs: importance of technological change. *Global Change Biology Bioenergy*, 7, 468–478. DOI: 10.1111/gcbb.12163.

COOPERBIO – Cooperativa Mista de Produção, Industrialização e Comercialização de Biocombustíveis do Brasil (2011) Avaliação da viabilidade da produção de álcool a partir de cana-de-açúcar e mandioca em pequenas unidades camponesas de produção. 2° Relatório Parcial Referente à Meta 2. Universidade Federal de Santa Maria. 55 p. Santa Maria, Brazil, December 2008.

Daugaard T, Mutti LA, Wright MM, Brown RC, Componation P (2015), Learning rates and their impacts on the optimal capacities and production costs of biorefineries. *Biofuels, Bioproducts and Biorefining*, 9, 82–94. DOI: 10.1002/bbb.1513.

Dias MOS, Junqueira TL, Cavalett O, Cunha MP, Jesus CDF, Rossell CEV, Maciel Filho R, Bonomi A (2012) Integrated versus stand-alone second generation ethanol production from sugarcane bagasse and trash. *Bioresource Technology*, 103, 152–161. DOI: 10.1016/j.biortech.2011.09.120.

Dias MOS, Junqueira TL, Cavalett O, Pavanello LG, Cunha MP, Jesus CDF, Maciel Filho R, Bonomi A (2013) Biorefineries for the production of first and second generation ethanol and electricity from sugarcane. *Applied Energy*, 109, 72–78. DOI: 10.1016/j.apenergy.2013.03.081.

Dias MOS, Junqueira TL, Sampaio ILM, Chagas MF, Watanabe MDB, Moraie ER, Gouveia VLR, Klein BC, Rezende MCAF, Cardoso TF, Souza A, Jesus CDF,

Pereira LG, Rivera EC, Maciel Filho R, Bonomi A. Use of the VSB to Assess Biorefinery Strategies. In: Bonomi A, Cavalett O, da Cunha, MP, Lima MAP (Org.). Virtual Biorefinery: An Optimization Strategy for Renewable Carbon Valorization. Series: Green Energy and Technology. Springer International Publishing, 1st Ed, 2016, pp. 189–256. DOI: 10.1007/978-3-319-26045-7_7.

EPA – U.S. Environmental Protection Agency. Renewable Fuel Standard Program. Available online at: www.epa.gov/renewable-fuel-standard-program. Accessed on Oct 2016.

Fleck TL, Vaccaro GLR, Demartini FJ, Cabrera RS, Moraes CAM (2011) Simulação e análise de cenários em uma cooperativa de produção de etanol do Rio Grande do Sul: um estudo considerando cana, sorgo e subprodutos. In: *XXXI Encontro Nacional De Engenharia De Produção*. 15 p. Belo Horizonte, Brazil, October 2011.

Frischknecht R, Jungbluth N, Althaus HJ, Doka G, Dones R, Heck T, Hellweg S, Hischier R, Nemecek T, Rebitzer G, Spielmann M (2005) The ecoinvent database: Overview and methodological framework. *International Journal of Life Cycle Assessment* 10 (1), 3–9. DOI:10.1065/lca2004.10.181.1.

Goedkoop M, Heijungs R, Huijbregts M, de Schryver A, Struijs J; van Zelm R (2013) ReCiPe 2008. A life cycle impact assessment method which comprises harmonized category indicators at the midpoint and the endpoint level. www.lcia-recipe.net/ (accessed 11.07.16).

Goldemberg J, Coelho ST, Nastari PM, Lucon O (2004) Ethanol learning curve -the Brazilian experience. *Biomass and Bioenergy*, 26, 301–304. DOI: 10.1016/S0961-9534(03)00125-9.

Hamelinck CN, Faaij APC (2002) Future prospects for production of methanol and hydrogen from biomass. *Journal of Power Sources*, 111 (1), 1–22. DOI: 10.1016/S0378-7753(02)00220-3.

Hamelinck CN, Faaij APC (2006) Outlook for advanced biofuels. *Energy Policy*, 34 (17), 3268–3283. DOI: 10.1016/j.enpol.2005.06.012.

Hettinga WG Juginger HM, Dekker SC, Hoogwijk M, McAloon AJ, Hicks KB (2009) Understanding the reductions in US corn ethanol production costs: an experience curve approach. *Energy Policy*, 37 (1), 190–203. DOI: 10.1016/j.enpol.2008.08.002.

Howarth, R.W., S. Bringezu, M. Bekunda, C. de Fraiture, L. Maene, L. Martinelli, O. Sala. (2009) Rapid assessment on biofuels and environment: overview and key findings. Pages 1–13. In: R.W. Howarth and S. Bringezu (eds), *Biofuels: Environmental Consequences and Interactions with Changing Land Use*. Proceedings of the Scientific Committee on Problems of the Environment (SCOPE) International Biofuels Project Rapid Assessment, 22–25 September 2008, Gummersbach Germany. Cornell University, Ithaca NY, USA. (http://cip.cornell.edu/biofuels/).

Janssen R, Rutz DD (2011) Sustainability of biofuels in Latin America: Risks and opportunities. *Energy Policy*, 39 (10), 5717–5725. DOI: 10.1016/j.enpol.2011.01.047.

Jonker JGG, van der Hilst F, Junginger HM, Cavalett O, Chagas MF, Faaij APC (2015) Outlook for ethanol production costs in Brazil up to 2030, for different biomass crops and industrial technologies. *Applied Energy*, 147, 593–610. DOI: 10.1016/j.apenergy.2015.01.090.

Kubota, A. M., Dal Belo Leite, J.G., Watanabe, M., Cavalett, O., Leal, M.R.L.V., Cortez, L. 2017. The role of small-scale biofuel production in Brazil: Lessons for developing countries. *Agriculture* 7, 61.

Lynd LR, Sow M, Chimphango AF, Cortez LAB, Brito Cruz CH, Elmissiry M, Laser M, Mayaki IA, Moraes MA, Nogueira LAH, Wolfaardt GM, Woods J, van Zyl WH (2015) Bioenergy and African transformation. *Biotechnol Biofuels*, 8, 18–35. DOI: 10.1186/s13068-014-0188-5.

Maroun MR, La Rovere EL (2014) Ethanol and food production by family smallholdings in rural Brazil: Economic and socio-environmental analysis of micro distilleries in the State of Rio Grande do Sul. *Biomass & Bioenergy*, 63, 140–155. DOI: 10.1016/j.biombioe.2014.02.023.

Mayer FD, Brondani M, Hoffmann R, Silva Lora EE (2015a) Environmental and energy assessment of small-scale ethanol production. *Energy & Fuels*, 29 (10), 6704–6716. DOI: 10.1021/acs.energyfuels.5b01358.

Mayer FD, Feris LA, Marcilio NR, Hoffmann R (2015b) Why small-scale fuel ethanol production in Brazil does not take off?. *Renewable & Sustainable Energy Reviews*, 43, 687–701. DOI: 10.1016/j.rser.2014.11.076.

Morais ER, Junqueira TL, Sampaio ILM, Dias MOS, Rezende MCAF, Jesus CDF, Klein BC, Gómes, EO, Mantelatto PE, Maciel Filho R, Bonomi A. Biorefinery Alternatives. In: Bonomi A, Cavalett O, da Cunha, MP, Lima MAP (Org.). *Virtual Biorefinery: An Optimization Strategy for Renewable Carbon Valorization*. Series: Green Energy and Technology. Springer International Publishing, 1st Ed, 2016, p. 53–132. DOI: 10.1007/978-3-319-26045-7_4.

Oliveira LC (2009) *Indústria de etanol no brasil: uma estrutura de mercado em mudança. Master thesis submitted to the Economic Development post-graduation program*. Federal University of Paraná. 182 p. Curitiba, Brazil, 2009.

Ruan R, Chen P, Hemmingsen R, Morey V, Tiffany D (2008) Size matters: small distributed biomass energy production systems for economic viability. *International Journal of Agricultural & Biological Engineering*, 1 (1), 64–68. DOI: 10.3965/j.issn.1934-6344.2008.01.064-068.

Santos RER (2011) *Análise da viabilidade energética e econômica da produção de etanol em microdestilarias*. Master thesis submitted to the Energy Engineering post-graduation program. Federal University of Itajubá. 128 p. Itajubá, Brazil, 2011.

Seabra JEA, Macedo IC, Chum HL, Faroni CE, Sarto CA (2011) Life cycle assessment of Brazilian sugarcane products: GHG emissions and energy use. *Biofuels, Bioproducts and Biorefining*, 5 (5), 519–532. DOI: 10.1002/bbb.289.

Tsiropoulos I, Faaij APC, Seabra JEA, Lundquist L, Schenker U, Briois JF, Patel MK (2014) Life cycle assessment of sugarcane ethanol production in India in comparison to Brazil. *The International Journal of Life Cycle Assessment*, 19 (5), 1049–1067. DOI: 10.1007/s11367-014-0714-5.

Van den Wall Bake JD, Junginger M, Faaij A, Poot T, Walter A (2009) Explaining the experience curve: cost reductions of Brazilian ethanol from sugarcane. *Biomass and Bioenergy*, 33, 644–658. DOI: 10.1016/j.biombioe.2008.10.006.

Wang L, Quiceno R, Price C, Malpas R, Woods J (2014) Economic and GHG emissions analyses for sugarcane ethanol in Brazil: Looking forward. *Renewable and Sustainable Energy Reviews*, 40, 571–582. DOI: 10.1016/j.rser.2014.07.212.

Watanabe MDB, Pereira LG, Chagas MF, da Cunha MP, Jesus CDF, Souza A, Rivera EC, Maciel Filho R, Cavalett O, Bonomi A. Sustainability Assessment Methodologies. In: Bonomi A, Cavalett O, da Cunha, MP, Lima MAP (Org.). *Virtual Biorefinery: An Optimization Strategy for Renewable Carbon Valorization*. Series: Green Energy and Technology. Springer International Publishing, 1st Ed, 2016, pp. 155–188. DOI: 10.1007/978-3-319-26045-7_6.

27 Vinasse, a new effluent from sugarcane ethanol production

Carlos Eduardo Vaz Rossell

Physical and chemical properties of vinasse

Physical and chemical composition data of sugarcane vinasses as reported by Prada (1998) for Brazilian distilleries musts and by Nimbalkar (2003) for heavily exhausted blackstrap molasses are presented in Tables 27.1 and 27.2.

Results show high polluting potential resulting of chemical (COD) and biological oxygen demand (BOD), low pH and high buffering strength, soluble organic and inorganic solids. Vinasse contains a large portion of soluble inorganic material originally present in sugarcane juice as well as those used in their processing to sugar or ethanol. It is important to remark that the

Table 27.1 Physical and chemical characteristics of vinasse from molasses, cane juice and mixed musts.

Parameter	Molasses	Cane juice	Mixed musts
pH	4.2–5.0	3.7–4.6	4.4–4.6
Temperature (°C)	80–100	80–100	80–100
BOD (mg/L O_2)	25,000	6,000–16,500	19,800
COD (mg/L O_2)	65,000	15,000–33,000	45,000
Total solids (mg/L)	81,500	23,700	52,700
Volatile solids (mg/L)	60,000	20,000	40,000
Fixed solids (mg/L)	21,500	3,700	12,700
Nitrogen (mg/L N)	450–1,610	150–700	480–710
Phosphorus (mg/L P_2O_5)	100–290	10–210	9–200
Potassium (mg/L K_2O)	3,740–7,830	1,200–2,100	3,340–4,600
Magnesium (mg/L MgO)	420–1,520	200–490	580–700
Sulfate (mg/L SO_4)	6,400	600–760	3,700–3,730
Carbon (mg/L C)	11,200–22,900	5,700–13,400	8,700–12,100
C/N ratio	16.0–16.27	19.7–21.07	16.4–16.43

Source: Adapted from Prada (1998).

Table 27.2 Physical and chemical characteristics of vinasse obtained from blackstrap molasses from different fermentation processes

Characteristics of vinasse from different fermentation processes

Parameter	Batch fermentation	Continuous multistage	Biostil process
pH	3.7–4.5	4.0–4.3	4.0–4.2
COD(mg/L O_2)	80,000–110,000	6,000–16,500	19,800
BOD (mg/L O_2)	45,000–50,000	55,000–65,000	60,000–70,000
Total solids (mg/L)	90,000–120,000	130,000–160,000	160,000–210,000
Volatile solids (mg/L)	60,000–70,000	60,000–75,000	80,000–90,000
Fixed solids (mg/L)	60,000–70,000	3,700	12,700
Inorganic dissolved (mg/L)	30,000–40,000	35,000–45,000	60,000–90,000
Phosphorus (mg/L P_2O_5)	200–300	300–500	1,600–2,000
Potassium (mg/L)	8,000–12,000	10,000–14,000	20,000–22,000
Sulfates (mg/L)	4,000–8,000	4,500–8,500	8,000–10,000
Sodium (mg/L)	400–600	1,400–1,500	1,200–1,500
Calcium (mg/L)	2,000–35,000	4,500–6,000	1,200–1,500
Total nitrogen (mg/L)	1,000–1,200	1,000–1,400	2,000–2,500

Source: Adapted from Nimbalkar (2003).

high content of potassium turns vinasse, a fertilizer agent, when applied in cane crops. Phosphorum, calcium, magnesium and sulfate are also present. Nitrogen is present in low quantity when compared to carbon.

Processing of vinasse

The use of vinasse in natura or preconcentrated, fertirrigation of sugarcane with vinasse, and the case of Brazilian distilleries

Vinasse fertirrigation in sugarcane plantations is a widespread practice among Brazilian mills and distilleries. This technology was wholly developed in Brazil, as no other country generates such a huge volume of this kind of residue. Mutton et al. (2014) report that vinasse promotes improvements in the agricultural yield of sugarcane, as well as chemical, biological and physical benefits to the soil, in addition to savings in fertilizer costs.

Several authors studied the effect of stillage on soils and, over time, showing that stillage increases pH and cation exchange capacity (CEC) in soils, improves soil structure, increases water retention and enhances biologic activity, promoting a large number of small animals (earthworms, beetles, etc.), bacteria and fungi.

Vinasse, when applied in proper doses for sugarcane nutrition, using the potassium fertilization recommendation, does not cause any salinization or ion lixiviation problems that may compromise the environment. Soil enrichment occurs in the 0–200 mm (soil surface) and 200–400 mm (soil subsurface) layers.

In order to regulate the fertirrigation procedures and avoid uncontrolled application, which can damage groundwater and sugarcane quality and lead to soil salinity, the Environmental Sanitation Technology Company (CETESB) (2006), São Paulo, Brazil, established criteria for fertirrigation of sugarcane crops prescribing the volume of application relating it to a potassium balance. The maximum concentration of potassium in the soil may not exceed 5% of the CEC of the soil. If this limit is reached, the application of stillage is restricted based on the average potassium extracted by the sugarcane, which is 185 kg K$_2$O per hectare per harvest. The following formula defines the maximum dose of vinasse in m^3/hectare to be applied:

$$\text{Vinasse dosing (m}^3/\text{hectare)} = [(0.05 \times \text{CEC} - K_{soil}) \times 3.744 + 185]/K_{vinasse}$$

where
Vinasse dosing = vinasse doses in m^3 per hectare per season;
0.05 = 5% of CEC;
CEC = cationic exchange capacity expressed in cmolc/dm^3 at pH 7, determined by soil analysis by a certified laboratory;
K$_{soil}$ = potassium concentration expressed in cmolc/dm^3, at a depth between 0 and 0.8 m, obtained by soil analysis (cmolc = centimole equivalent);
3.744 = constant for converting the analytical results of soil fertility to kg of potassium contained in a volume of 1 hectare (10,000 m^2) and 0.8 m depth;
185 = mass (kg) of K$_2$O extracted by cane culture per hectare per season;
K$_{vinasse}$ = potassium concentration in vinasse (kg K$_2$O/m^3) annually averaged from weekly results or an average obtained from semiannual data.

This regulation established by CETESB is being adopted in other states of the Brazilian federation, and it should be adopted as a starting point by the other countries that will adopt fertirrigation of vinasse.

The disposal cost of vinasse in natura in the sugarcane crops increases according to the distance from the distillery, while the cost of mineral fertilizers does not depend on distance. This implies that there is an economic distance that represents the distance where the application of vinasse has a lower cost than chemical fertilizer use (*Orlando* et al., 1980). For concentrated vinasse, this cost is a function of the final concentration, a higher concentration represents a lower volume to be transported and results in a greater economic distance, as reported by Carvalho (2010). However, the economic distance does not increase materially for values of concentration higher than 25–30 Brix, and above these values there is no economic advantage.

Transportation of vinasse to plantations is done by tank trucks or by a network of impervious channels and polymer pipes.

Fertirrigation is done by tank trucks fitted with a distribution system that sprays the vinasse in several rows of ratoon cane. The discharge is done by gravity, which does not provide uniform flow, or by a pump, which delivers pressurized vinasse with better performance. This system has the following disadvantages: it requires intensive labor; it is not applicable on sloping land; it gives a low daily yield; it leads to nonuniform flow in case of gravitational discharge, high operational costs and short lifetime of the vehicles; and it demands for roads and carriers to be in good condition, vehicle fleet availability, a need for planning to allow vinasse delivery 24 h/day, soil compaction and restricted application on rainy days. Therefore, the use of tank trucks moving in sugarcane plots has given way to alternative systems. For vinasse in natura, this system is being substituted by a station of a centrifugal pump powered by a diesel engine mounted on a four-wheeler truck, which directly sucks vinasse from the distribution channel and sprays on field by the use of a hydraulic cannon system. An improvement of this application system is the hydro-roll system, a mobile device with a winding reel with rotor-type hydraulic cylinder at the end of a flexible hose of HDPE (high-density polyethylene) mounted on a supported two-wheeler truck capable of operating at distances up to 300 m from the motor pump unit.

Fertirrigation with vinasse concentrated up to 20–25 Brix is done by tank trucks modified for application simultaneously in seven rows of sugarcane. Concentrated vinasse provides a significant volume reduction, enabling transport for longer distances and for discontinuous areas. Another advantage is the increased potassium and nitrogen content. In addition, it minimizes the risk of groundwater contamination.

Reduction of vinasse final volume

Reduction of vinasse volume is important from environmental viewpoint, also in terms of reduction of water use in ethanol production and reducing the cost of vinasse final processing or fertirrigation.

Some actions such as indirect heating during distillation stage, operation of the fermentation process at higher ethanol strength and vinasse recycling have a strong impact on vinasse volume reduction.

The use of indirect heating of distillation columns instead of steam sparging diminishes vinasse volume to near 8.6% (Fernandes, 2003). Separation of vinasse from spent lees (flegmasse) in ethanol rectification step diminishes volume to near 10%. The operation of fermentation process at higher ethanol strength has a strong impact on vinasse volume reduction, for example, increasing ethanol strength from 8.0°GL to 10.0°GL promotes a reduction from 10.0–11.9 to 7.8–9.4 L vinasse/L ethanol (Carvalho, 2010). The recycling of vinasse as water for molasses or syrup dilution is another alternative for vinasse volume reduction.

Recycling of vinasse at the fermentation process

Recycling of vinasse in the fermentation process is an alternative to contribute to the reduction of volume of final vinasse. The procedure consists in the use of vinasse as dilution water for must preparation. Initially, it was developed for the dilution of blackstrap molasses and later with the growing of fuel ethanol production extended to partially exhausted molasses and sugarcane syrup. Vassard (1982) reports that vinasse recycling has been a common practice in distilleries that processes molasses or juice of beet. He developed a correlation to determine the non-sugar-soluble solids content in wine after recirculation attains its threshold point. Later, Navarro (2000) proposed a mathematical model, equation below, to describe final soluble solids concentration according to the number of fermentation cycles and the ratio of vinasse recycling:

$$w_{DS} = \frac{w_{DS,0}\left(1-r^{N+1}\right)}{1-r}$$

where w_{DS} is the final dissolved solids concentration, $w_{DS,0}$ is the initial solids concentration, N is the number of cycles and r is the recycle ratio, which is the relationship between the amount recycled and total vinasse produced. Allard and Miniac (1985) studied vinasse recirculation in sugar beet mills and distilleries in France, they distinguished two procedures of recycling: batch fermentation with the preparation of a new yeast foot each run and fedbatch with yeast recycling (Melle-Boinot procedure), concluding that the recycling if conducted in well-established and controlled conditions was technically feasible and impacted favorably fermentation process, cost and environmental concerns. Fermentation time, yield and final ethanol strength are related to soluble solids content in vinasse due to osmotic pressure and organic acids content. The use of vinasse as dilution water on yeast foot preparation or dilution was not recommended.

The Biostil process. The case of Indian and other eastern distilleries

Biostil process, launched in 1977, is a process devoted to the reduction of final vinasse volume by recycling it as dilution water for molasses during must preparation. It is a continuous process in which fermentation and distillation are integrated. The actual version of Biostil as described by Guidoboni (1984), Wunsch (1989) and Chematur licensor (2018) operates in a single continuous stirred fermentation vat at high yeast density and with yeast recycling. The yeast used in this process is *Schizosaccharomyces pombe*, which is tolerant to high osmotic pressure and organic acids. The ethanol content in the fermented mash is around 5.7–7.6°GL. During fermentation, the yeast losses its activity with time, due to the heavily adverse conditions of high osmotic pressure, ethanol titer, organic acids and low sugar concentration, being periodically replenished by a new yeast mass. The mash from molasses/syrup is fed continuously to the aerated fermenter and wine is continuously withdrawn and pumped to the centrifuge separators. The yeast cream is recycled, and the clarified mash is pumped to mash column where an ethanol vapor phlegm of 40°GL is obtained at the top and sent to a conventional rectifying column where hydrous ethanol is obtained. The weak vinasse of mash column which is not totally depleted of ethanol is divided into two streams: one part is recycled to molasses dilution after cooling, and the rest is sent to a stripper column in which the remaining ethanol is removed. Biostil process had a good acceptation in eastern countries, mainly in India in autonomous blackstrap molasses distilleries. Deodhar and Shinagare (1990) claim a vinasse recycle ratio of 0.6, 5–6 liters of vinasse/liter of alcohol, a fermentation yield of 91%–92% and a distillation yield of 99%.

Thermal concentration of vinasse

The concentration of vinasse by evaporation reduces its volume. In addition, the water evaporated can be recovered for reuse, reducing freshwater demand; however, the quality of the condensed vapors is poor due to the appearance of volatiles like organic acids and others and high COD content (Couallier et al., 2006) requiring final treatment. As an example, condensed vapors are not recommended for dilution water of yeast foot or must. Two conditions are distinguished: evaporation to an intermediary soluble solids content of 20%–25% with the aim of reducing fertirrigation transport, and application costs and concentration to near 65% s.s. as an initial step for other processes such as composting, combustion or dry solids fertilizer preparation. The cost of fertirrigation of sugarcane crops with vinasse in natura increases according to the distance from the distillery, while the cost of mineral fertilizers does not vary in this way. This implies that concentration reduces this cost. Comparing with chemical fertilizers, there is an economic distance that represents the distance where the application of vinasse is more attractive. For concentrated vinasse, this cost is a function of the final concentration, a higher concentration represents a lower volume to be transported and results in a greater economic distance. However, the economic distance does not have significant increase for values of concentration higher than 25%–30% s.s., and above these values there is no economic advantage (Carvalho, 2010).

Thermal concentration is done in a cocurrent multistage evaporation system designed with the aim of optimizing heat integration, recovering as much secondary steams as possible and optimizing its use. The system operates under a final vacuum equivalent to an absolute pressure of 90–130 mmHg, in order to favor the use of low-pressure steam.

Evaporation equipment has high capital cost due to the need of corrosion-resistant materials for construction and be designed to withstand high-soluble solids content, severe scaling tendency and high viscosity. Most common designs of evaporator systems are as follows: falling film followed by forced circulation evaporators as described by GEA (2000), thermally accelerated short-time evaporation (TASTE) as described by Carvalho (2010) and Marengo (2011) and fluidized bed forced circulation (FLUBEX), a system that employs fluidized metal particles inside the heat exchange tubes, which promote the cleaning of tube walls, as reported by Godbole (2002).

Thermal concentration requires the optimization of steam production and use in the whole sugarcane mill in order to attain the extra requirement of steam for the concentration of vinasse. Scale formation during concentration is critical, and compared to cane juice evaporation, the frequency of evaporators cleaning is much higher and scale is more difficult to remove.

Composting of vinasse with other by-products of sugarcane processing

The aerobic composting of vinasse with other by-products and wastes or residues from sugarcane processing renders an organic product suitable for use as a solid fertilizer as reported by Gomes (2012). As described by Mehmood (2000), vinasse preferentially thermally concentrated is mixed with filter mud, ashes and soot from boiler and residue from cane yard, seeded with a fraction of already composted material as a microbiological starter culture and piled in rows. Windrow machines periodically revolve the piles with the aim of mixing and aerating. The compost losses water due to the combined effect of heat evolved during microbiological reactions and the drying action of unsaturated air. Periodically, more vinasse is fed to the compost. After 60–90 days, the compost is stabilized, the organic material content is reduced and the C/N ratio diminishes, which improves nutrient absorption. The moisture content drops to 70%, in this condition it can be applied as a solid fertilizer to sugarcane crops, eventually other crops. Vinasse provides K, N and P and Fe, Mn, Zn, Cu and B in minor quantities and yeast debris and cells. Filter mud provides mainly phosphates and organic matter, and ashes provide potassium and other minerals in minor quantities. Aerobic composting accelerates the microbial reactions, and *Trichoderma* genus is the more representative microorganism.

Typical parameters of a vinasse composting procedure (Mehmood, 2000) are as follows:

Mass of compost obtained: 2.5%–3.5% per TC
pH: 6.5–7.2
Electrical conductivity (1/10 dilution): 40–60 in dSm^{-1}
Total nitrogen: 1.45%–1.70%
Total phosphorum: 1.35%–1.55%
Potassium: 3.3%–3.8%
Organic matter: 55%–65%
Conil (2012) described a composting processing unit coupled with a sugar mill (10,000 TCD) with annexed distillery for ethanol production (ethanol:

90,000 LD and a vinasse ratio of 13 L/L ethanol) using final molasses as unique substrate in which a part of the vinasse generated is applied in composting, being the rest applied by fertirrigation. The composting processing is fed with a daily rate of 400 m^3 of vinasse, 400 tons of filter mud and 65 tons of ash and soot from the boilers, and produces 120 tons of compost.

This plant requires an area of 4 hectare in which the mixture is piled in 45 rows, requiring one windrow composting machine and two loaders. In the case in which the milling season coincides with the dry season, the composting plant can be built roofless, which diminishes significantly the capital cost.

Aerobic composting provides a stabilized efficient solid fertilizer. Its efficiency has been proved in sugarcane crops fertilizing. It has the advantage that transportation and dosing cost are reduced compared with fertirrigation with vinasse. It is an environmental friendly process, and it avoids obnoxious gas generation and insects' proliferation.

Recycling of vinasse, thermal concentration and composting. The case of Colombian distilleries

Colombian ethanol distilleries adopted a processing system that is devoted to attend vinasse regulations which ban the fertirrigation with vinasse in natura and involves recycling of vinasse in fermentation substituting the use of water for molasses dilution, concentration of vinasse in evaporators followed by a composting process which removes additional water and organic matter rendering a solid fertilizer. Distilleries in general are units annexed to sugarcane mills that produce sugar. They produce ethanol mainly from final and B molasses with a small percentage of juice or syrup. Ethanol production is done by employing an industrial process and equipment developed by Praj Industries Ltd. As reported by Figueroa-Benítez (2006) and Cenicaña (2008), the fermentation is done in continuous in a cascade of three or four mechanically stirred vats in which must is fed in parallel in different proportions. After fermentation step, final wine is sent to a continuous settler in which the flocculent yeast strain is separated.

The yeast cream recovered in the bottom of the decanter is submitted to acid wash and aeration and returned to the first fermentation vat. A propagator-aerated vat replenished the yeast lost during the settling step. The wine recovered at the top of the clarifier containing a fraction of unsettled yeast is sent to distillation which is done in a set of three columns (deflegmation, volatile removal and rectifying columns). Ethanol at 96°GL, vinasse and spent lees are obtained. The ethanol stream is dehydrated in molecular sieves to render anhydrous fuel ethanol at 99.8°GL. The vinasse stream separated at the bottom of distillation column is partially recycled to fermentation process and used to dilute molasses instead of water during the formulation of the must. As reported by Rueda-Ordoñez (2008), near 60% v/v of vinasse is recycled to the fermentation process increasing its soluble solid content to 15%–20%.

Fermentation process belongs to the class of osmotic fermentations considering soluble solids coming from molasses and from the recycle of vinasse requiring special osmophilic yeast strains.

The remaining vinasse is submitted to thermal concentration done in two steps in a forced circulation evaporator system named Flubex developed by Praj Industries Ltd. as described by Godbole (2002).

Vinasse after concentration reaches 30%–55% s.s. It is sent to a composting plant, a windrow aerobic composting unit in which it is blended with filter mud, ashes and soot from the boiler and wastes from the sugarcane yard. A solid fertilizer is obtained after a 70-day composting process, which is used in sugarcane cultivation as fertilizer in the planting groove with an application rate of 10–20 tons per hectare per year, depending on the potassium requirements of the soil.

The process shows advantages such as significant reduction of vinasse volume that attains 3.8–4.5 L/L ethanol after distillation and 2 L after thermal concentration, low process steam consumption of 2.75 kg/L ethanol, reduction of water used in the process and production of a solid fertilizer with much less cost of handling, storage and application. Final effluents (condensates and spent lees) after biological treatment are allowed to be used for irrigation or disposed in water streams.

Anaerobic biodigestion of vinasse

Anaerobic digestion is one of the options to process vinasse due to high level of COD and the opportunity for generating biogas or methane after biogas fractionation and obtaining additional energy from sugarcane as fuel for electrical power generation and as primary fuel as household gas for internal combustion engines. Biodigestion is a primary treatment considering that COD and BOD reduction is not enough to fulfill standards of water treatment environmental legislation to be dumped into water courses. An option for vinasse after biodigestion is fertirrigation as reported by Moraes et al. (2015). The process of anaerobic digestion occurs in two sequential steps: an acidogenic phase where bacteria convert the stillage components (glycerol, sugars, higher alcohols and esters, and nitrogenous organics from yeast) of the complex organic matter in short chain organic acids (acetic, propionic and butyric) reducing the level of COD. In the following phase, methanogenic microorganisms convert organic acids to methane and carbon dioxide. The UASB (upflow anaerobic sludge blanket) reactor has been successfully used in anaerobic digestion of vinasse. The UASB operates with a high content of active sludge that reduces residence time. This type of reactor has an internal biogas separator and a sludge decanter that maintains high level of active sludge in the biodigester. This type of reactor operates up to 30 kg of $COD/m^3/day$ with thermophilic microorganism (55°C) producing 10 Nm^3 of biogas/day. This performance is double compared to data reported in the literature for mesophilic process (35°C). The biogas produced has 65% of

CH4 and 35% of CO_2, a lower heating value of 19.5 MJ/m^3, a higher heating value of 23.4 MJ/m^3 and a specific gravity of 1.2 kg/m^3. Due to this characteristic, it is recommended to increase the content in methane removing the CO_2 by adsorption which increases its heating value and then compresses it to use in combustion engines (Freire and Cortez, 2000).

Lucas Jr. et al. (1997) presented an equation that estimates the methane production at standard temperature and pressure conditions (0°C and 1 bar) as a function of COD of vinasse:

$$COD(g) = 0.286 \times V_{CH_4}(L)$$

It has been a large research and development effort in biogas production from vinasse, and commercial units have been installed, but results obtained show that the process is not economically feasible yet and requires the financial support of government agencies.

The processing of vinasse combined with other by-products of sugarcane to a dry fertilizer. Dedini BiOFOM process

Sugarcane annexed or autonomous distilleries generate 28–40 kg/TC of filter mud, and 6.25 kg/TC of ash and soot resulting from bagasse combustion in boilers. Olivério et al. (2010) proposed the production of a solid and granular organomineral fertilizer called BIOFOM by the combination of these by-products with vinasse, and supplemented with nitrogen and phosphorus chemical fertilizers, which it can be formulated according to the specific needs of the soil and crop.

The production process of the organomineral fertilizer starts by the thermal concentration of vinasse rising its content of soluble solids from 3%–6% to 45%–60% in multiple effect evaporators. A fraction of this concentrated vinasse, near 10%, is directed to a furnace in which it is burned with bagasse and straw providing hot gases for drying a mixture containing the filter mud and the ashes and soot from the bagasse boilers and the furnace. In the remaining vinasse, urea, single superphosphate (SSP) and potassium chloride are dosed as nitrogen, phosphorus and potassium source in order to supplement and adjust composition formula. Then this stream of concentrated vinasse is blended with the dried filter mud and ashes, and granulated and dried by the furnace hot gas to obtain the final organomineral BIOFOM fertilizer. Typical parameters of a BIOFOM process unit integrated into a sugarcane distillery are as follows:

Ethanol production: 75 L/TC
BIOFOM organomineral fertilizer (15% moisture): 50.28 kg/TC
Water recovery: 663.24 kg/TC
Fertilizer doses on cane crops: 4,300 kg/hectare/year
Filter mud (60% moisture) used: 29.12 kg/TC

Ashes and soot from steam boilers and furnace: 11.3 kg/TC

Supplements of chemical fertilizer are as follows:

- Urea (45% N): 1.74 kg/TC
- SSP (21% P2O5): 0.37 kg/TC
- Potassium chloride (KCl): 0.054 kg/TC

Economic evaluation of the incorporation of a BIOFOM processing unit in a sugarcane ethanol distillery showed very attractive results as a discounted payback time of 2–3.5 years depending of the average distance between the mill and the sugarcane plantation.

BIOFOM is an alternative for turning vinasse in a competitive and sustainable organomineral fertilizer. It can be formulated according to the specific needs of the soil, minimizing the requirements of chemical fertilizers. Additional gains result because its high content of organic matter improves the soil physical–chemical properties such as the cation exchange capacity and porosity, facilitating the absorption of nutrients and reducing the losses caused by leaching. In addition, the use of a dry fertilizer allows a significant reduction of the infrastructure required to distribute fertilizer, vinasse, filter cake and ashes. It also diminishes the losses in nutrients, N, P, K and organic matter contained in the by-products due to a more efficient dosing.

By the use of this solid fertilizer, it is possible to reduce or even eliminate water catchment during sugarcane processing, by recovering the water contained in vinasse and filter mud.

BIOFOM was submitted to greenhouse experimental tests by Gurgel (2012), and results obtained using corn culture cycle proved its good performance in substituting chemical fertilizer.

Combustion of concentrated vinasse

Vinasse has a large proportion of organic matter soluble and suspended which after concentration to 50%–65% of soluble solids attains a heat of combustion value that makes able its combustion in boilers. Direct combustion was reported by Nilsson (1981) and Spruitenberg (1982) done in boilers similar to those used in the burning of black liquor from pulp and paper. Boiler units were installed in the 1980s by Alfa Laval AB and Hollandse Constructie Groep BV, in Europe and Thailand, producing steam from the combustion of concentrated vinasse and an ash by-product was recovered. Cortez (2000) did an extensive work focused on the combustion of vinasse in conventional sugar mill boilers reporting problems related to the high viscosity of concentrated vinasse which resulted in inefficient dispersion in the atomizing nozzles and the direct burning of vinasse due to its low heating value. He suggested using vinasse concentrated up to 40% s.s. blended with fuel oil to provide a lower viscosity and a higher heating value.

More recently, Avram et al. (2005) presented an optimized process for simultaneous sugar, ethanol and cogeneration that includes the thermal concentration of vinasse to 65% s.s. and its combustion in a boiler designed for vinasse combustion, generating additional electrical power and near 80% of the steam required for vinasse concentration.

Combustion of vinasse is a "zero-effluent" final treatment process, which can be integrated with other vinasse processing options. It should be more explored considering that it is an ecofriendly alternative which recovers the ash content as a potash dry fertilizer. It provides additional steam and electrical power coming from the heat content of organic matter contained in vinasse. Although being an interesting technology, it has not progressed yet as it may be expected due to the high investment related to the boiler.

Conclusion

We presented and discussed the processing of vinasse as a by-product which can be used as a fertilizer considering its content in potassium and other essential elements to the sugarcane growth and the recycling of these elements to soil.

The alternatives presented and discussed focus on the transformation of vinasse in a fertilizer.

In general, processing alternatives are orientated to the reduction of final volume, diminish fertirrigation or fertilization costs. Most of them improve the properties of vinasse as a fertilizer and recover water, reducing the quantity of water required in the sugarcane processing. An exception is biodigestion which does not reduce final vinasse volume.

It is expected that this chapter gave to existing ethanol distilleries as well as to new projects an understanding of the physical and chemical properties of vinasse, the environmental concerns and local regulations and advise them in the decision of the route selected to process vinasse.

It is important to continue the development of new alternatives to process vinasse, recover and recycle nutrients to soil and water, and continue optimizing the existing ones.

Abbreviations

BOD:	biochemical oxygen demand
CEC:	cationic exchange capacity expressed in cmolc/dm^3 at pH 7, determined by soil analysis by a certified laboratory
COD:	chemical oxygen demand
dSm^{-1}:	Decisiemens per meter
kg/TC:	kg per tons of cane
kg/hectare/year:	kg per hectare per year
K_{soil}	= potassium concentration expressed in cmolc/dm3, at a depth between 0 and 0.8 m, obtained by soil analysis (cmolc = centimole equivalent)

K_vinasse = potassium concentration in vinasse (kg K$_2$O/m^3) annually averaged from weekly results or an average obtained from semiannual data
L/L ethanol: liters per liter of ethanol
LD: liters per day
L/TC: liters per tons of cane
N: number of fermentation cycles
r: recycle ratio
s.s. %: percentage of soluble solids
TCD: tons of cane per day
TCH: tons of cane per hour
wDS: dissolved solids concentration,
wDS,0: initial solids concentration,

References

Allard, G., and Miniac, M. (1985): Recyclage des vinasses ou de leurs condensats d'évaporation en fermentation alcoolique des produits sucries lourds (melasses et égouts). *Ind. Aliment. Agric.* 102, 9, 877–882.

Avram, P., Morgenroth, B., and Seeman, F. (2005): Benchmarking concept for an integrated sugar, ethanol and cogeneration plant. *Proc. ISSCT*, V25, 7–12.

Carvalho, T. C. (2010): *Redução do volume de vinhaça através do processo de evaporação.* MSc Thesis. Universidade Estadual Paulista, Bauru, São Paulo, Brazil.

Cenicaña (2008): *Processo de obtenção de etanol.* www.cenicana.org/pop_up/fabrica/diagrama_etanol.php.

Conil, P. (2012): *Bioetanol de caña. El desafío ambiental de las vinazas.* Pistas para Tucumán y futuro de la agroindustria. Congresso Atalac-Tecnicaña. Septiembre, 10–14. Cali. Colombia.

Cortez, L. A. B. in: Freire W. J.; Cortez, L. A. B. (2000): *Vinhaça de Cana-de-Açúcar,* ed. 1, Agropecuária, Vol. 1, pp. 203, RS, Brasil. https://chematur.se/process-areas/bio-chemicals/biostil-ethanol/.

Couallier, E. et al. (2006): Recycling of distillery effluents in alcoholic fermentation: Role in inhibition of 10 organic molecules. *App. Biochem. Biotech.* 133, 3, 212–237.

Deodhar, A. S., Shinagare, K. K. (1990): *Biostil – a continuous fermentation-distillation system.* Technical Papers of the Fortieth Annual Convention of the Deccan Sugar Technologists' Association, Part 1, B25–B37.

Fernandes, A. C. (2003): Cálculos na Agroindústria da Cana-de-açúcar Fernandes, A. C. Cálculos na Agroindústria da Cana-de-açúcar. STAB Sociedade dos técnicos Açucareiros e Alcooleiros do Brasil, p. 162. Piracicaba, SP, Brazil.

Figueroa-Benítez, M. (2006): Producción de Alcohol Anhidro a partir de Caña de Azúcar. Destilería de Incauca S.A (personal communication).

Freire W. J., Cortez, L. A. B. (2000): Vinhaça de Cana-de-Açúcar, ed. 1, Agropecuária, Vol. 1, p. 203, RS, Brasil.

GEA Wiegand (2000): *Evaporators for stillage concentration.* GEA Wiegand GmbH, P14E0109, Etlingen, Germany.

Godbole, J. (2002): *Ethanol from Cane Molasses.* DOE+BBI Hawaii Ethanol Workshop, November 14, Honolulu, Hawaii.

Gomes, T. C. A. (2012): *Reciclagem de Vinhaça por Meio do Processo da Compostagem*. Embrapa. Boletim de pesquisa e desenvolvimento. N 74. ISSN 1678–1961, dezembro.

Guidoboni G. E. (1984): Continuous fermentation systems for alcohol production. *Enzyme Microbiol. Technol.* 6, 194–200.

Gurgel, M.N.A. (2012): *Tecnologia para aproveitamento de resíduos da agroindústria sucroalcooleira como biofertilizante organomineral granulado*. Dr. Sc. Thesis. Faculty of Food Engineering. Unicamp. Campinas. Brasil

Lucas, Jr.; Cortez, L. A. B.; Silva, A. (1997): Biodigestão. In: Cortez, L. A. B., Silva-Lora, E. (Coord.) *Tecnologias de conversão energética de biomassa*. Manaus: Ed. Universidade do Amazonas, cap. 10, 527 p.

Marengo, G. (2011): *Concentração de Vinhaça: Consumo de Vapor "ZERO"*. XII Seminário Brasileiro Agroindustrial. 26/10/2011.

Mehmood, Z. (2000): *Biocomposting of stillage and filter cake*. Shakarganj Mills Limited. Jhang, Pakistan.

Moraes, B.S.; Zaiat, M., Bonomi, A. (2015): Anaerobic digestion of vinasse from sugarcane ethanol production in Brazil: Challenges and perspectives. *Renew. Sustain. Energy Rev.* 44, 888–903.

Mutton, M.A., Rossetto, R., Mutton, M.J.R. (2014). In: Cortez, L.A.B. editor. Agricultural use of stillage. In: Sugarcane bioethanol-R&D for Productivity & Sustainability. Editora Edgard Blücher, pp. 423–440. São Paulo, Brazil.

Navarro, A. R., Sepúlveda, M. del C., Rubio, M.C. (2000): Bio-concentration of vinasse from the alcoholic fermentation of sugarcane bagasse. *Waste Manage.* 30, 581–585.

Nilsson, M. (1981): Energy recovery from distillery wastes, from Alfa-Laval A. B. *Int. Sugar J.* 83(993), 259–261.

Nimbalkar, S., Gunjal, B.B. (2003): *High rate biphasic anaerobic treatment for distillery spentwash*. Vasantdada Sugr Institute, Pune, India. Technical Report.

Olivério, J.L. et al. (2010): *Proc. Int. Soc. Sugar Cane Technol.* 27, 1–9.

ORLANDO FILHO, J.; SOUSA, I. C. de; ZAMBELLO JUNIOR, E.. Aplicação de vinhaça em soqueiras de cana-de-açúcar: economicidade do sistema caminhões tanque. Piracicaba: PLANALSUCAR, 1980. 35 p. (PLANALSUCAR. Boletim Técnico, 2, n. 5).

Prada, S., M., Guekezian, M., Suárez-LHA, M., E., V. (1998): Metodologia analítica para a determinação de sulfato em vinhoto. *Química Nova*, 21(3), 249–252.

Rueda-Ordoñez, D.A., Leal, M.R.L.V., Bonomi, A., Cortez, L.A. B., Cavalett, O., Rincon, J. M. (2018). Environmental and economic aspects in the Colombian sugarcane industry: evaluation of coal utilization and the compost production. (to be published)

Spruitenberg, G. P. (1982): Vinasse pollution elimination and energy recovery from Hollandse Constructie Groep *B.V. Int. Sugar J.* 73–74, Mar.

Vassard, O (1982): Recyclage des vinassses de betteraves. *Ind. Aliment. Agric.* 99(7–8), 526–528.

Wunsch, W. (1989): New continuous fermentation process in the production of ethanol. *Zuckerindustrie*, 114(Special edition), 39–44.

Part V

Assessing the LACAf Project findings

28 Final remarks on production model alternatives

*Manoel Regis L.V. Leal and
João Guilherme Dal Belo Leite*

Introduction

In developing countries, there is an overall agreement that the ideal production model favors rural development through an active participation of local communities (Ambali et al., 2011; von Maltitz and Stafford, 2011; Gasparatos et al., 2015; RISE, 2016). Therefore, job opportunities, food and energy security, health and education services and the engagement of smallholder farmers are decisive aspects for the project sustainability. Although Chapters 22, 23, 25 and 26 discuss many of these sustainability issues, in this chapter we explore in more detail how mechanization, scale and the engagement of smallholder farmers affects business viability in developing regions, such as Latin America, the Caribbean and sub-Saharan Africa (SSA).

Mechanization

In many sugarcane-producing regions, there seems to be a cost-effective ratio between manual and mechanical operations. For instance, it is common to lay cane seeds manually while the furrow is done with the help of a tractor. Manual harvest is another example of activity complemented with mechanical support for loading and transportation. In this case, pre-burning is necessary to improve both labor safety and productivity. Pre-burning, however, is not mandatory for mechanical harvest. Hence, green cane harvest leaves a blanket of residues on the ground that has positive (soil protection against erosion, increase in soil organic matter, soil moisture retention and nutrient recycling) and some negative impacts (risk of fire, pest proliferation and increased difficulty in fertilizer application). A fraction of the straw can be sustainably recovered from the field to supplement bagasse in power generation, thus boosting power surpluses.

Tables 28.1–28.4 compare harvesting alternatives, both manual and mechanical, with the creation of job opportunities, wages and production costs for an ethanol distillery crushing 2 million tons of cane per season under average Brazilian conditions.[1]

Table 28.1 Jobs in agriculture operations

Activities	Harvesting alternatives (no. of jobs)			
	Manual	Mechanical		
	Burned	Burned	Green	Green/Bales[a]
Manual operations	1,179	77	49	49
Mechanized operations	196	152	161	189
Transport	188	90	90	102
Total	1,562	319	300	339

[a] 50% of straw recovered by baling.
Source: Data from Chapter 25.

Table 28.2 Average wages in agricultural and industrial areas

Wage source	Harvesting alternatives (US$ month^{-1})[a]			
	Manual	Mechanical		
	Burned	Burned	Green	Green/Bales[b]
Agriculture	323	340	351	360
Industry	434	434	434	434
Average	348	395	400	402

[a] Exchange rate 1 US$ = 2.30 BRL (July 2014).
[b] 50% of straw recovered by baling.
Source: Data from Chapter 25.

Table 28.3 Sugarcane production costs under different harvesting alternatives

Production costs	Harvesting alternatives (US$/ha)[a]			
	Manual	Mechanical		
	Burned	Burned	Green	Green/Bales[b]
Agricultural operations	811	703	785	876
Inputs	412	434	497	572
Transport	360	248	249	275
Total	2,139	1,938	2,084	2,276
Stalks cost (US$/t)	27.57	25.50	27.41	27.41
Straw (US$/t$_{db}$)	–	–	–	36.66

t$_{db}$: t dry matter delivered at the mill.
[a] Exchange rate 1 US$ = 2.30 BRL (July 2014).
[b] 50% of straw recovered by baling.
Source: Data from Chapter 25.

Table 28.4 Industrial and economic performance

Factory performance and economic data	Harvesting alternatives			
	Manual	Mechanical		
	Burned	Burned	Green	Green/Bales[b]
Ethanol yield (Ltc^{-1})[a]	84.9	84.9	84.9	84.9
Surplus electricity (kWhtc^{-1})[a]	92	92	92	195
CAPEX (capital expenditure) (US$ mi)[c]	188.90	188.90	188.90	200.82
IRR (internal rate of return) (% year^{-1})	13.25	14.43	13.37	14.33
NPV (net present value) (US$ mi)[d]	17.8	35.5	19.6	36.2
Ethanol cost (US$ L^{-1})[e]	0.50	0.47	0.49	0.48

[a] tc = tons of cane.
[b] 50% of straw recovered by baling.
[c] Total investment in the industrial plant with an exchange rate 1 US$ = 2.3 BRL (July 2014).
[d] Considering a 12% minimum acceptable rate of return per year.
[e] Total ethanol production costs considering both operating and capital costs.
Source: Data from Chapter 25.

Table 28.4 presents electricity and ethanol yields, as well as some economic indicators.

Tables 28.1 and 28.2 indicate that under Brazilian conditions, mechanization affects jobs opportunities and average wages. On the other hand, Tables 28.3 and 28.4 show that sugarcane production costs do not differ largely between manual (burned) and mechanical green cane harvesting alternatives. There is, however, a significant difference between manual and mechanical harvest for burned cane, in favor of mechanical harvesting. Nevertheless, green cane harvesting allows straw collection to extend surplus electricity production from 92 to 195 kWh tc^{-1} with significant economic gains (Table 28.4).

The main drivers for mechanization of cane harvesting in most countries, like Brazil, are environmental laws prohibiting cane burning and/or labor shortage. On the other side, manual harvesting creates many more jobs than the mechanical case though with lower average wages. For developing countries and poorer regions, the burned cane manual alternative may be better suited due to social reasons. Manual green cane harvesting is possible, but not desirable because of the adverse working conditions for the cutter and the much higher costs. Manual burned cane, therefore, is likely to be the most suited option for smallholder farmers, where the logistics of mechanically harvesting cane, in small and disperse properties, becomes economically challenging.

Scale

In general, large scale[2] favors the use of more modern and efficient technologies (fluidized bed boiler for bagasse firing, reaction steam turbines, higher steam pressure and temperature, regenerative steam cycle, electric drives in the mills, fully automated processes, to name a few), with impacts on ethanol and sugar production costs. On the agricultural side, however, large scale implies large cane fields with detrimental impacts over logistics, as cane and production inputs need to travel longer distances.

As shown in Chapter 26, agricultural investment varies from US$ 17.5 t^{-1} $year^{-1}$ to US$ 19.5 t^{-1} $year^{-1}$ for a mill of 1 million tons per year to 6 million tons per year, respectively. At the same mill size range, industrial investment varies from US$ 85 t^{-1} $year^{-1}$ to US$ 62 t^{-1} $year^{-1}$. The balance between agricultural and industrial investments encourages mills to grow larger. Figure 28.1

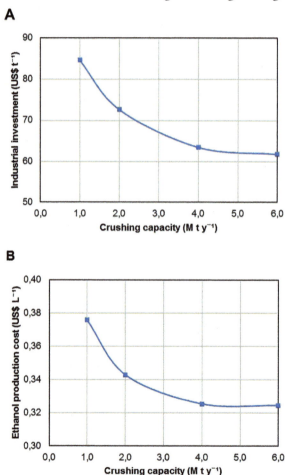

Figure 28.1 Impacts of plant size on industrial investments and ethanol production costs. Source: Data from Chapter 26.

shows the compounded effect of the agriculture and industrial sectors. Below 2 million tons of cane per season, ethanol production costs increase at a much higher rate. Therefore, in Brazil, 1 million tons is the minimum threshold for economic viability; below this level, investment costs (US$ t^{-1} $year^{-1}$) and ethanol production costs significantly hamper business competitiveness. Economies of scale lose momentum above 4 million tons per year capacity, thus discouraging the industry to go beyond such level due to the overwhelming increase in the complexity of harvesting logistics and availability of investment capital.

There are claims that microdistilleries (i.e. processing capacity ≤15,000 tons per season) are the sustainable model for sugarcane ethanol production. The reasons account for the easy inclusion of smallholder farmers, simpler and lower transportation costs, less expensive process technologies and food/fuel integration (e.g. cattle feed production) (Nogueira, 2006). Yet microdistilleries have not played an important role in ethanol production worldwide. High production costs derived from lower yield, industrial inefficiencies and the inability to meet ethanol fuel quality specifications are among the main limitations (Nogueira, 2006). Kubota et al. (2017) showed that other products of sugarcane, such as brown sugar and rum, produced in association with or without the production of fuel ethanol, are key to improve economic feasibility (for more details, see Chapter 26).

Smallholder inclusion

There are authors that consider small-scale sugarcane production a threat to the competitiveness of the sugar/ethanol industry (Chudasama, 2016). Key to such claim is the overall low productivity of agricultural systems under smallholder farmers. Collier and Dercon (2014) point out that to increase smallholder income, there is a necessity to increase labor productivity, which means more produce per labor hour. Although contentious, the inclusion of smallholder farmers in the biofuel production chain is a great opportunity to promote rural development (FAO/PISCES, 2009; Von Maltitz and Stafford, 2011) in areas where poverty and hunger prevails. There are several issues that complicate this inclusion: traditional land tenure systems, low education level, low access to agricultural technology, unskilled labor, poor infrastructure and political instability. In many regions where smallholder farms contribute to sugarcane production, such as India, China, Thailand, South Africa and Mexico, government intervention is necessary to sustain economic viability.

In SSA, a number of sugarcane projects aim at the engagement of smallholder farmers. Two well-known initiatives are the Bagamoyo Sugar project in Tanzania and the Addax Bioenergy project in Sierra Leone.

Agro EcoEnergy Tanzania Limited proposed the Bagamoyo Sugar Project in a joint venture with the government of Tanzania. The initiative foresaw an investment of US$ 569 million to produce sugarcane in 7,800 ha under the mill management and an additional 3,300 ha from independent outgrowers (AfDB, 2015). The production system accounts for the most advanced and modern agricultural methods: irrigation, genetics and mechanization. And high cane yields levels, i.e. 90–110 tons per hectare, were estimated. Bagamoyo would process 1 million tons of cane per year with a production of 125,000 tons of sugar, 8 million liters of ethanol and 100,000 MWh of surplus electricity. Unfortunately, the lack of agreement between government, farmers and company put an end to Bagamoyo project. Land tenure and farmers' relocation were the most contentious issues.

The Addax Bioenergy Project had similar aims of Bagamoyo. Addax and Onyx Group, a Swiss-based energy corporation, selected Makeni, Central Sierra Leone, to implement the project with funds from European Development Finance Institutions and the African Development Bank (Addax, 2013; BEFSCI, undated). To meet EU sustainability criteria, Addax implemented best practices for sugar and biofuel production as defined by the Roundtable on Sustainable Biofuels (RSB) and Better Sugarcane Initiative (BSI) (RSB, 2013).

The project covered 14,300 ha of land including 10,100 ha of irrigated sugarcane-producing estates, areas required for infrastructure (roads, power lines, industrial facilities, irrigation and resettlement areas) and 2,000 ha for the Farmer Development Plan (FDP). FDP accounted for 60 community fields established by Addax for local communities to produce staple food. Ecological corridors and buffer areas received the allocation of 1,800 ha. The sugarcane ethanol distillery design allowed the production of 90 ML of ethanol per year and 32 MW of electricity, of which 15 MW would be injected into the national grid and an investment of €267 million (Addax, 2013).

Undermined by land tenure disputes with local communities, agricultural and industrial challenges to maintain competitive production of bioenergy and the recent Ebola outbreak in 2015, Addax halted its activities leaving behind the promise to promote food and energy security through job opportunities and development services in Sierra Leone (Anane and Abiwu, 2011; Peyton, 2017).

Addax and Bagamoyo are the cases that demonstrate the challenges in developing large agroindustrial projects in SSA due to, mainly, land tenure issues, unskilled labor and limited access to technology. There are, however, successful cases for the engagement of smallholder farmers, such as the Xinavane mill in Mozambique by Tongaat Hewlett, the South African group that controls the mill. In Xinavane, independent growers and smallholder associations respond for up to 25% of the processed cane, with plans to reach 40% in some 5,000 hectares (Leite et al., 2016). The mill provides irrigation water, training, extension and some agricultural operations like harvesting (paid by the smallholders) and, in some cases, irrigated plots for food production. Farmers also have access to infrastructure (roads, power grid, and potable water) and social services such as school, clinic and cultural activities.

Illovo, another South African group, is developing a similar project in the KwaZulu-Natal province in South Africa that fosters sugarcane production in 3,000 ha under smallholder farmers. The group is also redistributing 28,086 ha of its own land to more than 50 black farmers, as part of the land reform program (Sugar Industry, 2017). The main difference between these two latter cases to Bagamoyo and Addax is that it accounts for reforms in the production model of existing companies with ample expertise in sugarcane production and processing.

Starting sugarcane ethanol production

There are two main alternatives to kick off sugarcane ethanol production: (1) introduce sugarcane and build a new processing plant (greenfield distillery) or (2) in countries that already produce sugar, start annexing an ethanol distillery to the sugar factory (brownfield distillery) and produce ethanol from final molasses. In this latter alternative, ethanol production is possible without compromising sugar output. Further increase in ethanol production comes at the expense of either sugar production or planting more cane and retrofitting the sugar/ethanol-producing plant to increase the crushing capacity.

In the greenfield alternative, cane production should increase in steps to avoid having all cane at the same age with yields going down continuously. The first step is to install the cane nursery to produce seeds of the selected cane cultivars under strict phytosanitary management and then propagate in areas close to the future mill in a process that takes normally three to four years. The construction of the processing plant should follow the sugarcane-increasing production. When done, the industrial plant can produce only ethanol and electricity or sugar and electricity (e.g. Addax) or a mix of ethanol, sugar and electricity. Bhardwaj et al. (2007) present a case for the installation of a sugar greenfield project in Tanzania and provide valuable details to a successful implementation.

For the brownfield option, which we consider more adequate in terms of investments and risks for those developing counties with a solid sugar industry and experience in sugarcane cropping (selected varieties, good management practice in place, trained personnel), the ethanol production can increase progressively. Leal et al. (2016) present a case for the introduction of ethanol starting by annexing a molasses distillery to the sugar factory and increasing ethanol production at the expenses of sugar output through the expansion of the distillery capacity. There is also scope for energy surpluses (i.e. electricity) enabled by the installation of modern retrofit systems. Table 28.5 presents a range of investments from the baseline (only sugar) to greenfield (ethanol + electricity).

Results in Table 28.5 show that ethanol can be produced at significant scale by retrofitting an existing sugar mill, with modest investment costs compared to a new greenfield distillery of the same cane crushing capacity. Upgrades to increase power generation also increase costs, which may be justifiable depending on the power market.

Table 28.5 Sugarcane mill investment under different scenarios

Mill outputs	Baseline	Step 1	Step 2	Step 3	Greenfield
Sugar (t year^{-1})	200,000	200,000	100,000	200,000	0
Ethanol (ML year^{-1})	0	16	67	16	136
Surplus electricity (kWh tc^{-1})	0	0	0	90	296
Investment (MUS$)	0	4.4	22.3	54.9	141.4

Note: Exchange rate considered 1 US$ = 3.22 BRL. Baseline: A sugar mill crushing 1.6 Mt year^{-1} and producing sugar and molasses. Step 1: Install an annex ethanol distillery to process all final molasses to ethanol. Step 2: Increase the distillery capacity to process the molasses of the sugar factory (from 50% of cane juice) and 50% of the cane juice. Step 3: Retrofit the energy sector (install high-pressure boiler and turbine generator to generate surplus electricity for sale. Greenfield processing capacity 1.6 Mt year^{-1} to produce ethanol and surplus electricity.
Source: Leal et al. (2016).

Sugarcane ethanol production model: recommendations for developing countries

In the following paragraphs, we suggest some basic benchmarks as to the design of appropriate sugarcane production models in developing countries, particularly in SSA.

Plant scale: 1–2 million tons of cane (Mtc) per year is compatible with the African mills sizes. Although there are economic benefits of larger scales (i.e. above 2 Mt), it remains restricted to regions where sugarcane, both agricultural and industrial sectors, is well developed such as Brazil and Thailand. On the other hand, small-scale mills, i.e. below 1 Mtc/year, may become poorly competitive at the national and international markets.

Sugarcane processing: modern technology and mixed production of sugar, ethanol and electricity and, possibly, other coproducts to improve competitiveness and reduce business risks.

Sugarcane production: integrated system with the engagement of outgrowers, particularly smallholder farmers, that seem to be key in promoting rural development and food security, and improving project acceptability. Land tenure is the most sensitive and delicate issue, followed by farmers relocation, extension services, access to irrigation and cane payment system.

Mill location: areas with adequate soil and climate for rainfed sugarcane production, suitable for mechanization and sparsely populated to minimize the displacement of local communities. Water availability for irrigation, where necessary, is an important asset.

Infrastructure: roads and energy (electricity and clean cooking fuels) access, as well as social responsibility projects including health clinics, schools, potable water and sanitation infrastructure. The cost of implementing these projects should be shared between the enterprise and the government in a negotiated way.

Last, but not least, adequate and timely public policies are a main driver for sustainable biofuels production, with implication to the type, volume (scale size) and sustainability criteria. Public policies are also necessary to create an enabling environment for investors through the setting of medium- to long-term conditions for biofuel programs, such as legal frameworks, price structure and market size.

Concluding remarks

Optimizing the production model for sugarcane ethanol is not a trivial task. It accounts for a complex balance of sustainability indicators and, sometimes, conflicting stakeholder priorities. UN Sustainable Development Goals and sustainability certification systems can provide valuable guidance to business, extension agents and farmers, as well as assurance to national and international investors.

In this chapter, we explored nuances in the selection of the production model that can be useful for brand new initiatives (i.e. greenfield projects) or ongoing projects (i.e. brownfield projects). The greatest challenge is to balance economic competitiveness with the engagement of smallholder farmers and their inherited low-productivity agricultural systems.

Despite failures, such as Bagamoyo Sugar and Addax Bioenergy projects, there seems to be cases where the association of smallholder farmer is working better (e.g. Mozambique and South Africa). These experiences deserve attention and lessons carefully assessed to serve as indication of the main causes for success and failure. No project development should be started without careful evaluation of the lessons learned elsewhere; it will save time, money and headaches.

Notes

1 80 tons of cane per hectare, five harvests per cane production cycle, 25 km average cane transport distance and 12 and 20 tons of seed cane per hectare for semi-manual and mechanical planting, respectively.
2 ≥2 million tons of processed cane per year under Brazilian conditions.

References

Addax, 2013. *Addax Bioenergy sugarcane ethanol project in Makeni, Sierra Leone.* Freetown: Addax Bioenergy SA.

AfDB, 2015. *Bagamoyo integrated rural infrastructure development sugar project (BIRIDESP).* Available at: www.afdb.org/en/projects-and-operations/project-portfolio/project/p-tz-aag-004/. Accessed in March 2015.

Ambali A, Chiwa PW, Owen C, van Zyl WH, 2011. A review of sustainable development of bioenergy in Africa: An outlook for the future bioenergy industry. *Scientific Research and Essays*, 6(8), 1697–1708.

Anane M and Abiwu CY, 2011. Independent study report of the Addax bioenergy sugarcane-to-ethanol project in the Makeni region in Sierra Leone. Sierra Leone

Network on the Right to Food (SiLNoRF), Bread for All, Switzerland, Bread for the World and Evangelischer Entwicklungsdienst (EED), Germany.

BEFSCI, undated. Bioenergy and Food Security Criteria and Indicators: Addax Bioenergy. Food and Agriculture Organization of the United nations (FAO). Rome.

Bhardwaj V, Alloo EY, Alloo IY, 2007. Setting up of a greenfield sugar Project in Africa A practical approach. *Proceedings of International Society of Sugar Cane Technologists*, 26. 1759–1766.

Chudasama A, 2016. Will the inefficiency of small-scale cane growers continue to mar industry? *International Sugar Journal*, 115(1376), 2013 ed. 1.

Collier P and Dercon S, 2014. African agriculture in 50 years: Smallholder in a rapidly changing world? *World Development*, 63, 92–101.

FAO/PISCES, 2009. Small-scale bioenergy initiatives brief description and preliminary lessons from studies in Asia, Latin America and Africa. Food and Agriculture Organization of the United Nations (FAO) and Policy Innovation Systems for Clean Energy Security (PISCES). Final Report.

Gasparatos A, von Maltitz GP, Johnson FX, Lee L, Mathai M Oliveira JAP, Willis KJ, 2015. Biofuels in sub-Saharan Africa: Drivers, impacts and priority policy areas. *Renewable and Sustainable Energy Reviews*.

Kubota AM, Leite JGDB, Watanabe M, Cavalett O, Leal MRLV, Cortez LAB, 2017. The role of small-scale biofuel production in Brazil Lessons for the Developing Countries. *Agriculture*, 7, 61. doi:10.3390 agriculture7070061.

Leal MRLV, Leite JGDB, Chagas MF, da Maia R, Cortez, LAB, 2016. Feasibility assessment of converting sugar mills to bioenergy production in Africa. *Agriculture*, 6, 45. doi:10.3390;agriculture6030045.

Leite JGDB, Leal MRLV, Langa FM, 2016. Sugarcane outgrower schemes in Mozambique findings from the field. *Proceedings of International Society of Sugar Cane Technologists*, 29, 434–440.

Nogueira LAH, 2006. Ethanol as Fuel in Brazil; Technical Report. Banco do Povo—Projeto GAIA, Belo Horizonte, Brazil.

Peyton N, 2017 Africa land projects promise much, leave locals hungry: report. REUTERS. Available from: www.reuters.com/article/us-landrights-africa/africa-land-projects-promise-much-leave-locals-hungry-report-idUSKBN1D82LV. Accessed in 16 March 2018.

RISE, 2016. *Regulatory Indicators for Sustainable Energy – A Global Scorecard for Policy makers*. International bank for the Reconstruction and Development; The World Bank. Washington, DC.

RSB, 2013. Addax bioenergy earns first African certification by the RSB. Available at: www.addaxbioenergy.com/uploads/PDF/February_%202013_Addax_Bioenergy_RSB_Certification_Press_Release_FINAL.pdf. Accessed in June 2015.

Sugar Industry, 2017. Illovo secures Treasury support to develop 3000 ha in KZ-Natal Sugar Economics and Business. *Sugar Industry* 142(42), 678.

Von Maltitz G and Stafford W, 2011. *Assessing opportunities and constraints for biofuel development in sub-Saharan Africa*. Working Paper 58, CIFOR, Bogor, Indonesia. 66p (available at www.cifor.cgiar.org).

29 Alcohol and LACAf project in Colombia

José Maria Rincón Martínez, Jessica A. Agressot Ramirez and Diana M. Durán Hernández

Introduction

Colombia is located in the northwestern corner of South America, and it has a surface of 1.1 million km^2. Its climate conditions go from warm tropical at sea level to the perpetual snows at 4,000 meters at the sea level (msl). Moreover, most of the country has a tropical and subtropical weather, which is suitable for the growing of sugarcane.

Sugarcane was brought to Colombia by Pedro de Heredia in the year 1538, who introduced it to the city of Cartagena. In addition, it arrived to the city of Cali with Sebastián de Belalcázar, the founder of this city, in the year 1540. From that moment, this crop was spread through the basin of the Cauca River valley (Cenicaña, n.d.). Also, with the first crops, it was found that in the Valle del Cauca the sugarcane can be produced throughout the whole year, without limiting it to harvest periods. On the other hand, the first sugar mill "Ingenio Manuelita" was inaugurated at the beginning of the 20th century, and by 1930, there were two other sugar mills: "Providencia and Riopaila"; since those years, the sugar industry has expanded throughout the region.

The LACAf project is carried out in order to diagnose the country's energy and food situation; determine the physical potential for the production of ethanol based on sugarcane and verify that an expansion of cane crops can be carried out; and determine which are the current production models and which could be modified in the future without affecting food security by determining social factors and economic development.

Bioalcohol in Colombia

Since the Spanish colonial times, alcohol was processed from sugarcane honey by fermentation either for the production of beverages or as an antiseptic for medicinal use. The first distilleries produced alcohol of 95% purity, and its main purpose was the production of beverage known in Spanish as "aguardiente" with varying alcohol concentration according to the distillation cut; this alcohol is also used in the pharmaceutical and cosmetic industry.

During the 1990s, Bogotá, the capital city of Colombia, which is located at 2600 m above sea level, had a huge problem due to its contaminated atmosphere; the Japan International Cooperation Agency (JICA) conducted a study on the "Air pollution control plan in the area of the city of Bogotá", revealing that gas emissions from motor vehicles were responsible for 96% of the pollution produced during that period of time, the air had high concentrations of SOx, NOx, CO and unburned hydrocarbons (HCs) emissions. This study recommended the use of oxygenated fuels such as MIBK (methyl isobutyl ketone), which were used at that time in the United States, Japan and Canada,

Due to the problems mentioned above, Rincon and collaborators (Rincón, Romero, Camacho, & Montenegro, 1996) conducted a study at the height of Bogotá using a mixture of oxygenated products, in different proportions, and gasoline; they found that the use of a mixture of anhydrous ethanol with gasoline was an option to reduce emissions of monoxide and HCs. Figures 29.1 and 29.2 correspond to the variations of carbon monoxide and carbon dioxide, respectively, found in the mentioned study.

The results of the tests showed that the greater the proportion of alcohol in the gasoline, the lesser the emissions of carbon monoxide and HCs, without affecting the performance of the engine at the altitude of Bogotá; the work concludes that the addition of anhydrous alcohol as oxygenates to gasoline is an effective way to improve the quality of emissions from the exhaust in automotive. Prior to this work, other researchers had already proposed the

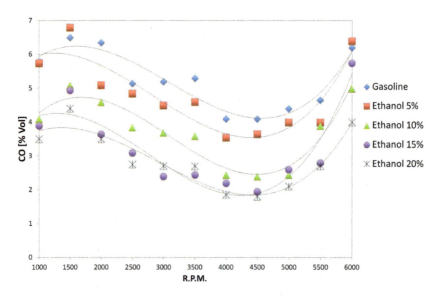

Figure 29.1 Variation of carbon monoxide emissions by adding ethanol to gasoline: speed characteristic test.

Source: Rincón, Romero, Camacho, & Montenegro (1996).

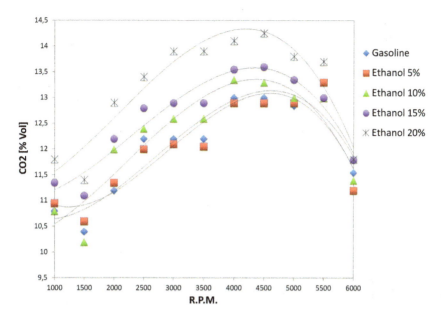

Figure 29.2 Variation of carbon dioxide emissions by adding ethanol to gasoline: speed characteristic test.
Source: Rincón, Romero, Camacho, & Montenegro (1996).

production of anhydrous alcohol (Araujo & Torres, 1986) for the production of alcohol fuels in Colombia.

Later, in 2001, it was established by the Law 693 in Colombia, according to which vehicular the gasoline used in the urban centers of the country with more than 500,000 inhabitants should contain oxygenated components such as fuel alcohols, in the quantity and quality established by the Ministry of Mines and Energy, and in accordance with the regulations on the control of emissions and the environmental sanitation requirements established by the Ministry of the Environment for each region of the country (Ministerio de ambiente, vivienda y desarrollo terriorial, 2003).

As a consequence of this law, since the end of the year 2005, the sugar mills such as Mayagüez, Providencia, Incauca, Risaralda and Manuelita established distilleries to produce anhydrous ethanol. Anhydrous ethanol or fuel alcohol is produced from sugarcane in five different stages: adequacy of the raw material, fermentation, distillation, dehydration and concentration of vinasse. Figure 29.3 shows the actual process of obtaining anhydrous ethanol in Colombia. The main difference of the Colombian production process is the use of vinasse residues in composting with other residues of the sugar mill such as boiler ash and solid residual of the clarification process. Moreover, the compost product is used as a biofertilizer of the sugarcane avoiding the use of half of the chemical fertilizers.

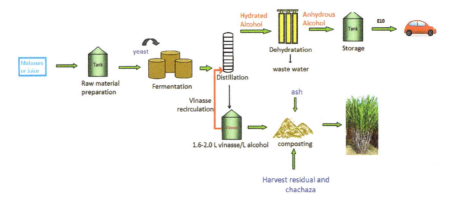

Figure 29.3 Anhydrous alcohol obtaining process.

In 2011, the distilleries were expanded, bringing the installed capacity to 1,250,000 liters per day. In August 2015, a new distillery belonging to the Riopaila Castilla sugar mill began operating with an installed capacity of 400,000 liters per day and increased the installed capacity of anhydrous bioethanol production to 1,650,000 liters per day (Asocaña, 2017a). Currently, Colombia is the third ethanol producer in Latin America, after Brazil and Argentina.

Bioethanol in Colombia has become a source of renewable and sustainable energy. According to this, the production of alcohol in the dual system sugar–alcohol becomes strategic to the country: generating formal employment, benefiting the environment, enabling it to reduce its energy dependency and maintaining reserves of nonrenewable energy sources, such as gasoline or other fossil fuels derived from petroleum.

The Sugar Sector generates more than 188,000 direct and indirect jobs. According to a study of life cycle analysis conducted by the Ministry of Mines and Energy, Colombian bioethanol production observes the standards for human rights, labor rights, social and environmental impact; there are no tenure problems, and there is no negative effect on food security (Asocaña, 2017).

The Colombian Sugar Sector operates under the principle of prioritizing demand and internal sugar consumption over sugar exports. This way, in 2016, 459,000 tons of sugar was used for alcohol production and was no longer exported, without affecting the national market sugar consumption. In order to increase the percentage of bioethanol mixture in gasoline within the national market and develop a bioethanol export industry, the current availability of land should be better exploited (Asocaña, 2017).

There is only one example of sugar crops used exclusively for the production of bioethanol, that is, the company BIOENERGY of Ecopetrol, which lies in the Colombian Eastern Plains, where land has been adapted in which no agricultural activities were carried out, in such a way that the growth of raw materials for the production of bioethanol will not diminish the national production of available food. Bioenergy has produced roughly/around 15 million liters of ethanol, for which it has ground around 262,000 tons of sugarcane. In parallel, it has generated 31,440 MWh of energy, out of which 18,542 MWH has been sold to the National Electricity Network and the remaining has been supplied to the operation of the plant (Bioenergy, 2017).

Cogeneration with sugarcane bagasse – as bioalcohol and sugar by-product

The country's geographic and climatic characteristics help generate hydroelectric energy, as the main source of primary generation thanks to constant rains and conditions for water storage. This high hydroelectric storage capacity makes the energy system vulnerable during long dry periods as it happened, in 2015, when there was an increase in the intensity of the ENSO (NIÑO) phenomenon that caused a significant decrease in the water supply and prompted the rise of electricity rates due to the higher cost of the thermal energy source (Rincón, Vera, Guevara, & Duarte, 2017). In these circumstances, cogeneration and anhydrous alcohol productions becomes important.

Cogeneration of energy is a process in which electrical and thermal energies are simultaneously produced and used. According to Asocaña (2017),

> As it was identified by national and international studies, the sugar sector has the greatest potential to perform cogeneration in Colombia due to the availability of biomass, especially bagasse, a by-product derived from cane harvesting and milling processes, which is the primary source of energy for cogeneration.

Since its inception, the sugar mills have used sugarcane bagasse as fuel to power their boilers and use steam as energy for the execution of their processes. At the present time, cogeneration uses high-pressure steam to generate electrical energy through turbogenerators and exhaust gases that are used for the heat supply for the productive process. In Colombia, part of the bagasse is also used in the paper industry as a source of fiber, hence some amount of coal is used in cofiring as a substitute of the bagasse that is used in the paper industry. Figure 29.4 shows a diagram of the cogeneration process in a sugar mill.

In Colombia, cogeneration has been stimulated since the issuance of Law 788 of 2002. On July 16, 2008, Law 1215 came into force, which exempts cogenerates from paying the 20% contribution on energy that is generated for consumption. Subsequently, from December 2011, this exemption would be extended to all industrialists who request it. Resolution CREG 005 of February 2010 regulates cogeneration, differentiating it from other types of generation (Asocaña, 2017).

Given the importance of cogeneration for the country, especially for the summer time, and in order to obtain safe sources of energy, the Commission for Energy and Gas Regulation (CREG) established the Resolution 153 of October 31, 2013, for awarding contracts to supply Agricultural Fuel (COA),

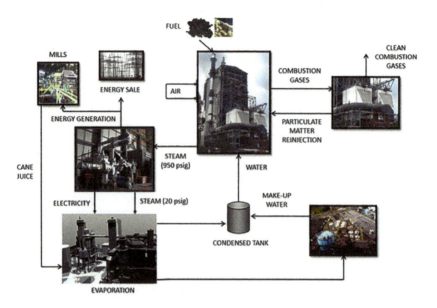

Figure 29.4 Process diagram of cogeneration in the sugar industry.
Source: Adapted from Paredes and Bermúdez (2016).

in such a way that the sugar mills can access to the reliability charge through a technical evaluation.

On May 13, 2014, the Law 1715, better known as the "Renewable Energy Law", was established. This law regulates the integration of nonconventional renewable energies into the national energy system.

Thanks to the advances in energy politics, the sugar mills have been trying to make projects for the use of cogeneration. In 2016, the installed capacity of cogeneration was 253 MW. Moreover, right now 93.6 MW of the installed capacity is sold (Asocaña, 2017). Figure 29.5 shows the installed capacity of cogeneration in the Colombian sugar mills.

Also, it is expected that with the implementation of the Law 1715, the Colombian sugar mills will have a cogeneration capacity of about 284 MW per year and that 123 MW could be sold to the main energy network.

The cogeneration gives firmness to the system since the Colombian matrix depends on about 78% of hydraulic generation. This in turn depends on the rainfall and flow of the rivers that feed the reservoirs. During the summer, the availability of water is reduced. In the case of the sugar and energy sector, summer is when the sugarcane crops have a greater aptitude for harvesting, so there is a greater supply of cane in the mills and therefore there is an abundance of fuel for cogeneration.

Figure 29.6 shows the complementarity between the production of sugarcane and rainfall per month.

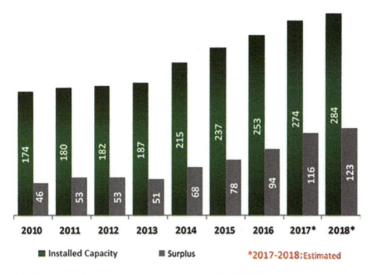

Figure 29.5 Cogeneration capacity in sugarcane mills (MW).
Source: Asocaña (2017).

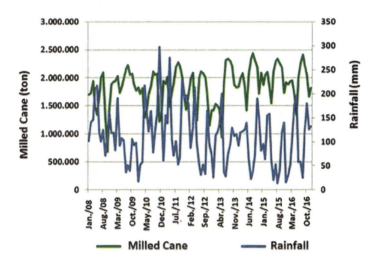

Figure 29.6 Complementarity between the production of sugarcane and rainfall.
Source: Asocaña (2017).

The worldwide importance of energy through cogeneration and the linked projects to carry it out, have experienced an important development. Among the causes of this change, can be considered, the rise of fuel prices, the progress of the technology in renewable energies and policies to reduce greenhouse gas (GHG) emissions that are reflected in the goals of the different countries agreed in COP 21 and COP 23.

Potential for bioethanol expansion in Colombia

With the current bioalcohol production, including that used for bioenergy generation, it is possible to make bioalcohol mixtures E10. However, due to the Colombian policies to increase the percentage of bioalcohol used in gasoline and to export it, it is important to find new ways for the development of sugarcane crops, in such a way that they can be used in the production of bioalcohol as a fuel for domestic consumption and export.

These kinds of projects seek the expansion of the agricultural frontier, optimizing the use of the available land without compromising the food security in Colombia. In fact, a recent study by FAO (United Nations Organization for Food and Agriculture) concluded, for Colombia, that

> In general terms, so far there is no conflict between the development of biofuels and food security since to meet the domestic demand for biofuels, only raw sugar exports have been substituted for the production of bioethanol. The analysis of the behavior of sugar availability and per capita consumption confirms that the effects of the production of biofuels on food consumption have been imperceptible and that, in effect, domestic supply has been protected from direct human consumption at low cost.

Estimations were made by the Instituto Geográfico Agustín Codazzi (IGAC) in 2012, and Minminas (Ministerio de Minas y energía – CUE, 2012) show that there are more than 34 million underutilized hectares for extensive livestock with less than 0.5 animals per hectare; the majority of this areas are suitable for the production of different crops, which tells us that there is a big portion of land that can be used to grow appropriate crops for biofuels and still maintain the food security. The main goal of the CUE study was "to provide a first filter of potentially suitable areas for sugarcane and oil palm nationwide. The ability to grow raw materials for biofuels is evaluated using a set of biophysical, legal, environmental and socioeconomic variables, addressing key sustainability issues" (Ministerio de Minas y energía – CUE, 2012).

In the case of sugarcane, the study first evaluated climatic and biophysical conditions in order to identify the areas in which sugarcane can be grown, and subsequently areas with prior legal restriction, such as national parks or Indian reservations, were excluded. Afterward, it identified the areas with a high impact on biodiversity, water scarcity and GHG emissions, and then considered socioeconomic criteria (based mainly on existing research). Finally, they obtained aptitude maps in all possible categories, which were the key to discuss the sustainability of the potential areas.

Most of the soils in the tropical and subtropical zones of Colombia are considered to be moderately suitable for sugarcane crops. Nevertheless, the alluvial plains in the Andean valleys of northern Colombia are considered more fertile and therefore more suitable for cultivation.

Additionally, current GHG studies show the importance of considering changes in the use of land with respect to the environmental performance of agricultural biofuels. According to Fargione et al. (2008), the change in land use caused by the production of biofuels can lead to a "carbon debt" due to the release of large amounts of CO_2, which were stored in the soil and above it. Therefore, the CUE study also evaluated the "carbon debt" for potential sugarcane plantations (Ministerio de Minas y energía – CUE, 2012).

Finally, the study showed that the possible areas of sustainable expansion are reduced to the northern plains, some areas in the Andean valleys and the eastern non-forested area as shown in Figure 29.7.

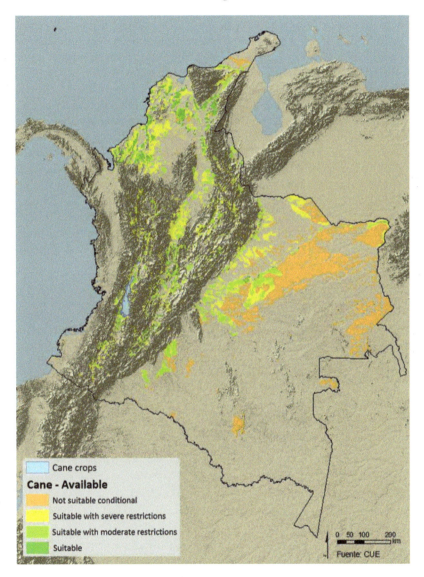

Figure 29.7 Sugarcane suitability.

Note: Sugarcane suitability excludes biophysically unsuitable and protected areas, areas with less than 40% GHG savings, high priority areas for biodiversity, areas currently used for agriculture and access limited areas (orange areas).

Currently about 40,000 ha of sugarcane crops are dedicated to the production of ethanol, and there is a potential for an expansion of up to 4,918,000 ha, of which 1,518,000 are highly suitable and 3,400,000 are moderately suitable (Ministerio de Minas y energía – CUE, 2012).

Most of the moderately suitable hectares are located in the eastern of the Andes mountain range (the Piedemonte) in the Meta department and partially in the Caquetá department. In addition, the majority of the hectares with high potential for growing sugarcane were found in the north, more specifically in the departments of Cesar, Córdoba and Magdalena. The study also showed that the areas in the Vichada department proved to be moderately suitable for the cultivation of raw materials for biofuels, if the transportation infrastructure is significantly improved (Ministerio de Minas y energía – CUE, 2012).

The results of the CUE study provide a scientific knowledge base for strategic planning (i.e., sustainable land use planning) and could draw attention to sustainable investment in biofuels. However, more detailed analysis is required based on higher resolution maps, which are frequently updated and refine the set of criteria, in order to plan specific biofuel projects (Ministerio de Minas y energía – CUE, 2012).

Based on the study, we can finally recommend that the northern areas of the country can be preferentially used for the production of bioethanol for export, given its proximity to the coast and good transportation infrastructure; while the zones of the center and east of the country near the consumption sites (Tolima, Huila, Meta, Casanare and Vichada) are used in alcohol production for internal consumption.

Alcohol perspectives

A comparison of the production of sugarcane, sugar and alcohol between Brazil and Colombia shows that although Brazil leads in terms of production, Colombia has higher productivity. Table 29.1 shows some comparative data between the two countries.

Table 29.1 Comparison between Colombia and Brazil

	Brazil[a]	*Colombia*[b]
Sugar production (tons)	38,724,993	2,271,926
Sugarcane production (tons)	657,572,586	21,598,667
Productivity average sugar (ton/ha)[c]	10.7	15.5
Machining index (%)	84.80	51
Electricity generation	32.3 TWh	1,381 GWh

[a] Data extracted from UDOP (2017).
[b] Data extracted from Asocaña (2017).
[c] Data extracted from Asocaña (2016.

Additionally and due to some financial crisis such as the 1973 one, the main purpose for the Brazilian bioalcohol production has been its use as a gasoline substitute, Proálcool program (Regis, 2008), whereas the Colombian alcohol production was aimed for the bioalcohol to be used as an additive, solving environmental concerns about pollution caused by combustion engines.

In Brazil since 1931, the addition and mixture of a 5% of alcohol in imported gasoline led to the development of technology production from then on. Brazilian cars work with flexi-fuel engines, capable of using anhydrous alcohol and hydrated alcohol without any trouble; in the Colombian case, E8 blend is already used to solve the pollution problem. In both cases, the goals had been achieved.

On the other hand, recently the European Commission argues that by 2020, there can be only a 7% of first-generation or conventional biofuels (bioalcohol and biodiesel) on the fuel consumption of the European vehicles (EU-biofuels policy, 2018), and that by 2030, this percentage has to fall to 3.8%, according to the energy plan proposal presented by the EU Commission. In its plan, Brussels maintains in its plan – known as "winter package" – that a "cap" must be imposed to "minimize the indirect impacts of land use change", which contributes to climate change by transforming forest masses – which they retain CO_2 – in crops (bioenergyinternational, 2018).

On the other hand, the entire world has a special interest in climate change as it is proved by the creation of new regulations against polluting emissions. In countries like France and the United Kingdom there are plans to ban sales of internal combustion engine vehicles from 2040 onward. Members of German government have just sent a resolution to forbid, by the year 2030, sales of internal combustion engines across the European Union and only allow in the market zero emissions vehicles, according to the resolution.

The advances in battery technologies (Graphene-info, 2017) and energy regeneration systems made it possible that in the year 2017, large companies produced a significant number of electric vehicles, making this year the starting point of a new energy era, where electric cars are the future all over the world (Rincón, 2017).

The foregoing shows that we are close to having commercial electric vehicles with better characteristics than the current fossil fuel competitor. So the questions posed are as follows: how can sugarcane bioenergy take part in the new zero emission policy and take a role in the electric vehicles development? Do we need a new route for the bioenergy use from sugarcane?

Both Colombia and Brazil have a great capacity to double the production of sugarcane and consequently the production of alcohol and sugar; for this reason, it must be evaluated whether this alcohol generated in excess can be useful for the sector or whether, on the contrary, an alternative should be found for the use of the agro-industrial waste generated during the production of sugar.

Therefore, one of the viable suggestions for the future would be to switch from alcohol production to biogas production, as a source of energy for the bioelectricity and biomethane production. Biogas is produced from the anaerobic digestion of biomass waste, like molasses and crops; it contains about

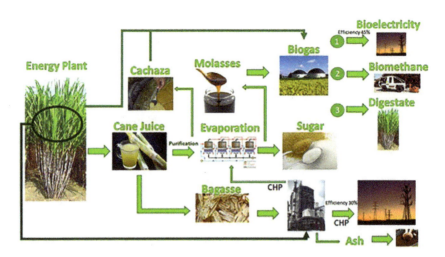

Figure 29.8 Future processing of sugarcane.
Source: Ministerio de Minas y Energía – CUE (2012).

60% CH_4 and 40% CO_2 with other impurities. Proven as 100% renewable energy, biogas is mainly used for the bioelectricity production and one of its components, biomethane, is used to replace natural gas in all its applications including the production of Fischer-Tropsch fuels.

Following this path, the future route processing of sugarcane is shown in Figure 29.8.

From this figure, the future of sugarcane as energy crops is clear, the main products will be alcohol, until there are internal combustion vehicles, harvest residues and molasses can be used to generate biogas and biomethane that will be the new biofuels. Meanwhile and given the international uncertainty about the future of mobility, due to the different interests of the developed countries, it is important to establish research and development programs to show the environmental and economic benefits of this new route proposed for the new biorefineries from sugarcane crops.

In this way, we can show that having enough bioenergy to meet a substantial fraction of future demand for energy services (≥25%) of global mobility, as proposed by the IANAS Energy program and the GSB (Global Sustainable Bioenergy) initiative, is possible.

References

Araujo, A., & Torres, C. (1986). *Investigaciones realizadas en Colombia sobre la producción de alcohol carburante*. Comité Nacional de Sucroindustria.

Asocaña. (2016). *Aspectos Generales del Sector Azucarero Colombiano 2015–2016*. Retrieved from www.asocana.org/documentos/252016-B27E6326-00FF00,000A000,878787,C3C3C3,0F0F0F,B4B4B4,FF00FF,FFFFFF,2D2D2D,A3C4B5,D2D2D2.pdf

Asocaña. (2017a). *Balance Azucarero Colombiano Asocaña 2000–2017.*
Asocaña. (2017b). *Cogeneración.* Retrieved from www.asocana.org/documentos/562017-BC7B477D-00FF00,000A000,878787,C3C3C3,0F0F0F,B4B4B4,FF00FF, 2D2D2D.pdf
Asocaña. (2017c, June). *Las cifras del sector azucarero colombiano y la produccion de bioetanol a base de caña de azúcar.* Retrieved from www.asocana.org/modules/documentos/10396.aspx
Bioenergy. (2017, October). *Más de 15 millones de litros de etanol producidos.* Retrieved from Bioenergy.com: www.bioenergy.com.co/SitePages/Noticia.aspx?IdElemento=38
Bioenergyinternational, 2018. https://bioenergyinternational.com/opinion-commentary/ecs-winter-package-an-insoluble-blend-for-transportation-biofuels
Cenicaña. (n.d.). *Centro de Investigación de la Caña de Azucar de Colombia.* Retrieved from Fechas históricas de la agroindustria de la caña en Colombia: www.cenicana.org/quienes_somos/agroindustria/historia.php
Graphene-info. (2017). *Graphene batteries: Introduction and Market News.* Retrieved from www.graphene-info.com/graphene-batteries
EU-biofuels policy- www.europarl.europa.eu/RegData/etudes/BRIE/2015/545726/EPRS_BRI(2015)545726_REV1_EN.pdf), 15-1–2018
Ministerio de ambiente, vivienda y desarrollo terriorial. (2003). *Resolución 0447 de 2003.* Retrieved from www.alcaldiabogota.gov.co/sisjur/normas/Norma1.jsp?i=15720
Ministerio de Minas y energía – CUE. (2012). *Evaluación del ciclo de vida de la cadena de producción de biocombustibles en Colombia.* Medellín.
Paredes, J., & Bermúdez, L. (2016). Aprovechamiento capacidad instalada de cogeneración de energía. *Seminario internacional de cogeneración y bioenergía.* Cali, Colombia.
Regis, M. (2008). Evolução Tecnológica da Produção de Etanol: O Passado. *60ª REUNIÃO ANUAL DA SBPC.* Campinas.
Rincón, J. M. (2017). Sustainable Energy From Sugarcane and Its Perspectives In L.A.C. *Workshop Ethanol in the Americas.* Gainesville, Florida.
Rincón, J. M., Romero, G., Camacho, A., & Montenegro, E. (1996). *Combustible de Transporte y medidas para reducir la contaminacion: I - Uso de Combustibles oxigenados a la altura de Bogotá.* Bogotá.
Rincón, J. M., Vera, M. A., Guevara, P., & Duarte, S. (2017). Cofiring in the sugar mills industry in Colombia. *VGB PowerTech.*
UDOP. (2017). *Produção Brasileira.*

30 Sugarcane ethanol in Guatemala

Aida Lorenzo, Mario Melgar and Luiz A. Horta Nogueira

Introduction

Sugar in Guatemala has been for many years the main export product, along with coffee, bananas, vegetable oils and cardamom, making this country as the fourth largest sugar exporter in the world and the third in global productivity (ASAZGUA, 2017). In fact, Guatemala has been producing sugar since the colonial times and during the last decades has improved notably its production practices, not only in the agricultural area but also in the sugar manufacturing area and diversifying the production mix towards energy and ethanol (industrial, as fuel and other uses). The weather of Guatemala is favorable for the production of sugarcane, and a relatively small area of national territory is cultivated with this crop.

Guatemala has been moving towards a renewable energy matrix in the electricity generation sector; however, the transportation sector still depends totally on imported fossil fuel. Guatemala exports more than 80% of its ethanol production to countries such as Europe and the United States because it does not have implemented ethanol blending with gasoline. The use of fuel ethanol can replace imported fuel and improve the trade balance, as well as help to meet the goals of reducing CO_2 emissions in the country.

The following paragraphs describe sugar industry history, its efficiency, the electricity generation, the current production of ethanol and the potential that the country has to make a blend of fuel ethanol in gasoline.

The sugarcane industry in Guatemala

Sugarcane began to be cultivated in Guatemala in 1536, the first Guatemalan "trapiches" (old sugar mills) were founded in the central valley of Guatemala and in the Salama Valley, during the 16th century. In the 17th century, the number of "trapiches" increased, the most important were in hands of religious orders. It was until the middle of the 19th century that Guatemala began to export sugar in small amounts.

In 1957, the Guatemalan Sugar Association, ASAZGUA was founded, and in 1960, when the total production of sugar was 68,000 metric tons, the country received its first quota from the United States. The year 1960 is taken as a starting point for the modern history of sugarcane. In the world, the industrial era was highly developed and changes in the world dynamics were foreseen; it was then that sugar mills defined their modernization and growth strategy. Sugar factories were evolved from local to exporting industries, becoming one of the most important agro-industrial activities of the country. When Guatemalan sugar exports expanded, ASAZGUA started to develop a series of projects and strategies that were the driving force of the national sugar agro-industry. In order to increase sugarcane production, the sugar mills introduced improvements in agronomy, harvest, factory, distribution and product commercialization, as well as better life conditions for the workers of the sugarcane agro-industry (Melgar et al., 2012). Currently, the sugar agro-industry of Guatemala is made up of 12 sugar mills, which are located in the coastal plain of the Pacific Ocean of Guatemala (Figure 30.1). The harvest is carried out in an average of 180 days beginning in November and ending in May.

In 2017, sugar was the second export product of Guatemala and the first among agricultural exports. In the 2016–2017 harvest, Guatemala harvested almost 25.8 million tons of sugarcane and produced 2,719 million metric tons of sugar, of which 83% was exported. Representing US$ 979 million income to the country, sugar is the second most important tradable product. Sugar agro-industry generates more than 82,000 direct jobs and 410,000 indirect jobs, which represents 5% of the total formal jobs in the country. The planted area with sugarcane in that harvest was 255,859 ha, equivalent to 3.3% of

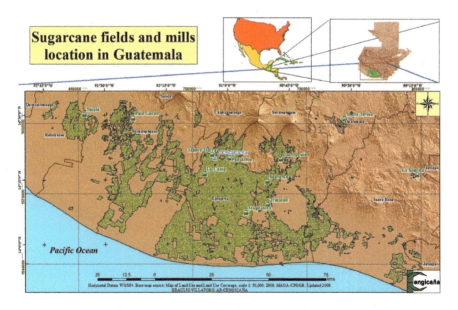

Figure 30.1 Guatemala sugarcane geographical location (Melgar et al., 2012).

whole national territory and 5.9% of agriculture area, which includes pastures and perennial and annual crops (Table 30.1).

Sugar production in Guatemala has been growing steadily since 1960 as a result of the increase in area and mainly due to the increase in productivity. In the 1950/1960 harvest, 12,534 ha was harvested and 65,163 tons of sugar was produced, with a productivity of 5.24 ton/ha. For the 2016/2017 harvest, these values reached, respectively, 255,859 ha, 2,719,231 tons of sugar production and 10.63 ton/ha. Thus, due to productivity gain, more than 100%, the area currently planted with sugarcane is about the half of that would be needed without productivity gains. The evolution of sugarcane area, sugarcane processed and sugar production in Guatemala is presented in Table 30.2, and the

Table 30.1 Land use in Guatemala in 2016 (ASAZGUA, 2017)

Land use in Guatemala	Percentage (%)
Forest (natural forest and natural plantations)	36.0
Pastures (natural pastures and bushes)	22.8
Annual agriculture (corn, beans, rice, vegetables, etc.)	13.6
Cultivated pastures	10.5
Perennial agriculture (coffee, palm, banana, etc.)	8.9
Other (infrastructure, bodies of water, wetlands, etc.)	4.9
Sugarcane	3.3
Total	100.0

Table 30.2 Area cultivated with sugarcane, milled cane and sugar production in Guatemala

Harvest season	Area (ha)	Milled cane (tons)	Sugar production (tons)
1970–1971	30,633	2,075,293	197,717
1980–1981	78,000	5,485,805	447,896
1990–1991	120,000	9,934,918	974,798
2000–2001	179,471	15,174,029	1,711,832
2010–2011	231,505	19,219,653	2,048,142
2011–2012	252,871	21,562,263	2,252,954
2012–2013	263,056	26,747,489	2,782,461
2013–2014	270,178	27,733,769	2,806,080
2014–2015	271,313	28,267,605	2,975,801
2015–2016	268,735	27,987,308	2,822,590
2016–2017	255,859	25,835,487	2,719,231

Source: ASAZGUA (2017).

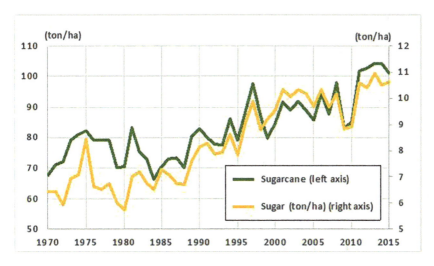

Figure 30.2 Sugarcane and sugar yields in Guatemala (Meneses et al., 2017).

increase in sugarcane and sugar productivity is depicted in Figure 30.2. As can be observed, the sugarcane yield, affected directly by climatic conditions, is the main drive of sugar productivity, which is affected also by sugarcane quality and industry efficiency.

The electricity generation in the Guatemalan sugar mills during the 2016/2017 harvest season represented 31% of the total generation to cover the national demand and exports of the National Interconnected System (NIS). For the 2016/2017 harvest, the effective power available to the NIS was 698.4 MW. The total generation of the sugar mills is made up of the generation used to feed their own consumption (factory processes of the sugar mills and auxiliary generation services) and the generation that they inject into the NIS. In this harvest period 2016/2017, there was a total production of 3,160 GWh and an injection of 2,220 GWh to the national grid. The fuel consumption in sugar mills was shared in 79.96% bagasse, 19.97% coal and 0.07% bunker (ACI, 2017).

Fuel ethanol in Guatemala

The first distillery was installed during the 1980s in Guatemala, and it was not until the year 2000 that other four distilleries were installed; currently Guatemala has an installed capacity of 253.6 million liters per year (fuel and other uses). Table 30.3 presents the installed capacity of each of distillery. Three of these distilleries operate during the harvest and the other two during most of the year. All distilleries use molasses as feedstock, exporting 80% of ethanol production to Europe and the United States, accomplishing independent sustainability certification such as ISCC (International Sustainability and Carbon Certification) and BONSUCRO (Better Sugar Cane Initiative).

Table 30.3 Guatemala sugarcane ethanol production capacity

Distillery	Capacity (L/day)	Days of operation	Production capacity (L/year)	Year
Palo Gordo	120,000	155	18,600,000	1984
Servicios Manufactureros	120,000	300	36,000,000	2001
DARSA	250,000	300	75,000,000	2006
Bio Etanol	600,000	155	77,500,000	2006
Mag Alcoholes	300,000	155	46,500,000	2007
Total			253,600,000	

Source: APAG (2017).

Today, all fuel consumed by Guatemalan vehicles is imported, but adopting a 10% ethanol blend (E10) in the gasoline, significant savings in fuels imports and reduction in the emissions in the transport sector can be achieved. The amount of ethanol required for this blend is about 200 million liters of ethanol per year, less than the current installed annual production capacity, 250 million liters.

Several studies, sponsored by international cooperation institutions such as Organization of American States, Inter-American Development Bank, United Nations Economic Commission for Latin America and the Caribbean, have assessed and confirmed the potential for promoting the use of ethanol blends in Guatemala, however with few concrete results. The obstacles remain essentially in the legal and regulatory side. In 1985, the Fuel Alcohol Law was created, Decree 17–85, stating in Article 13 that all gasoline must be blended with at least 5% of fuel ethanol. The program lasted less than a year and did not have the necessary political support to continue. When this law was approved, the Guatemalan fuels market was directly controlled by government through the Ministry of Energy and Mines, which established the price of all fuels and blends. More than 30 years passed, this market today is liberalized, with the prices set freely by the economic players, but that law remained the same, hampering the blend gasoline with local ethanol and imposing a revision in the regulatory framework.

In order to clarify any doubts about the reliability and efficiency of ethanol blends, in 2015 the Ministry of Energy and Mines with the support of the Organization of American States made a pilot plan with 25 vehicles and 5 motorcycles representative of the national vehicle fleet. They used E5 (5% ethanol and 95% gasoline), E7 (7% ethanol and 93% gasoline) and E10 (10% ethanol and 90% gasoline) blends. The results showed that emissions in the exhaust pipe decreased, with the greatest impact on the carbon monoxide (CO), which presented in average a 79% reduction. By using E10, an increase in the octane rating was found, improving the power and engine torque according to the tests performed. As expected, there was no mechanical or operation problem related to the use of blends.

According to calculations based on the GREET –(the Greenhouse Gases, Regulated Emissions, and Energy use in Transportation) model, developed by the Argonne National Laboratory (Wang et al., 2014), it was estimated that, considering the current (2017) situation, the adoption of E10 blend mitigates the emission of 433,000 tons of CO_2e per year, about 2.3% of the total emissions of the country or 4.7% of the emissions from the transport sector. By this way, Guatemala would mitigate annually the equivalent to 8% of 5.92 million tons of CO_2e, the emission reduction pledge (11% of 53.8 million tons of CO_2e) presented as the Nationally Determined Contributions (NDC), in the context of the UNFCC and derived agreements (Guatemala, 2015).

Perspectives and final comments

In recent years, when many countries in the world have bet on the ethanol blend in gasoline, the government role has been essential in setting and implementing public policies to encourage the required procedures and their follow-up; aiming to reduce fuel imports, save foreign currency and promote economic activity, income and employment; and improving the air quality. Possibly the Guatemalan decision makers still are not enough convinced about this perspective, to go ahead implementing blending mandates, an absolutely necessary step. Guatemala for many years has had the installed capacity to make an E10 blend, but the lack of public policies and coordination between the public and private sectors has not allowed this country to use the ethanol it produces and exports. Although there are policies, laws, regulations, agreements, strategies and national agendas that promote the blend of ethanol in gasoline, it has not been implemented because a national dialogue between public and private entities is necessary to establish robust and long-term public policies.

It is fundamental that public and private sectors, academia, civil associations and environmentalists come together to propose a country strategy for the early use of the ethanol blend, initiating at least with E5 blend and get to a E10 for a maximum period of 5 years.

References

ACI, 2017. *Cogeneration Data 2016–2017*, Cogeneration Association of Guatemala, Guatemala.
APAG, 2017. *Ethanol Production Capacity*, Alcohol Producers Association of Guatemala, Guatemala.
ASAZGUA, 2017. *2016 Yearbook*, Sugar Association of Guatemala, available at www.azucar.com.gt/
Guatemala, 2015. *Contribución Prevista y Determinada a Nivel Nacional* (INDC/UNFCCC), República de Guatemala.

Melgar, M.; Meneses, A.; Orozco, H.; Pérez, O.; Espinosa, R., 2012. *Sugarcane Crop in Guatemala*. Centro Guatemalteco de Investigación y Capacitación de la Caña de Azúcar, Guatemala.

Meneses, A.; Melgar, M.; Galiego, M. 2017. Series históricas de producción, exportación y consumo de azúcar en Guatemala. *Boletín Estadístico* 18(1). Guatemala, CENGICAÑA. 8 p.

Wang, M., Sabbisetti, R., Elgowainy, A., Dieffenthaler, D., Anjum, A., Sokolov, V., et al., 2014. *GREET Model: The Greenhouse Gases, Regulated Emissions, and Energy Use in Transportation Model*. Argonne National Laboratory.

31 Methodology of assessment and mitigation of risks facing bioenergy investments in sub-Saharan Africa

Rui Carlos da Maia

Introduction

The last 40 years of agriculture policy in Africa have been under the discourse of the role of subsistence agriculture in fighting hunger in Africa. The panacea of subsistence farming is explained in terms of it being the major rural employer or showcasing the poor farmer as the major food producer for Africa. In the cases where subsistence agriculture didn't succeed in feeding urban elites, African governments started a new discourse on mechanization or, at least, increasing usage of tractors in subsistence agriculture; this is bringing a movement of transfer of rural wealth to mechanization operators, under sponsorship of urban elites, developing modern ways of still exploring the rural power. Be it with by hand hoe or with tractors, subsistence agriculture is still incapable of paying a real salary in sub-Saharan Africa. Even under contract farming projects, per capita earnings from subsistence agriculture will be eroded by high transportation costs to deliver the farming goods to urban markets or dealers will charge unsustainable fees for transporting products. The negative effects of reliance in subsistence agriculture are even more pronounced for the future of Africa since earnings of rural farmers do not match the established minimum requirements for having them to pay the mandatory government taxes. Subsistence farming keeps 70% of African farmers outside of the tributary systems, and, in this way, it represents a heavy burden to Africa's development. This means that having African farmers continuously to rely on subsistence agriculture is in fact a form of massive, authorized and irresponsible tax evasion strategy. At the same time, African ruling elites, by disseminating continuously the strategic role of subsistence agriculture to Western audiences, succeed well in aligning their agriculture and rural development interests with the paradigms and mind setup of international donors, very interested in developing value chain mechanisms to improve production of rural farmers in Africa. So, money will flow to be allocated to African governments as part of large basket of "Funds for Agriculture Development". This money will certainly end up in perpetuating the role of subsistence agriculture in Africa stagnation and increasing the relative power of the ruling elites in African countries. The possibility of direct taxation of earnings of subsistence agriculture in Africa is remote, and since we are living in a globalized world, it would certainly trigger a broad discussion on human rights at continental level. It is also important to mention that young Africans have no interest in working at the subsistence level. Thus, subsistence agriculture is perceived and has an inferior economic activity by African youth.

We consider that new strategies are needed to reduce drastically the percentage of African farmers involved in subsistence agriculture, while commercial agriculture is to find its adequate place and take over agriculture production in Africa and bring back rural areas to be part of the tributary base of African countries. Because of their very well-structured value chains, sugarcane bioenergy projects could be a starting point for the introduction of massive commercial agriculture in Africa.

African comparative advantages for commercial agriculture and bioenergy are as follows: huge size of countries, huge amount of unused land and lands under extensive pasture activities, immense water resources for irrigation and more than 100 years of experience of sugarcane cultivation.

Background

The study of pathways to bioenergy in Africa was the subject of at least six seminars of the LACAf Consortium in Africa and Brazil in 2014; Maputo, Mozambique, in 2015; Kruger Park, South Africa; Piracicaba, Brazil; São Paulo/FAPESP (2016) and Stellenbosch (2017). These seminars discussed the main research questions: "why bioenergy in Africa", "where to implement bioenergy in Africa" and "how to implement it." Those questions are not independent from each other and have culminated in one publication (Leite et al., 2016). In answering these questions, a sort of general model was followed, and several actors had suggested the Brazilian bioenergy model as inspiration for the development of matured solutions for launching bioenergy projects in Africa and Latin America (LACAf). The resort to the Brazilian bioenergy model is justified with the allegations that there are climatic, soils, phenotypic and ecologic similarities between sugarcane formation in Brazil and the tropical areas of sub-Saharan Africa with high potential for bioenergy commissioning. Some other colleagues argued that although there are similar geographical formations in both Brazil and Africa, the context is very different in terms of nations rearrangements, history, politics, institutional context, level of industrial development and consumer matrix. Moreover the "why question" has been lately aligned in Brazil and Western countries with the discourse of emission reductions (Seabra et al., 2011) under the Paris Agreement (2015).

In trying to answer the "why question" for the case of sub-Saharan Africa, a comprehensive intervention logic was developed. It is based on the need of Africa to achieve two desirable African goals and six objective and tangible targets. The goals are as follows: establish connection of African agriculture with international trade and develop a supply driven agriculture in Africa. The tangible objectives are as follows: elimination of the prevalence of food insecurity; expansion of the establishment of acceptable quality rural jobs where rural farmers contribute to the expansion of tributary base of African countries; creation of logistical, roads, water and energy infrastructure; development of rural markets and transformation of the quality of human capital at the local level. For the global world, the answer to the question, "why to invest in bioenergy in Africa" is based on the need to achieve two objective targets: to increase the worldwide access to liquid fuels at reduced costs and to reduce emissions of carbon dioxide from liquid fuels.

The question, "where to install bioenergy systems in Africa", has also been sorted out by selecting the following countries: Angola, Burundi, DR Congo, Ethiopia, Liberia, Uganda, Mozambique, Sierra Leone and South Sudan. This was done using as base their huge geographic size and amount of unused lands coupled with the need of improvement in the socioeconomic status of people in rural areas through job creation and real need of achieving growth in their agriculture settings, after several decades of internal military conflicts.

There are very few references in the literature concerning common challenges in investments in commercial agriculture in sub-Saharan Africa. This paper concentrates in contributing to finding answers to the question of risks to implement bioenergy systems in the political, economic and social context of sub-Saharan Africa. The main objective of this paper is listing major causes, problems, barriers and possible interventions and ways to mitigate those risks.

Literature study on risks of biomass to electricity projects

The potential of sugarcane to complement the energy matrix of sub-Saharan countries has been studied by Souza et al. (2016) in terms of the possibility of replacing the use of traditional biomass and fossil fuel, reduce GHG emissions and enlarge electricity access in southern Africa. This paper takes in consideration the technical scenarios of expansions of sugarcane cultivation in southern Africa (Angola, Malawi, Mauritius, Mozambique, South Africa, Swaziland, Tanzania, Zambia and Zimbabwe). This paper did also mention, in short, the part of the context in which African commercial agriculture is performed, namely challenges in implementing sugarcane energy; appointing some economic, social and environmental challenges; poor logistics and lack of human capital and legal issues related to land use and land acquisition. In a recent paper by Othieno and Awangue (2016) referring to the state of energy resources in Kenya, Uganda and Tanzania, it is stated that

> although…governments recognize that the success of socioeconomic and industrial developments will depend on the performance of the energy

sector...very little support has been given to energy initiatives. The major challenge is how to deal with weak generation, transmission and distribution infrastructure and policies.

This statement points for the need of development of massive awareness programs concerning the potential synergies or linkages between bioenergy and rural electrification, at the government levels of sub-Saharan countries. As a matter of fact, and with the exception of Mauritius, an emerging country that has succeeded in developing an energy mix based on coal and sugarcane electricity, very few African governments do have a working strategy and policies to achieve rural electrification through the use of bioenergy programs. Since we believe that rural electrification can be achieved in most African countries through sugarcane bioenergy, we have developed a robust methodology of problem identification and problem solution, with the support of the Zopp intervention family of participative methods (Toni, 2005).

Research boundaries

In this paper, we use the term bioenergy to refer mostly to integrated production of biomass, sugar, ethanol and electricity. Simple sugar factories claim to be energy self-sufficient but most of them, in Africa, rely on a major supply from national electricity grids, because they are designed to produce mostly sugar and some electricity or, because there are subsidies on the cost of electricity for agriculture projects. Bioenergy investments in Africa will need to be focused in the production of electricity, ethanol and some sugar, and new business models will have to accommodate that purpose, which in real terms mean a need of investing much more money on the types of boilers and their capacity in the sugar factories.

It is also important to delimitate our discussions only to sugarcane. We do not include biodiesel since very few African countries have produced a surplus of nonedible vegetable oil from respective commercial farms. Moreover, we do not include the possibility of using surplus production of cassava or maize for ethanol production in Africa in accordance with FAO frameworks and discussions related to the use of food crops to bioenergy purposes. We do not include other sources of biomass for energy, namely wood, charcoal, agriculture residues or plant and animal wastes.

Methodology

The listing of barriers and enabling mechanisms for investments in bioenergy in Africa was made possible by employing the Metaplan/Zopp and logical framework methods of building "problem tables" and "objective tables" (Toni, 2005). We have identified four layers of risks and their causes. From these four layers, we have proceeded in identifying the priority programs for removing investment barriers. These priority programs are the ones to be

Table 31.1 Problem table – listing barriers to development of bioenergy projects in Africa

International barriers caused by	Continental barriers caused by	Regional barriers caused by	Country barrier caused by
(Common to world economic system in relation to Africa)	(Common to Africa as a continent)	(Common to regions of Africa)	(Common to countries)
Exclusion of Africa from international trade	No demand of biofuels from Africa	No regional demand/markets of biofuels from region	No local demand of biofuels from local business
Conflicting colonial power interests	Wars and conflicts around natural resources	Military and common border confrontations	Ethnic and social and electoral conflicts
Conditional foreign aid	Huge deficit in national budgets	Lack of investments in integrated power systems	Weak generation transportation and consumption of electricity
No investments in infrastructure	Poor logistics	Lack of regional railway systems	Lack of national and local roads
Deterioration of commercial exchange terms	Poverty increase • Massive unemployment rates	Low agriculture productivity • Conflicting land tenure issues	Food insecurity and recurrent Outbreak of tropical diseases and • Massive emigration
Weak education in commercial agriculture	Lack of agriculture management skills	Lack of integration in agriculture policies	Reliance of subsistence agriculture
Non sustainable foreign debt	Enrichment of African government elites	Bad governance and corruption	Weak social policies
Lack of financial markets in Africa	Lack of rural collaterals and insecurity	Lack of trained managers in agriculture finances	No access to finances to agriculture projects
Bad press and risk communication on role of biofuels versus food security	Main international NGO's lobbying against biofuels	Commercial agriculture/biofuels project delayed	Legal processes against biofuels at country level
Biophysical vulnerabilities associated with negative interventions/forest fires	Lack of unified programs of disaster risk reduction	High economic and social losses	Low bankability of investments in agriculture
Human factors and "divide and rule" politics	Nondemocratic/military regimes	Lack of mutual trust in regional programs	Isolated decision-making processes

Source: Developed from data on metrics of development economics referred by Hoeffler (1999).

Table 31.2 Interventions table – enabling actions to help success of bioenergy projects in Africa

International barriers removed by	Continent barriers removed by	Regional barriers removed by	Country barriers removed by
(From world economic system toward Africa)	(In sub-Saharan Africa)	(Region by region)	(Country by country)
Continuous dialogue at the United Nations, Africa Union and International Trade Organizations to present Africa as a future hub of bioenergy	Increased demand of biofuels from Africa	Creation of markets of biofuels in the region	Developing new products from existing sugarcane projects
Aligning with former colonial power strategic interests	Resolving major conflicts with inclusion of former colonial powers	Increasing levels of regional integration	Increasing inclusion in governance and access to wealth by rural people
Transforming foreign aid into investment in commercial agriculture	Long-term investments in commercial agriculture and bioenergy	Investments in integrated power systems	Development of programs of accessing energy for all through the usage of bioenergy at local level
No investments in infrastructure	Investment in rural infrastructure	1. Establishment of regional railway systems where needed to support bioenergy projects. 2. Building water retention and irrigation infrastructure	Investments in building up regional roads, local power lines, rural bridges
Improvement of commercial exchange terms	Poverty reduction programs	Allowing local processing of finished products	Increasing agriculture productivity and creating jobs throughout grower schemes in sugarcane projects
	Development enabling guidelines for land tenure for commercial agriculture		

Establishing education in commercial agriculture	Training technicians in agriculture management skills	Integration in of agriculture polices in university curricula	Introduction of teaching commercial agriculture in schools
Controlling the accumulation of foreign debt through international organizations (IMF, World Bank)	Surveillance to avoiding occult debts at continental level	Increasing programs of good governance	Reinforcing policies and funding of education health and welfare at country levels
Allowing creation of financial markets for agriculture in Africa	1. Sorting out insecurity in rural Africa by solving electoral conflicts 2. Involving World Bank and Africa Development Bank in funding projects	Training people in agriculture financing	Creating competitive funding sources for commercial agriculture and bioenergy
Improving risk perception and communication by involving International NGOs, FAO, UNIDO and African Union Commissions in developing guidelines for integration between production of bioenergy and reducing food security	Improving risk perception and communication by involving NGOs in dissemination of good practices and guidelines on integration between biofuels and food security	Training regional NGOs on the implementation of guidelines on integration of biofuels projects with food security projects	Strengthening countries in avoidance of land grabbing incidents and monitoring of risk of bioenergy negatively affecting food security
Supporting Africa with science and technology transfer programs against hydrometeorological risks	Systematic mapping of biophysical and ecological risks at continental level	Selection of regions with lower biophysical and ecological risks for bioenergy projects	Construction of dams, reservoirs and development of disaster management education programs
Win-win approaches in ways of "doing business"	Improved coordination in bioenergy policies at continental level	Development of regional frameworks for investments in bioenergy	Systematic demonstration at local level of the role of bioenergy in rural electrification and social development

continuously checked by investors in the process of commissioning bioenergy projects in sub-Saharan Africa.

a   Layer 1: International barriers (e.g., Africa exclusion from international trade);
b   Layer 2: Continental barriers (e.g., lack of biofuels markets in sub-Saharan Africa);
c   Layer 3: Regional barriers (lack of regional integration and investments in regional blocks of Africa: southern Africa-SADC, West Africa and East Africa country blocks);
d   Layer 4: Country barriers (e.g., barriers identified for Mozambique – lack of qualified human resources for bioenergy and so on).

Under each layer, we have listed the main barriers in Table 31.1. This table is to be worked out from left to right. At the left side we list the probable causes, and at the right side we list the effects at continental, regional and country levels.

The information summarized in Table 31.1 is a list of "negative statements" that could impact bioenergy projects at different times, before, during and after commissioning these systems in sub-Saharan Africa. Some items represent root causes, and other items do represent effects. The same list is considered as valid for justifying the lack of capacity of Africa feeding its own people and, in general, all together are contributors to the main truth: economic weakness of sub-Saharan Africa.

Statements in Table 31.1 can now be transformed into "positive declarations" and are used as input data for development of a table of interventions (Table 31.2), which can provide us with steps that have to be dealt with at international, continental, regional and country levels to assure the success of implantation of bioenergy projects in sub-Saharan Africa. It is obvious that we are dealing here with development economics. Bioenergy systems will have to contribute to the development of rural sub-Saharan Africa in order to succeed.

Risk evaluation and mitigation process

After development of the list of barriers as well as the list of actions/interventions that are to be taken into consideration, we will now list the steps for organizing the risk assessment process. A risk assessment platform for investing in bioenergy in sub-Saharan Africa can be made up of five steps (BHP Billiton, 2015):

- mapping of probabilities of investments going wrong country by country and region by region in considering the statements listed in Table 31.1 and the possibility of removing the barriers that was stated in Table 31.2;
- quantifying the risk in terms of magnitude and frequency of political instability incidents, institutional and economic strength country by

country, social indexes, environmental criteria, security issues and human rights situation, country by country and region by region;
- determining the level of acceptable risks for investors and assuring highest political commitment at continental and country levels;
- designing a plan of risk management, mitigation or reduction;
- developing a sort of review and monitoring plans to watch progress country by country before commissioning bioenergy projects.

In doing these exercises, we will have to set a list of "goals" (political and social, intangible outcomes, complex to quantify) and a list of objectives (concrete steps, measurable outputs). Goals are to be dealt politically by integrating high-level international organizations dealing with development issues to facilitate the development of a general framework for investments in bioenergy in Africa.

Dealing with "objectives" is a question of bilateral talks between investors and specific countries in a way to use comparative advantages for establishing bioenergy projects as soon as the desired mutual conditions are met. These objectives should be prioritized in accordance with weighing criteria that will be established in accordance with project size and country specificity (logistics and infrastructure). In order to succeed, bioenergy investments should be preceded by political dialogue at the African Union, making use of expertise in economic risk assessment of United Nations Industrial Development Organization (UNIDO), using also agriculture recommendations from Food and Agriculture Organization (FAO) and involving finances of the African Development Bank. Special emphasis should be given to avoiding investments in countries with undemocratic governments since they are prone to long-term economic, social and internal military conflicts.

Bioenergy projects use billions of dollars and will last for decades (100 years). Adequate care should be taken to use the time to have a mature negotiation with African governments before embarking into the venture. Further and taking into consideration that bioenergy projects are in essence energy projects, it is also important to verify the following list of concerns:

- Is the resettlement of populations in the selected areas secured for the next 30 years?
- Is the bioenergy project going to have adverse effects on fishing or fishing biodiversity in water formations, especially in landlocked countries?
- Is the land lease secured for at least 60 years?
- Are the proposed technologies environmentally approved and safe?
- Is the project logistics planned for the next 30 years?
- Are finances available to support the project for the next 30 years?
- Are there markets of bioenergy products for the next 30 years?
- Are there fail-safe contracts for the next 30 years of operation of the bioenergy consortium?

- Is there a plan for selecting, training and maintaining local human capital for the next 30 years?
- Is there a framework for conflict management and avoidance of grievance and strikes at the sugar plantations and factories?
- Are there provisions for inclusion of mandatory and regular training and simulation for fatal risk control protocols, human rights, hygiene health, safety, environment and community consultation as well as industrial risk assessments in the proposed project?
- Are there any signs of discontent or mal-satisfaction with the mechanisms that were used to allocate land to the bioenergy projects?
- Is there any framework to monitor progress and evaluate if the bioenergy project is improving the quality of jobs created in Africa especially the long discussion on manual or automated sugarcane collection process?
- Is the whole project economically, socially and environmentally sustainable?

Conclusion and final remarks

The paper summarizes the potential benefits of bioenergy projects in Africa as being the high potential of transforming Africa's agriculture from an archaic system of reliance of subsistence to a phase of commercial agriculture. We consider that sugarcane bioenergy projects (biomass to electricity) can be leapfrogging mechanism to start or reinforce the penetration of commercial agriculture systems in countries of sub-Saharan Africa in order to achieve, locally, the six major goals of African rural development: elimination of the prevalence of food insecurity, expansion of the establishment of acceptable quality rural jobs, creation of logistical and energy infrastructure, development of rural markets and transformation of the quality of human capital at local level as well as to increase the worldwide access to liquid fuels at reduced costs and reduce emissions of carbon dioxide from liquid fuels. But installing bioenergy systems in Africa will be very challenging since Africa is in process of stagnation because of several factors ranging from the international economic power relationships, continental and regional context as well as local military, ethnic, electoral and border conflict situations. We discussed in this paper the methodology and form of identification of challenges and major risks that must be well considered before, during and after commissioning bioenergy systems in Africa, and we concluded that several developmental barriers will have to be removed at various levels. The paper recommends a list of steps of risk assessment: mapping risks, quantifying the acceptable level of risk in accordance with statements listed in Tables 31.1 and 31.2, developing mitigations plans country by country and dealing with or involving the support of international organizations, case by case and country by country during the whole process of commissioning bioenergy investments.

References

Hoeffler, A. (1999), Challenges of infrastructure rehabilitation and reconstruction in war-affected economies. *African Development Bank*, economic research papers, No. 48. www.afdb.org/fileadmini/uploads/afdb/Documents/Publications/00157630-EN-ERP-48.PDF

BHP BILLITON HSEC Controls (2015). *Health Safety Environment and Community.* 141114_suppliers-petroleum-phse00c01petroleumhseccontrols.pdf

Leite, G.D.B., Leal, M.R.L.V., Nogueira, L.A.H., Cortez, L.A.B., Dale, B., da Maia, R.C., and Adjorlolo, C. (2016).Sugarcane: a way out of energy poverty. *Biofuels, Bioprod. and Biorefining*, Wiley Online Library, http://onlinelibrary.wiley.com/doi/10.1002/bbb.1648/abstract

Paris Agreement (2015). United Nations Climate Change Conference. http://en.wikipedia.org/wiki

Seabra, J.E.A., Macedo, I.C. Chum, H.L., Faroni, C.E., and Sarto, C.A.(2011). Life cycle assessment of Brazilian sugarcane products: GHC emissions and energy use. *Biofuels Bioprod. Biorefining* 5, 519–532. doi:10.1002/bbb.289

Souza S.P., Nogueira, L.A.H., Watson, H.K., Lynd, L.R., Mosad, E., and Cortez, L.A.B. (2016).Potential of sugarcane in modern energy development in southern Africa, *Frontiers in Energy Research*, doi:10.3389/fenrg.2016.00039

Toni, de Jackson (2005). Zopp and Logical Framework methods. *National School of Public Administration (ENAP)*, Brazil.

32 Can sugarcane bring about a bioenergy transformation in sub-Saharan Africa?

Willem Heber van Zyl

Saharan Africa, opportunities and challenges

Africa, the second largest continent, has much to offer in resources but faces major challenges. The size of Africa is often not appreciated; with equal land area as that of the United States, Europe, India and China have about half of the world population, Africa has the most underutilised arable land of any continent. Of the about 3,000 million hectares of land in Africa, an estimated 840 million hectares are arable, compared to 890 million hectares in the Latin Americas. However, only about 27% of the arable land in Africa is utilised in formal agriculture, leaving about 600 million hectares underutilised (Ejigu, 2008). The primary energy consumption of Africa is very low compared to the rest of the world. If the land available in Africa is divided by the primary energy use, the bioenergy potential of 176 ha/TJ/year is more than twice that of the second continent, South America, at 74 ha/TJ/year and more than six times the world average of 28 ha/TJ/year. The potential for the production and harvest of biomass for food and bioenergy in Africa is thus enormous (Lynd et al., 2015).

Despite the large potential for bioenergy in Africa, current realities present major challenges for the continent. Although the percentage of people in extreme poverty declined in the past 20 years to about 43% in 2012, the total number increased to 330 million due to the rapidly expanding African population (Beegle et al., 2016). The United Nations projected that the world population will increase from 7.6 billion in 2017 to 11.2 billion in 2100. The largest expansion in the African population from 1.26 billion in 2017 to 4.47 billion in 2100, which represents a 3.5-fold increase and more than 80% of the growth in world population towards 2100, makes Africa almost the most populous continent (Table 32.1) (United Nations, 2017). The high growth rate, combined with a high extreme poverty index, will place tremendous pressure on the arable land to provide both food and bioenergy in the future.

Compounding the population expansion and accompanying extreme poverty projections, Africa has been suffering from poor infrastructure, lack of agricultural extension and historically inefficient use of biomass as energy source. In contrast to wind and solar energy, which are primarily capital

Table 32.1 Population of the world regions forecasted until 2100*

Region	Population (millions)			
	2017	2030	2050	2100
Africa	1,256	1,704	2,528	4,468
Asia	4,504	4,947	5,257	4,780
Europe	742	739	716	653
Latin America and Caribbean	646	718	780	712
Northern America	361	395	435	499
Oceania	41	48	57	72
World	7,550	8,551	9,772	11,184

* According to the medium-variant projection by the United Nations: DESA.
Source: United Nations (2017).

dependent, modern bioenergy production is much more complex and inextricably linked to agriculture, environmental factors and social development, left alone political will and support. Water availability, agricultural development, available transport infrastructure, financial backing, land tenure and good governance are but a few prerequisites for large-scale bioenergy production. Developing modern bioenergy as a major energy source with limitations in several of these prerequisites in Africa therefore remains a daunting task.

Why does Africa need modern bioenergy?

In Africa, nearly 730 million people still use firewood for daily household energy needs with more than 620 million people who do not have access to electricity (Figure 32.1). When excluding South Africa, 90% of the 280 Mt biomass used in sub-Saharan Africa is for household cooking and heating. Even with the rollout of electricity in Africa, more than 500 million people will remain excluded it as a result of population expansion (IEA, 2014), rapid urban growth accompanied by extreme urban poverty and resulting social problems (UN-Habitat, 2014). Thus, electrifying Africa will not totally address energy needs as often proposed by the developed world, and bioenergy will remain a major source of household energy in the larger part of rural areas. Therefore, the use of traditional bioenergy in the form of charcoal and firewood should be minimised by providing more efficient modern bioenergy alternatives.

About a third of sub-Saharan Africa is covered by forest, with the total forest biomass estimated to be about 130 billion tonnes in 2010. It should be noted that less than 1% should be used annually so as to not promote deforestation (IEA, 2014). It is therefore not surprising that about 48% of Africa's

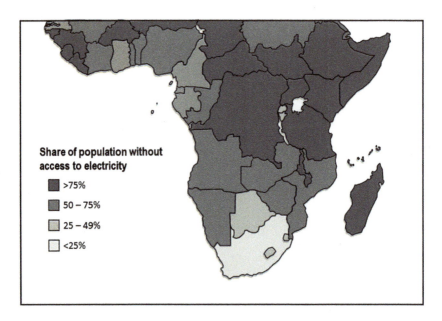

Figure 32.1 Over 620 million people in sub-Saharan Africa do not have access to electricity.
Source: Adapted from IEA (2014).

primary energy supply comes from biomass. However, the bulk of biomass is used in the form of firewood, which is rather inefficient and also hazardous when used in-house (IRENA, 2015). In urban areas, charcoal remains a preferred household fuel option, since it has a higher energy density than firewood and can be easily transported, stored and distributed. The bulk is still home-produced in traditional earth-mound kilns that have a conversion efficiency of 8%–12%, while industrial kilns are more than twice as efficient (IEA, 2014).

Wood as preferred biomass source, combined with the associated inefficiencies, results in significant deforestation (Allen and Barnes, 1985), which is a major environmental concern (Chidumayo and Gumbo, 2013). At the same time, significant amounts of biomass residues and invasive plants responsible for bush encroachment remain underutilised in terms of energy production (Smith et al., 2015). The use of modern bioenergy, which constitutes a switch away from deforestation towards sustainable biomass sourcing, will not only address these inefficiencies and changes of fuel mixes but can play a significant role in addressing socio-economic challenges in sub-Saharan Africa.

Modern bioenergy not only offers more efficient use of biomass, but, if well planned and governed, could also provide economic, social and environmental benefits. Modern bioenergy could include traditional biofuels, such as biogas, bioethanol, biodiesel, more efficient charcoal production than in primitive kilns and the rollout of advanced lignocellulosic conversion technologies to cellulosic ethanol, high-density pyrolysis oils, char and gases for the production of synthetic fuels using Fischer–Tropsch (Van Zyl et al., 2017). The particular value of advanced lignocellulosic conversion technologies is that they provide options for combining the production of traditional and advanced biofuels, and optimise biomass utilisation. These technologies also allow the integration of bioenergy production with existing bio-based industries, such as sugar industries, paper and pulp, waste streams valorisation and electricity production for local decentralised grids (Van Zyl et al., 2011).

Apart from being a more efficient source of energy, modern bioenergy can also provide economic, social and environmental benefits to particularly sub-Saharan Africa at large. Biomass production is inevitably linked to the farming community and can therefore improve the social well-being of rural communities. Reducing the time-consuming collection of firewood, particularly by women, could mean better caring for children and their education. At the same time, it could reduce health hazards associated with in-house smoke inhalation (Ejigu, 2008).

Modern bioenergy could also provide an additional income and decentralised energy (heat, electricity and biofuels for transport) to local communities, particularly those in landlocked countries. This could overcome the 'double penalty' of the higher cost of importing fertilisers and then transporting goods to seaports (Lynd et al., 2015). Food losses because of a lack of refrigeration and storage energy could also be reduced, which would help to overcome food insecurity and malnutrition. Furthermore, creating other markets for biomass parallel to food markets, and thus making food production less vulnerable to the fluctuation in international markets, would also aid food security (Lynd et al., 2015).

Furthermore, modern bioenergy would be more efficient than the use of traditional firewood and charcoal produced in traditional earth-mound kilns, because less carbon is released, leading to climate change mitigation (Ejigu, 2008). It is important that biomass for bioenergy does not fall into the same trap of large-scale raw material exports, as in the case of crude oil and precious minerals. Bioenergy production should first focus on empowering local communities by directing attention on smallholder production and processing, thus empowering farmers by generating new income streams that can help with rural economic transformation. Domestic markets should preferably be developed, rather than relying on international markets. More modern bioenergy technologies (e.g., biogas and biodiesel for households and farming equipment) are simple and easily transferable, and can be operated at village level (Ejigu, 2008).

Why does sub-Saharan Africa not have many success stories with biofuels?

Introspection is prudent to understand why sub-Saharan Africa, one of the subcontinent regions with the greatest potential, is not a major biofuels producer. Contributing to the problem is diverse and often wrong priorities, false expectations and non-committal governance. One of the most important drivers has been economic development; potential investors saw biofuels as a lucrative commodity that could serve international markets, e.g. EU biofuel markets without necessarily considering land ownership and involving local communities. Also, the lack of proper research into the resilience of crops in the African setting resulted in an overestimation of yields. Jatropha is such an example, rising as a 'miracle crop', but many projects collapsed when realism settled in and harvests did not meet expectations for various reasons, including the aridity of land used (von Maltitz et al., 2014; Kgathi et al., 2017).

Energy security is another such driver, mostly for poor landlocked countries such as Malawi, Zimbabwe and Zambia. In South Africa, the main driver has been to try and link the formal first and informal second economies through job creation in former homelands, primarily on small farmer scale. Until today, no major venture has taken off, primarily due to the restriction of feedstock sources and the focus on small-scale farmers in rural communities, effectively locking out large industries (Letete and Blottnitz, 2011). Eventually, the lack of compulsory blending, incentives to kick-start the vulnerable biofuels industry and low crude oil prices brought the biofuels initiative in South Africa in stalemate (Kohler, 2016). The development of biofuels in the sub-Saharan Africa region has thus been hampered by a combination of diverse factors and diverse drivers, but also the lack of formal regional strategy or overarching policy framework and political commitment, regional infrastructure, focus and commitment, which makes the establishing of a regional biofuels industry very challenging. It is therefore not surprising that biofuels/modern bioenergy has not come to fruition in sub-Saharan Africa, despite the huge potential (Gasparatos et al., 2015).

What does sugarcane offer as bioenergy crop to Africa?

Many countries in sub-Saharan Africa, notably South Africa, Swaziland, Malawi, Mauritius, Mozambique, Tanzania, Zambia, Zimbabwe and Madagascar, have had past experience working with sugarcane for both sugar and bioethanol production (FOASTAT, 2016). The regional focus has been on sugar production (47 Mt in 2014 and 44 Mt in 2016 when South Africa experienced droughts) with South Africa producing about 40% of the raw sugar in the region (about 2.2 Mt in 2014) (FOASTAT, 2014; Wright, 2018). Ethanol production in sub-Saharan Africa has been marginal at just above 600 ML/annum in recent decades, mostly converting sugarcane molasses to bioethanol (Munyinda et al., 2012). In South Africa, ethanol has been produced for the technical market and not for biofuels (Ambali et al., 2011).

Watson (2011) found that 6 million hectares of suitable land (excluding primary food production land or conservation areas) are available in sub-Saharan

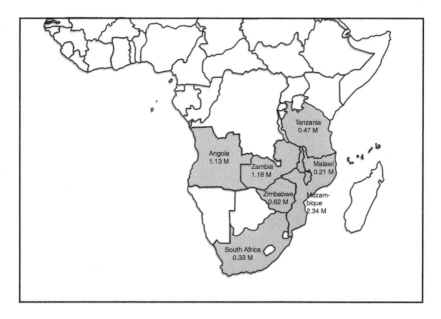

Figure 32.2 Areas available and suitable for sugarcane cultivation in southern Africa.
Source: Adapted from Watson (2011). The study excluded environmentally sensitive protected areas, closed canopy forests and wetlands. It also excluded areas currently under food and cash-crop production.

Africa for sugarcane production (Figure 32.2), eight times more than currently in production in the aforementioned nine countries (FOASTAT, 2016). When considering the efficiency in South Africa to produce sugarcane (65 t/ha) and processed sugar (8 t/ha) in 2014, the 6 million hectares of land could yield 390 Mt sugarcane or 48 Mt processed sugar. If 50% of the sugar is converted to ethanol, almost 14 BL of ethanol can be produced from the available land.

It is thus clear that sugarcane is well known as a major crop in Africa, but it has not been utilised optimally. Sugarcane thus presents interesting and unique opportunities, particularly in sub-Saharan Africa, to address not only agricultural production and income generation, but also access to bioenergy, either as ethanol fuel or electricity.

Brazilian experience and LACAf findings

Sugarcane has been cultivated in Brazil since the 16th century, but it was the onset of the National Alcohol Program (Proálcool) in 1975 that really set off the development of the very successful bioethanol industry in Brazil. In 2016, Brazil had 10.2 Mha under sugarcane cultivation that yielded almost 770 Mt of sugarcane, 40% of the global production of 1,900 Mt (FOASTAT, 2016). Undoubtedly, Brazil is a major world player in sugarcane cultivation and second

Table 32.2 Lessons from Brazil experience for sub-Saharan Africa

No	Context	Experience
1	Expertise	It is essential that bioenergy feedstocks are well known and regional factors (environmental factors and agricultural extension) are taken into account. Good local breeding programmes and germ plasm are crucial.
2	Multiple markets	Selling into multiple markets (food, fuel and electricity) is advantageous and can mitigate unnecessary risk.
3	Life cycle indicators	Bioenergy chains should score well in terms of life cycle indicators – such as efficient use of land, water and energy.
4	Political will and governance	Governments have a key role in creating/monitoring/enforcing legal frameworks, fuel specifications, mandatory blending and tax regimes to enable new, vulnerable bioenergy projects to come off the ground.
5	Social benefits	Should not be an add-on but explicitly considered within the integrated frameworks that look at commercial viability.

Source: Lynd et al. (2015).

to the USA in ethanol production at 28 BL in 2016 (Statista, 2016). It is important to note that the soils and climate in sub-Saharan Africa and Brazil, the two continents with the largest potential for bioenergy production, are similar. At first glance, it is obvious that the Brazilian experience with sugarcane and bioethanol, as well as the geographical and climatic similarities between the continents, could be very relevant to sub-Saharan Africa (Table 32.2).

The study by Leite et al. (2016) is therefore very timely and provides interesting options for consideration, from simple extension of the existing sugar industry to utilise molasses for ethanol production with significant gains at low-investment capital costs, to advanced scenarios that consider the simultaneous production of electricity or ethanol (either 50% or 100% conversion of sugarcane juice). The latter, more advanced scenarios may yield more benefits in both energy security and combating environmental degradation and deforestation, but the authors rightly recognised that limited investment opportunities and poor regional political frameworks (including compulsory blending and incentives to provide economic viability to an upcoming biofuels/bioenergy industry) and infrastructure for efficient distribution or markets to sell to, remain some of the main obstacles. Initially, ethanol production from molasses as by-product of the existing industry would provide for better sustainability while ethanol and energy markets are developed, infrastructure secured and hopefully better enabling regional frameworks developed. This was eloquently demonstrated by Leal et al. (2016) that brownfield development of molasses-based ethanol plants annexed to existing sugar mills (called Ethanol-1 scenario) is the most logic starting point. If the obstacles mentioned above (including both local and international markets) are adequately addressed to ensure a regional biofuels industry, Ethanol-2 scenarios, where flexible management of sugar and ethanol production can respond to market opportunities, can be explored. Ethanol-2 scenarios would clearly yield better financial benefits and assist rural development. These scenarios have been confirmed by local studies at Stellenbosch University (Petersen et al., 2014).

When considering South Africa with an established sugar industry and several modern sugar mills, economic analysis not only promoted ethanol production from molasses or cane juice, but found that the mild treatment of bagasse on-site for the production of ethanol with cogeneration of electricity from residual bagasse present very attractive scenarios with very competitive ethanol production costs (Petersen et al., 2014). Furthermore, the production of lactic acid from the bagasse and leave fraction is much more environmentally friendly than fossil-based lactic acid production, allowing the industry to further diversify its markets (Daful et al., 2016).

Priorities for sustainable biofuels production in Africa

For the sustainable development of biofuels in Africa, certain important aspects have to be considered (Gasparatos et al., 2015). First, biofuels policies must be aligned with regional needs and priorities, and it should be very clear why certain biofuels options are considered and what energy needs biofuels will serve. It is also imperative that southern Africa should develop regional policy frameworks to ensure unity in biofuel drivers. Ideally, national and local energy security should be considered with rural development and poverty alleviation as priorities. However, strong regional agricultural and biofuel markets are essential, and the lack therefore has been one of the key stumbling blocks for biofuel expansion in southern Africa. Furthermore, many countries in southern Africa still have relatively small vehicle fleets, and for significant expansion of biofuels production, flexi-fuel vehicles have to be promoted. This is unlikely to happen, except if it is a regional initiative in southern Africa. For that reason, international markets might be crucial to get a sustainable biofuels industry from the ground. However, proper policy frameworks and commitments from national governments to promote biofuels production are crucial, otherwise investors will remain hesitant to invest in a biofuels industry with too many uncertainties (Fundira and Henley, 2017).

Feedstock selection should be based on proper agronomic knowledge, and in this regard, sugarcane offers many advantages as an internationally recognised biofuel feedstock and having ample experience in producing sugarcane in southern Africa. However, if a significant biofuels industry is foreseen, local research and development would be essential to ensure suitable feedstocks for southern Africa and continual improvement of sugarcane varieties, for example, when energy cane is considered in future. Southern Africa cannot afford the mistakes made with the large-scale plantation of Jatropha in the recent past (von Maltitz et al., 2014; Singh et al., 2014). Fortunately, the South African Sugar Research Institute (SASRI, www.sasa.org.za/sasri/) and the Sugar Milling Research Institute (SMRI, www.smri.org/) have a good track record in supporting the sugar industry in southern Africa and should also respond to the expansion of the sugar industry into bioenergy (Braude, 2014).

Regional policies and frameworks should safeguard against speculative and predatory corporate behaviour as observed during large-scale international investments in Jatropha production with detrimental consequences for the local rural communities. Sugarcane expansion in underutilised land may still lead to dispossession, displacement or disruption of local communities. It is thus imperative that the land rights of particularly rural communities should be entrenched, and land grabbing avoided at all costs. Any potential food-fuel competition should be minimised, and synergies between food and fuel production sought to ensure that biofuel production positively impacts on food and agricultural security. Ideally, a proportion of earnings from the biofuels industry should be reinvested in the agricultural sector, notable to advance food security, as well as social and education programmes in rural communities.

Often sugarcane expansion necessitates the conversion of natural ecosystems to production land. Harmful environmental practices, such as deforestation and excessive use of fertilizers, should be avoided as to not erode unnecessary top soils. There should also be a concerted effort in developing household biofuels that are less harmful to replace paraffin, charcoal and firewood. One such approach would be to focus on safe energy-efficient ethanol stoves at competitive prices relative to the use of paraffin stoves and ensure necessary supply chains that could also reach remote rural communities. The main impeding factors to the use of ethanol stoves are energy density, the availability of standard stoves, as well as the availability of ethanol in rural areas. It is thus not surprising that paraffin is still preferred, despite the health and fire hazards (Kimemia and Van Niekerk, 2017). At the same time, it might be important to actively seek alternative and more sustainable sources for firewood and charcoal production to minimise the deforestation of natural woodlands and forests (Mohammed et al., 2015).

Concluding remarks

Africa, specifically southern Africa, is facing major challenges with poverty, food security, weak infrastructure and a fast-expanding population. Traditional biofuels, primarily firewood and charcoal, remain the main source of household energy, with detrimental environmental and health risks. At the same time, Africa has ample underutilised arable land that can provide both food and modern bioenergy. South African countries are also well acquainted with sugarcane cultivation and sugar production, but to a lesser extent bioethanol production. The potential expansion of the Brazilian experience in sugar and bioethanol production to southern Africa is thus logical and much needed. If 6 Mha estimated by Watson (2011) is used for sugarcane cultivation and bioethanol production, the region could become a significant player in the biofuels arena as currently the case for Brazil with all the accompanying socio-economic benefits.

However, trying to simply apply the highly successful Brazilian sugar and ethanol model to southern Africa might not be ideal, considering inherent difference, both cultural, political and socio-economical. Particular attention should be devoted to develop proper political frameworks, aligning biofuels/bioenergy policies with regional and national policies and objectives, involving the local sugar industry to buy in and take ownership in expanding the sugarcane industry to produce biofuels and electricity, where appropriate. It may take time, but it will eventually pay off to not only involve the local industry but also involve research institutions and universities to optimise sugarcane varieties and cultivation, diversify the industry and ensure continual support and improvements as the industry expands and matures. Furthermore, sensitive aspects such as land ownership, social development in local communities and actively replacing the inefficient use of firewood and charcoal with better bioenergy options. Thus, sugarcane clearly has the potential to transform bioenergy use and assist in socio-economic developments in Africa, but it has to be carefully planned and all necessary stakeholders should be involved from day 1, starting with governments and industry, but not excluding research/tertiary institutions and rural communities.

References

Allen, J. C. and Barnes, D. F. (1985) 'The Causes of Deforestation in Developing Countries', *Annals of the Association of American Geographers*, 75(2), pp. 163–184. doi: 10.1111/j.1467-8306.1985.tb00079.x.

Ambali, A., Chirwa, P. W., Chamdimba, O. and van Zyl, W. H. (2011) 'A review of sustainable development of bioenergy in Africa: an outlook for the future bioenergy industry', *Scientific Research and Essays*, 6(8), pp. 1697–1708. doi: 10.5897/SRE10.606.

Beegle, K., Christiaensen, L., Dabalen, A. and Gabbis, I. (2016) *Poverty in a rising Africa*. Washington, DC. doi: 10.1596/978-1-4648-0723-7.

Braude, W. (2014) *Towards a SADC fuel ethanol market from sugarcane : Regulatory constraints and a model for regional sectorial integration*. Available at: www.tips.org.za/files/sadc_ethanol_market_constraints_and_intgration_model.pdf.

Chidumayo, E. N. and Gumbo, D. J. (2013) 'The environmental impacts of charcoal production in tropical ecosystems of the world: A synthesis', *Energy for Sustainable Development*, 17(2), pp. 86–94. doi: 10.1016/j.esd.2012.07.004.

Daful, A. G., Haigh, K., Vaskan, P. and Görgens, J. F. (2016) 'Environmental impact assessment of lignocellulosic lactic acid production: Integrated with existing sugar mills', *Food and Bioproducts Processing*. Institution of Chemical Engineers, 99, pp. 58–70. doi: 10.1016/j.fbp.2016.04.005.

Ejigu, M. (2008) 'Towards energy and livelihoods security in Africa: Smallholder production and processing of bioenergy as a strategy', *Natural Resources Forum*, 32, pp. 152–162. doi: 10.1111/j.1477-8947.2008.00189.x.

FOASTAT (2014) FOASTAT, *Production – Crops processed*. Available at: www.fao.org/faostat/en/#data/QD (Accessed: 3 February 2018).

FOASTAT (2016) FOASTAT, *Production– Crops*. Available at: www.fao.org/faostat/en/#data/QC (Accessed: 3 February 2018).

Fundira, T. and Henley, G. (2017) *Biofuels in Southern Africa: Political economy, trade, and policy environment*. 2017/48. Available at: http://hdl.handle.net/10419/161612.

Gasparatos, A., Von Maltitz, G. P., Johnson, F. X., Lee, L., Mathai, M., Puppim De Oliveira, J. A. and Willis, K. J. (2015) 'Biofuels in sub-Sahara Africa: Drivers, impacts and priority policy areas', *Renewable and Sustainable Energy Reviews*, pp. 879–901. doi: 10.1016/j.rser.2015.02.006.

IEA (2014) *Africa Energy Outlook. A focus on the energy prospects in sub-Saharan Africa, World Energy Outlook Special Report, International Energy Agency Publication*. doi: www.iea.org/publications/freepublications/publication/africa-energy-outlook.html.

IRENA (2015) *Africa 2030: Roadmap for a Renewable Energy Future, REmap 2030 programme*. doi: 10.1017/CBO9781107415324.004.

Kgathi, D. L., Mmopelwa, G., Chanda, R., Kashe, K. and Murray-Hudson, M. (2017) 'A review of the sustainability of Jatropha cultivation projects for biodiesel production in southern Africa: Implications for energy policy in Botswana', *Agriculture, Ecosystems and Environment*, 246(July 2016), pp. 314–324. doi: 10.1016/j.agee.2017.06.014.

Kimemia, D. and Van Niekerk, A. (2017) 'Cookstove options for safety and health: Comparative analysis of technological and usability attributes', *Energy Policy*, 105, pp. 451–457. doi: 10.1016/j.enpol.2017.03.022.

Kohler, M. (2016) 'An economic assessment of bioethanol production from sugar cane: The case of South Africa', *Economic Research Southern Africa*, (August), p. 14.

Leal, M., Leite, J., Chagas, M., da Maia, R. and Cortez, L. (2016) 'Feasibility assessment of converting sugar mills to bioenergy production in Africa', *Agriculture*, 6(3), p. art 45. doi: 10.3390/agriculture6030045.

Leite, J. G. D. B., Leal, M. R. L. V., Nogueira, L. A. H., Cortez, L. A. B., Dale, B. E., da Maia, R. C. and Adjorlolo, C. (2016) 'Sugarcane: a way out of energy poverty', *Biofuels, Bioproducts and Biorefining*, 10(4), pp. 393–408. doi: 10.1002/bbb.1648.

Letete, T. and Blottnitz, H. von (2011) 'Biofuel policy in South Africa: A critical analysis', in Janssen, R. and Rutz, D. (eds) *Bioenergy for Sustainable Development in Africa*. Springer, Heidelberg, Germany, pp. 191–199.

Lynd, L. R., Sow, M., Chimphango, A. F. A., Cortez, L. A. B., Brito Cruz, C. H., Elmissiry, M., Laser, M., Mayaki, I. A., Moraes, M. A. F. D., Nogueira, L. A. H., Wolfaardt, G. M., Woods, J. and Van Zyl, W. H. (2015) 'Bioenergy and African transformation', *Biotechnology for Biofuels*, p. art 18. doi: 10.1186/s13068-014-0188-5.

von Maltitz, G., Gasparatos, A. and Fabricius, C. (2014) 'The rise, fall and potential resilience benefits of Jatropha in Southern Africa', *Sustainability (Switzerland)*, 6(6), pp. 3615–3643. doi: 10.3390/su6063615.

Mohammed, Y. S., Bashir, N. and Mustafa, M. W. (2015) 'Overuse of wood-based bioenergy in selected sub-Saharan Africa countries: Review of unconstructive challenges and suggestions', *Journal of Cleaner Production*. Elsevier Ltd., 96, pp. 501–519. doi: 10.1016/j.jclepro.2014.04.014.

Munyinda, K., Yamba, F. and Walimwipi, H. (2012) 'Bioethanol potential and production in Africa: Sweet sorghum as a complementary feedstock', in Janssen, R.

and Rutz, D. (eds) *Bioenergy for Sustainable Development in Africa*. Springer, Dordrecht, pp. 81–91. doi: 10.1007/978-94-007-2181-4_8.

Petersen, A. M., Aneke, M. C. and Görgens, J. F. (2014) 'Techno-economic comparison of ethanol and electricity coproduction schemes from sugarcane residues at existing sugar mills in Southern Africa', *Biotechnology for Biofuels*, 7(1), pp. 1–19. doi: 10.1186/1754-6834-7-105.

Singh, K., Singh, B., Verma, S. K. and Patra, D. D. (2014) 'Jatropha curcas: A ten year story from hope to despair', *Renewable and Sustainable Energy Reviews*, 35(2014), pp. 356–360. doi: 10.1016/j.rser.2014.04.033.

Smith, J. U., Fischer, A., Hallett, P. D., Homans, H. Y., Smith, P., Abdul-Salam, Y., Emmerling, H. H. and Phimister, E. (2015) 'Sustainable use of organic resources for bioenergy, food and water provision in rural Sub-Saharan Africa', *Renewable and Sustainable Energy Reviews*, 50, pp. 903–917. doi: 10.1016/j.rser.2015.04.071.

Statista (2016) *Statista: Ethanol fuel production in top countries 2016*. Available at: www.statista.com/statistics/281606/ethanol-production-in-selected-countries/.

UN-Habitat (2014) *State of African cities 2014, re-imagining sustainable urban transitions*. UN-Habitat. Available at: https://unhabitat.org/books/state-of-african-cities-2014-re-imagining-sustainable-urban-transitions/.

United Nations (2017) *World Population Prospects Key Findings and Advance Tables, World Population Prospects: The 2017 Revision, Key Findings and Advance Tables*. doi: 10.1017/CBO9781107415324.004.

Watson, H. K. (2011) 'Potential to expand sustainable bioenergy from sugarcane in southern Africa', *Energy Policy*, 39(10), pp. 5746–5750. doi: 10.1016/j.enpol.2010.07.035.

Wright, G. (2018) 'SADC Sugar Digest 2018', p. 70.

Van Zyl, W. H., Chimphango, a F. a, den Haan, R., Görgens, J. F. and Chirwa, P. W. C. (2011) 'Next-generation cellulosic ethanol technologies and their contribution to a sustainable Africa', *Interface Focus*, 1(2), pp. 196–211. doi: 10.1098/rsfs.2010.0017.

Van Zyl, W. H., Görgens, J. F. and Brent, A. C. (2017) 'The Science base for Biofuels', in Ruppel, O. C. and Dix, H. (eds) *Roadmap for Sustainable Biofuels in Southern Africa*. Konrad Adenauer Stiftung, Nomos Verlagsgesellschaft, Baden-Baden, Germany, pp. 11–46.

33 Sugarcane bioenergy, an asset for Latin America, the Caribbean and Africa?

Patricia Osseweijer

Evidence for opportunities

The book presents an abundance of technical knowledge on projected opportunities for sustainable energy provision and transport fuels made from sugarcane (SC), supported by the long-term successful experience in Brazil. Examples show that other Latin American (e.g. Colombia) and Caribbean countries are already implementing the increasing use of SC for their energy mix, and recommendations are made to set blending targets to help create a solid market. The opportunities for sustainable ethanol and linked electricity production also seem abundant for sub-Saharan countries. van Zyl calculated a SC ethanol production of 14 billion litres/year, and Cutz et al. saw Zimbabwe, Zambia, Mozambique, Malawi and Angola as potential exporters of SC ethanol. Both are in line with International Renewable Energy Agency (IRENA) that predicts a possible overproduction of biofuels (of diverse crops) to the needs of 2050 based on opportunities in five countries (Uganda, Mozambique, Nigeria, Ghana and South Africa) (IRENA, 2017). So, it should be possible to sustainably produce bioenergy from SC. Notable contributions by, among others, Leal, Oliverio, Moreira, Cardoso, Rossell and Leite et al., provide detailed answers on how this could be done, what investments are needed (e.g. gradual investments in existing refineries) and where (soil) opportunities are best. Many, however, mention social constraints, and while several considerations for improving conditions are given, this seems to be the biggest question.

This was also a leading question in a dedicated multidisciplinary workshop organised by Osseweijer, Lynd and Landeweerd in 2016, supported by the Netherlands Royal Academy and Lorentz Centre ("Bridging Technological and Social Innovation for a Biobased Economy"). Inspired by insights produced by LACAf authors and the Global Sustainable Bioenergy (GSB) project, experts from different disciplines and continents were invited to discuss the urgency and possible actions to accelerate the move to a sustainable bioeconomy. They produced a joint statement about the role of biomass-derived energy and products for sustainable development, addressing global challenges and offering a set of recommendations (https://nias.knaw.nl/news/lorentz-biopanel-releases-statement-on-biomass-and-sustainability/).

The expert group found solid evidence to support the expectation that biomass will have to play an essential role in the mitigation of greenhouse gas (GHG) emissions and has a unique position for the production of materials, which cannot be (easily) replaced by other alternatives, and that biomass also has a distinctive role in providing sustainable liquid fuels. They choose to spotlight on the bio-based component of renewable alternatives, as this provides some unique relations with sustainability pathways, including rural development, and energy and food security. However, to maximise sustainability benefits with respect to the environmental, economic and social aspects, chances in biomass utilisation should be carefully assessed and managed. The multidisciplinary input in the workshop provided insights on values and worldviews; ethics; transition management; options for bio-based technology development; economic consequences; sustainability issues; and accountability, responsibility and policy options. The group concluded that the positive contribution the biomass can provide in the mix of renewable resources deserves more attention. This book is therefore a very welcome contribution to help enabling the urgent transition.

Challenges and constraints

The joint statement of the expert group identified key enablers and action points for sustainable biomass utilisation in both developed and less developed regions. The paper starts with a vision, making the point that

> ...A bio-based economy calls for a fundamental paradigm shift, a change in our foundational values, aspirations, and worldviews, to transform current patterns of unsustainable production and consumption and change the way fossil fuels are engrained in social, economic and political mentalities....

This view demands the explicit consideration of non-technological factors, which may provide criteria for change and investment in SC production facilities, as also offered by several authors in this book. "...The debate thus should not be restricted to technological and economic potentials, but needs to include the social and political realities that constrain the pace of change...". Rui Carlos da Maia underlines this with his conclusion that bioenergy expansion should be preceded by political dialogue at an international level, including the World Bank, the ADB (African Development Bank) and FAO (Food and Agriculture Organization), while taking perceptions of local governments and youths into perspective. He points out that in fact the present support of aid organisations to subsistence farming is counterproductive, with funds enriching the elites and youths remaining uninterested in rural jobs. He suggests to put efforts in connecting sub-Saharan Africa (SSA) agriculture with international trade and developing a supply-driven agriculture.

The expert joint statement echoes this in stating that technological innovation and new markets could inspire young farmers to re-engage with agriculture, acknowledging that subsistence farming is not appealing to the young and is inherently unstable. The statement does however put more emphasis on the urgency of the required transition, in the acknowledgement that an economic crisis is emerging, with associated problems of financial uncertainty and the need for risk-taking (Worldbank, 2013 and 2015). Like Leite et al., the authors do also acknowledge the need to address poverty and energy security. While the statement underlines the crucial need for improved efficiency of biomass conversion and the need to involve public and private institutions in securing energy for social development which is expected to increase food security, this book further illustrates the complexities in this endeavour. Referring to the almost total absence of an energy grid infrastructure, Horta points to the difference between energy security, national concept and energy poverty, felt by households. Leal describes the UN SE4ALL program which tries to deal with this problem but Leite et al. conclude that it is not an energy access but the fact that households cannot afford energy that provides the problem. SC could help in providing excess electricity but is restrained by seasonal production. A further complication is the economy of scale that would be needed, as Chagas points out, small-scale (below 4 Mtons/annum) production of SC transport fuels and electricity makes ethanol too expensive. Investments in technical innovation and R&D may offer remedies as suggested by Cortez, but more is needed. Education is crucial to make poor countries self-supporting, not only to provide technology expertise and innovation both in agriculture and in production, but also to acquire knowledge of supply chains, markets and policy incentives.

The boxes below present the leverage points and enabling factors identified by the expert panel in 2016.

Box 33.1 Leverage points*

(a) Avoiding carbon lock-in through sustainable infrastructure consistent with bio-based economy (BBE), which involves lower cost due to the low level of existing fossil infrastructure and needs to be based on a longer-term perspective. Support agricultural extension and investment, and highlight the importance of the BBE as a driver of social enterprise at the local scale;

(b) Encouraging natural indigenous talent and entrepreneurship (especially information and communications technology (ICT)) to develop transformative ideas into practice. Identifying, finding and supporting transformative ideas/topics and activities (avoid reinventing the wheel);

(c) Drawing on existing markets, including traditional products, e.g. medicines, foods, artefacts);
(d) Identifying key stakeholders, e.g. decision makers/thought leaders/public sector/private sector/politicians/technologists, as change agents—can also be institutions, e.g. SEBRAE (www.sebrae.com.br) in Brazil;
(e) Reinforcing existing centres for excellence for research and education in developing countries or create them where they don't exist.

---

* **Leverage points** are places within a complex system (a corporation, an economy, a living body, a city, an ecosystem) where a small shift in one thing can produce big changes in everything.

Lorentz BioPanel expert Statement, 2016.

Box 33.2 Enabling factors*

(a) Be location-specific, identify barriers in infrastructure and build required infrastructure, take cultural and social issues and values that impact social development into account;
(b) Ensure the prerequisites for development: personal security, sufficient food and shelter – since people are in survival mode and need to get out;
(c) There are plenty of land and labour, but no market or technology – establish platforms to create technology and market channels;
(d) Upgrade basic infrastructure for education and communication for sustainable agriculture and innovation in BBE continuously and (more) equitably;
(e) Establish strong institutions and legal structures for land and property that are enforced, so people can work and invest in the future;
(f) Organise access to markets (distribution), access to water and access to information.

---

* **Enabling factors** are factors that make it possible for individuals or populations to change their behaviour or their environment include resources, conditions of living, societal supports, and skills that facilitate a behaviour's occurrence.

Lorentz BioPanel expert Statement, 2016.

This book offers a very welcome in-depth knowledge on the opportunities SC can offer to local and regional development, which seems very promising. It resonates the constraints identified by the expert panel workshop, providing insights on economies of scale (Chagas), environmental sustainability and economic impacts (da Cunha), adding suggestions for political incentives such as reinvestment of earnings in bioethanol production to agricultural R&D and education (Cortez). All agree that governance is an important factor, specific mention is made to aspire equity and organise land tenure.

Final considerations – what can we do?

Especially the transition challenges, highlighted in the expert statement, require further attention. Tuned (further) development of technology, resources, markets, logistics, financial systems and others is crucial. But, in addition to investment, other actions are needed related to issues of risk, functional and cross-sector governance and culture. This transition requires multi-actor networks defining guiding principles, values and pathways, and keeping the persistent problem of continuous use and dependency of fossil resources on the agenda. Here the Brazilian experience can be of only limited value, as cultures in SSA and other poor and emerging nations differ. Transdisciplinary networks help to identify and understand the problems and find solutions. It is necessary to understand that different cultures have different values and worldviews, which provide the basis for their acceptance, and willingness and directions of change to novel practices. A sufficiently powerful positive and persuasive vision for the BBE, including development of greater consensus and in some cases myth-busting, is needed in order to coalesce critical mass for action. The expert statement specifies a series of recommendations for action, with emphasis on collaboration with other renewable sectors and increased communication, involving credible actors and aligning with (inter)national organisations and local people, supported by independent advice on ethical dimensions.

Several authors in this book refer to the difficulties of public perceptions and the need to reconcile the persistent debate on food versus fuel (Rosillo-Calle et al.). While persistent beliefs on the lack of land, negative linkage to food prices and unsustainability are dismantled, it remains difficult to show the positive contribution that SC ethanol can bring to energy *and* food security. It seems that these debates are fought high above the heads of the people they refer too. Indeed, science communication studies show that rational argumentation does not convince those with emotionally based opinions (Sleenhoff et al., 2014; Osseweijer, 2006). Revisiting the problem definition may prove a better strategy for communication than rebutting myths: The world population is growing, not in the developed nations but especially in the poor areas, and most strikingly in SSA: where poverty is already a problem for the majority, where 1.8 billion are under 18 years of age, where 15 million enter the workforce every year, where unemployment reaches 80% and where urbanisation is a growing problem (Kline et al., 2016; Osseweijer et al., 2016, Souza et al., 2017). Yet, these are regions where suitable land is available for sustainable agriculture, that with technological (agriculture related) investment can deliver jobs which are liked by the youth, providing energy for households and social development, and creating a market for production of goods. This can (eventually) provide export opportunities, creating more stable governments. A sustainable planet will depend on this social and economic development. Such approaches can more easily align bio-based (business) opportunities with approaches by NGOs and other influential international organisations. The details provided here in this book may lead the way into how this can be done.

Must we?

As presented above and in more detail in the various chapters of this book, there are many challenges in finding leverage points and implementing enabling factors to improve the situation in less developed countries (LDCs). But as the expert workshop concluded, inaction on implementing the modern BBE is particularly unethical in the LDCs because there is so much potential to emerge from the current hybrid of traditional biomass used and the fossil economy that condemns a significant portion of the population to poverty and resource degradation. And, as Lynd suggests in Chapter 8, we have a global responsibility, not only to water, air and climate but also to land.

The overall outcome suggests that we should change our strategy in communication, and foremost join forces to help regional, national and local communities to implement sustainable and equitable land tenure and management systems and policies to sustainable transition paths. This should be accompanied by *joint* efforts in providing (online and free accessible) education and training to achieve self-supportiveness. Energy poverty reduction and job creation are key focus points for public–private initiatives to ensure successful (long-term) development. In the identified areas of appropriate conditions, SC-based ethanol and electricity production with its required scale and supply chain organisation can play a crucial role in helping to realise the required transition from subsistence farming and energy poverty to successful agricultural production for food, feed, energy and materials, contributing to rural, economic and national development.

> …Current political and institutional regimes at national and international levels remain largely aligned with existing fossil-based economic structures. This gives rise to an urgent need for action from leaders in public and private sectors as well as civil society in order to create a level playing field that can reverse the dirty development pathways of the past. The implementation of a sustainable biobased economy as part of a circular economy, that improves energy and food security is not only an ethical imperative, but also offers an inspirational and therefore mobilizing vision of a more just and higher-quality future for all.

References

IRENA (2017) Biofuel potential in Sub Saharan Africa: Raising food yields, reducing food wastes and utilising residues. www.irena.org/-/media/Files/IRENA/Agency/Publication/2017/Nov/IRENA_Biofuel_potential_sub-Saharan_Africa_2017.pdf (last accessed 28-3-2018).

Kline, K.L.; Siwa Msangi, Virginia H. Dale, Jeremy Woods, Gláucia M. Souza, Patricia Osseweijer, Joy S. Clancy, Jorge A. Hilbert, Francis X. Johnson, Patrick C. McDonnell, Harriet K. Mugera (2016). *Reconciling food security and bioenergy: priorities for action.* GCB Bioenergy.

Osseweijer, P. (2006). A new model for science communication that takes ethical considerations into account – The three-E model: Entertainment, Emotion and Education. *Science and Engineering Ethics*, 12(4), 591–593.

Osseweijer, P., Watson, H.K., Johnson, F.X., Batistella, M., Cortez, L.A.B., Lynd, L.R., Kaffka, S.R., Long, van Meijl, J.C.M., Nassar, A.M., and Woods, J. "Bioenergy and Food Security" (2015) *in* Souza, G. M., Victoria, R., Joly, C. Verdade, L. (Eds). (2015). *Bioenergy & Sustainability: Bridging the Gaps* (Vol. 72, p. 779). Paris. SCOPE. ISBN 978-2-9545557-0-6.Rosillo-Calle F (2012) Food versus Fuel: toward A New Paradigm-The Need for a Holistic Approach (Commissioned paper "Spotlight Article for ISRN Renewable Energy, Volume 2012, Article ID 954180

Sleenhoff, S.; E. Cuppen and P. Osseweijer (2014). *Unravelling emotional viewpoints on a bio-based economy using Q methodology.* Public Understanding of Science DOI: 10.1177/0963662513517071 published online 13 June 2014.

Souza et al. (2017) "The role of Bioenergy in a climate changing world", Journal of Environmental Development, Elsevier.

Worldbank reports: "Shock waves: managing the impact of Climate Change and Poverty" (2015); and "Turn down the heat" (2013): https://openknowledge.worldbank.org/bitstream/handle/10986/22787/9781464806735.pdf; https://openknowledge.worldbank.org/bitstream/handle/10986/22787/9781464806735.pdf; http://documents.worldbank.org/curated/en/2013/06/17862361/turn-down-heat-climate-extremes-regional-impacts-case-resilience-full-report.

34 Closing remarks

Luís A. B. Cortez, Manoel Regis L. V. Leal and Luiz A. Horta Nogueira

Socioeconomic development is certainly the final goal of emergent countries and societies. Developed countries are now focused on reducing their greenhouse gas (GHG) emissions while making the transition from fossil-based economy to a more sustainable model based on renewable energies. This transition will probably take several decades, and it can be expected an enormous effort and will to undertake the necessary work involved in this endeavor.

For the developing countries, particularly in Africa and South Asia, the crucial matter is to eliminate poverty in all its aspects, create a more equal society, with more opportunities and, at the same time, preserve and protect their environment. It is expected that until the end of the 21st century, Africa alone will have 4 billion inhabitations, so almost quadrupling its present population. This extraordinary increase in population will occur with great pressure on food and energy securities.

The first part of the book was devoted to understand the transition context and the difficulties involved in choosing bioenergy as a primary source of energy. It is becoming clear that bioenergy will have to be competitive, not with petroleum, natural gas or coal, but with the other renewable energies, such as wind and solar photovoltaics. It is quite impressive the speed with which these renewables are increasing their participation in the market. So, bioenergy will suffer great competition but, on the other hand, bioenergy offers comparative advantages that can last for several decades, may be a century, until it will be phased out.

In this book, we tried to understand if bioenergy, in particular from sugarcane, can contribute both to improve energy and food supply, while generating income and increasing the offer of new employment. Several chapters were devoted to understand the food vs bioenergy dilemma. Several sensitivity studies were conducted giving numbers, not opinions, to quantify the possibilities and try to generate conclusions regarding bioenergy projects. At a certain point, it was concluded by our researchers that for better understanding the options and possibilities we had to use countries context for the correct analysis. Projects are country dependent, and the best way to understand the advantages and disadvantages involved is to select a representative country for the analysis.

Mozambique was taken for a case study in Africa because the country still presents low human development index but has enormous natural resources, including large tracts of land almost without use. Several studies were made before exploring Mozambique case and still there is no clear consensus about the benefits of bioenergy in this kind of scenario. In those conditions, social impact was quantified for the implementation of sugarcane ethanol projects. We tried to evaluate if a country such as Mozambique, with so many energy possibilities, should or should not devote its resources to produce sustainable bioenergy. It is important to say that traditional bioenergy plays a very important role in the Mozambican reality but it doesn't seem to be giving any positive contribution to improve living conditions. On the contrary, traditional bioenergy seems to be perpetuating the suffering of people, submitting women and children to the hard work of gathering wood and cook in inefficient and smoky stoves.

In the course of presenting a systematic way to produce modern sustainable sugarcane bioenergy, we tried to demonstrate the advantages and positive aspects involved making a transition from underdeveloped to a more satisfactory society. We don't expect that by the simple adoption of sugarcane bioenergy (ethanol and bioelectricity), developing countries can make the way out of poverty, but it can help significantly.

In the case of Brazil, when the oil crisis hit heavily the Brazilian economy during the 1970s, it was sugarcane bioenergy that helped the country to alleviate oil dependence while improving food security. Yes, it is exactly right: the expansion of sugarcane plantation in Brazil helped, at the same time, to decrease fossil fuel dependence and to improve productivity not only of sugarcane but also of other crops helping the country to boost a robust agribusiness in a very competitive agricultural sector.

The book also presented a systematic way to map agricultural land for sugarcane. This is another important contribution, indeed. While mapping land with good potential to produce sugarcane, we can understand the aptitude of land for other crops. Organizing land use is a crucial aspect of agricultural development and environment protection. This has been done by developed countries in one way or another. It is almost impossible to develop a country without planning its land use. The mapping that was done and presented here can be considered as a first step for the implementation of a sustainable agriculture not only for bioenergy but for food and other uses.

Last, while trying to answer the "how" question, it was presented in several studies using the virtual biorefinery tool developed by the CTBE (Brazilian Bioethanol Science and Technology Laboratory), Brazil. With this tool one can understand, for example, if scale is or isn't an important factor to take into consideration. Many people still believe that "small is beautiful"... we tried to demonstrate that bioenergy hardly can be competitive at small scale. On the other hand, using the virtual biorefinery we tried to show how integrated systems can be advantageous, and why bioelectricity can be produced in a very competitive way helping small communities to insure a better quality of life.

While we come to an end of this book, we expect to show how bioenergy is model-dependent, how important it is to make right choices to conceive sustainable models. We certainly don't believe there is only one way to go, but we hope to have demonstrated that sustainable bioenergy can certainly play an important role in the development process.

Index

2°C increase scenario (2DS) 14, 110
6°C increase scenario (6DS) 14
2DS see 2°C increase scenario (2DS)
2G ethanol research 12–14

ABE see acetone-butanol-ethanol (ABE)
ABM see agent-based model (ABM)
aboveground net primary production (ANPP) 51
acetone-butanol-ethanol (ABE) 99, 100
ACP see Africa, Caribbean and Pacific (ACP)
acrisols, subsoil 224
Addax Bioenergy Project 352
advanced motor fuels (AMF) 20
Africa, bioenergy in: biofuels 391–2; Brazilian experience and LACAf findings 393–5; methodology 380–4; modern bioenergy 389–91; opportunities and challenges 388–9; overview of 377–9; research 380; risk evaluation and mitigation process 384–6; and rural electrification 379–80; sugarcane 392–3; sugarcane lignocelluloses in 98; sustainable development 395–6
Africa, Caribbean and Pacific (ACP) 178
Agave model 52
agent-based model (ABM) 202
agricultural system, yield gap of 50
agricultural zoning 218, 220, 228
agroecosystem modeling 54–5
agro-industrial technologies 73–4
alcohol production 366–8
AMF see Advanced Motor Fuels (AMF)
anaerobic digestion 339–40
anhydrous ethanol 10, 71
ANPP see aboveground net primary production (ANPP)
ASAZGUA (Guatemalan Sugar Association) 370–2

AspenPlus™ 98, 303
Average annual temperature (AAT) 218

Bagamoyo Sugar Project 351–2
bagasse-based power generation 148
bagasse, cogeneration with 361–3
"battery buying power" 94
Better Sugarcane Initiative (BSI) 352, 373
bioalcohol 357–61
bioenergy: biofuels 391–2; Brazilian experience and LACAf findings 393–5; challenges and constraints 401–4; considerations 404–5; development of 405; diplomacy 21–2; evidence for opportunities 400–1; feedstock production 110; methodology 380–4; modern bioenergy 389–91; opportunities and challenges 388–9; overview of 377–9; production system 266–8; research 380; risk evaluation and mitigation process 384–6; and rural electrification 379–80; sugarcane 135, 392–3; sustainable development 395–6; systems models, types of 287–8
Bioenergy and Food Security (BEFS) 265
bioenergy crop expansion: animal productivity 51; land, distribution of 47; pastureland 46; yield gap 47
bioenergy trade: bagasse-based electricity fed 121; barriers and opportunities 125–8; crude-based diesel fuel 123; government policies in 123–5; long-term markets 128–30; prospects for 128–30; sustainable trade 130; trade dimensions 121–3
bioethanol production 364–6
BIOFOM process 340–1
biofuels 391–2; markets 122; policy and strategy 227; production 219, 220, 221; sustainable development 395–6

biofuels-based food crops 37
biomass, bioenergy 29
biomass to liquids (BTL) 100
Biostil process 335–6
BONSUCRO sustainability principles 267
Brazil, bioenergy in: agricultural production in 136, 162; bioenergy production 30–2; biofuel output 162; biofuels, evolution of 122; biofuels industry in 111–12; cattle production 163; deforestation 162; ethanol-based fuels 10; ethanol fuel programme 37; ethanol industry in 62; ethanol production, evolution of 31; food security and deforestation in 155, 157–61; fuel ethanol in 9–12; landscape-scale process-based modeling 48; oil crisis of 9; pasture area and livestock performance 49; sugarcane ethanol in 156–7; sugarcane production systems 300
Brazilian Center for Research in Energy and Materials (CNPEM) 12
Brazilian Center for Sugarcane Technology (CTC) 229
Brazilian energy matrix 290
Brazilian experience and LACAf findings 393–5
Brazilian sucroenergy industry: bioelectricity 76; design development 75; engineering capability 79; ethanol production model 269–70, 317; financial capability and guarantees 78–9; future competitiveness 79; growth of 78; industrial/manufacturing capability 78; products, capacities, and technologies 80; sugarcane mills, profile of 75–6; sustainability 80–3; technological evolution of 76–7
Brazilian Sugarcane Zoning 215
Brazil's Agroecological Zoning 111
Brownfield project 266
BTL *see* biomass to liquids (BTL)

California Low Carbon Fuel Standard 126
CanaSoft model 318–19
capital expenditures (CAPEX) 309
carbon emissions 93, 111, 138, 141
cation exchange capacity (CEC) 222
CBP *see* consolidated bioprocessing (CBP)

cellulosic biomass 113
cellulosic ethanol technology 51–2
CGEE 221, 223
CHP *see* combined heat and power (CHP)
"climate-binning" method 50
climatic restrictions 221–2
CNPEM *see* Brazilian Center for Research in Energy and Materials (CNPEM)
coal-to-liquid (CTL) process 97, 101, 179
cogeneration capacity 361–3
Colombia, bioenergy in 215–16, 357; alcohol production 366–8; bioalcohol in 357–61; bioethanol production 364–6; case 218, 220, 221; cogeneration capacity 361–3; land cover maps 246–8; law 244; natural regions map of 244; production potential of 252–6; restrictions of 250–2; sugar-energy sector 244; sugar mills/sugarcane map in 245, 246; territory 218, 219, 246, 248, 249
combined heat and power (CHP) 98
Commission for Energy and Gas Regulation (CREG) 361–2
consolidated bioprocessing (CBP) 116
cooking ethanol 280
cooperatives, comparative analysis 293–5
crop-based biofuels 37, 38
crop irrigation 221
Cropland/natural vegetation mosaic 221, 223
crop-livestock systems 46
CTL process *see* coal-to-liquid process
CUE study 364–6
CVA *see* High conservation value areas (CVA)

DANE *see* National Administrative Department of Statistics (DANE)
DDGs *see* dried distillers' grains (DDGs)
DDGS *see* distillers' dried grains with solubles (DDGS)
Dedini Sustainable Mill (DSM) 80–3
deforestation 162, 166
Delphi method 207
Diesel 291
direito de uso e aproveitamento da terra (DUATs) 228
distillers' dried grains with solubles (DDGS) 127
dried distillers' grains (DDGs) 11, 37

Index

DSM *see* Dedini Sustainable Mill (DSM)
DUATs *see* direito de uso e aproveitamento da terra (DUATs)

econometric model 202
economic analysis 321
economic assessment 322–4
economic growth, energy consumption 274
Economic Research Service (ERS) 128
electric vehicles: in 1900 90–1; battery storage capacity 91; CO_2 emissions 90; electricity 88, 92; Lithium batteries 89
EM *see* equivalent-mill (EM)
energy consumption: Africa 388; economic growth 277
Energy Independence and Security Act (2007) 122
energy ladder theory 275
Energy Policy Act 10, 122
energy poverty (EP) 172, 274
energy security (ES) 172, 274; global view 173–6; Mozambique 227; sugarcane, improver of 176–7
environmental analysis 321
environmental assessment 324
environmental preservation 279
Environmental Sanitation Technology Company (CETESB) 333
EP *see* energy poverty (EP)
"EPZ – Ethanol Project in Zimbabwe" 83
equivalence ratio 277
equivalent-mill (EM) 73
ERS *see* Economic Research Service (ERS)
ethanol-based fuels 10
ethanol consumption 73
ethanol production model 265–6; Brazilian case 269–70
European Sugar Regime 178

FAO *see* Food and Agriculture Organization (FAO)
Farmer Development Plan (FDP) 352
FCEV *see* fuel cell electric vehicle (FCEV)
Ferralsols 223
Fischer-Tropsch (FT) 100–3
Fluvisols 224
Food and Agriculture Organization (FAO) 68, 247, 289, 290
"food-fuel" conflict 155
Food Price Index 41
food security: in Brazil 157–61; children, malnutrition 157; methodological approach 156; sugarcane ethanol 156–7
food *versus* biofuel dilemma: agricultural products 43; biofuels production 37–8; food prices and wastes 40–3; grain production 38–40; historical development of 36; historical perspective 36–7; land use 37
fossil-derived fuels 99
fossil energy: correct use of 290–1; emissions 14
FT *see* Fischer-Tropsch (FT)
fuel cell electric vehicle (FCEV) 15
fuel–engine interactions 19
fuel ethanol 373–5; bioenergy, markets for 18–19; biofuels costs comparison 20; in Brazil and US 9–12; energy consumption 19; ethanol diplomacy 21–2; future transportation 16–18; high-level ethanol blends 20; liquid biofuels 18; projects of 12; sustainable biofuels 20–1; technological considerations for 19–20; transportation markets 14–16; 2G ethanol research 12–14
fuelwood collection 285
"Future-Mills" design 73–4, 84

gasification-FT (G-FT) process 101
gasification-synthesis process 100, 106
gasoline additive 71
gasoline consumption 73
gasoline distribution networks 71–2
gas-to-liquid (GTL) 97
GBEP *see* Global Bioenergy Partnership (GBEP)
GBEP sustainability indicators 267, 268
GCH *see* green cane harvesting (GCH)
GDP *see* gross domestic product (GDP)
GEM *see* general equilibrium model (GEM)
general equilibrium model (GEM) 202
G-FT process *see* gasification-FT process
GHG *see* greenhouse gas (GHG)
Gleysols 224
Global Bioenergy Partnership (GBEP) 220
global-scale simulations 52
Global Sustainable Bioenergy (GSB) 48, 109, 215
Global Tracking Framework 173

Global Trade Analysis Project (GTAP) 182
grain production 38–40
Granbio Project 14
"Great Recession of 2008" 11
green cane harvesting (GCH) 98
greenhouse gas (GHG) 12, 14, 52, 103, 135, 219, 317
Greenhouse Gases, Regulated Emissions, and Energy use in Transportation (GREET) 375
gross domestic product (GDP) 173
GSB see Global Sustainable Bioenergy (GSB)
GTAP see Global Trade Analysis Project (GTAP)
GTL see gas-to-liquid (GTL)
Guatemala 370; fuel ethanol in 373–5; perspectives 375; sugarcane industry in 370–3

HDI see Human Development Index (HDI)
HEFA see hydro-processed esters and fatty acids (HEFA)
High conservation value areas (CVA) 220
High Impact Opportunities (HIOs) 173
High Octane Fuels (HOF) 19
HIOs see High Impact Opportunities (HIOs)
Histosols 224
HOF see High Octane Fuels (HOF)
Human Development Index (HDI) 89
hydropower 279
hydro-processed esters and fatty acids (HEFA) 103
hydrous ethanol 10

IA see integrated analysis (IA)
IC see inherent context (IC)
ICEs see internal combustion engines (ICEs)
IEA see International Energy Agency (IEA)
IGBP see International Geosphere-Biosphere Program (IGBP)
IIAM see Institute of Agricultural Research of Mozambique (IIAM)
ILO see International Labour Organisation (ILO)
ILUC see indirect land use changes (ILUC)
Improved Biomass Technologies (IBTs) 288; comparative analysis 293–5
inappropriate soil 217
indirect land use changes (ILUC) 155
Industrial Innovation Plan 13
inherent context (IC) 205
input-output analysis (I-O) 202
Institute of Agricultural Research of Mozambique (IIAM) 222, 233
Instituto Geográfico Agustín Codazzi (IGAC) 364
integrated analysis (IA) 199, 204
Integrated Food-Energy Systems (IFES) 265, 289
Inter-American Development Bank (IDB) 289
internal combustion engines (ICEs) 88, 93
internal rate of return (IRR) 309
International Energy Agency (IEA) 14, 27, 88, 144, 172
International Geosphere-Biosphere Program (IGBP) 229, 246
International Labour Organisation (ILO) 64
International Renewable Energy Agency (IRENA) 27, 29, 400
International Sustainability and Carbon Certification (ISCC) 373
investment capital 280
I-O see Input-output analysis (I-O)
IRENA see International Renewable Energy Agency (IRENA)
IRR see internal rate of return (IRR)

Japan International Cooperation Agency (JICA) 358
Jatropha curcas 52
JICA see Japan International Cooperation Agency (JICA)

Köppen classification 221

LACAf see Latin America, Caribbean and Africa (LACAf)
LACAf project 109, 142, 265, 357
Landeweerd, L. 400
landscape design 202
large-scale distilleries 319–20
Latin America and the Caribbean (LAC) 141, 177–8; bioelectricity supply 147–8; bioenergy infrastructure in 149–51; closing remarks 151–2; crop-livestock-forestry systems 144; ethanol production 145–7; GHG mitigation in 148–9; new framework

143–4, 146; sugarcane bioenergy production 143–5; sustainable bioenergy development 151–2
Latin America, Caribbean and Africa (LACAf) 216
Latin Hypercube method 305
LCA *see* life cycle assessment (LCA)
LDCs *see* least developed countries (LDCs)
LDV *see* light-duty vehicles (LDV)
least developed countries (LDCs) 178
legal restrictions, agricultural expansion 218–19
life cycle assessment (LCA) 202, 204, 321
light-duty vehicles (LDV): passenger, market of 15–16; technological considerations for 19–20
liquid biofuels 18, 93
Lithium batteries 89
Lixisols 224
low-carbon bioenergy production 46
LPG 290
Luvisols 224

macroeconomic objectives 1
Malawi 280
man-made GHG emissions 14
MAP *see* Market Access Program (MAP)
Market Access Program (MAP) 127
MCA *see* multi-criteria analysis (MCA)
MDGs *see* Millennium Development Goals (MDGs)
mechanization 347–9
methyl *tert*-butyl ether (MTBE) 10, 72, 122
microdistilleries technologies 320–1
Millennium Development Goals (MDGs) 172
minimum acceptable rate of return (MARR) 321
Moderate Resolution Imaging Spectroradiometer (MODIS) 54; images of 229; MCD12Q1 data 247
modern bioenergy (MB) 286, 389–91; comparative analysis 293–5; deforestation reduction 291, 293; in developed countries 287; in less developed countries 290; production costs 291, 292; social impacts 291, 293; *vs.* traditional bioenergy 289–90
"modern biomass" 27, 29
MODIS *see* Moderate Resolution Imaging Spectroradiometer (MODIS)

Mozambique, bioenergy in: agricultural sector of 194, 239; area without strong restrictions 234–6; bioenergy policy in 241; bioenergy sector 188; economy of 227; electricity production in 227; final comments 258–9; independence war 227; influence over HDI 194–6; I-O analysis 182–6; land cover map of 228–31; large-scale agriculture 228, 241; large-scale bioenergy production 215; mapping process 229, 239; output multipliers 188–91; overview of 196; production potential of 233–8; restrictions of 232–3, 237–8, 240; small scale agriculture 232; social accounting matrix 186; socioeconomic impacts 191–3; sugarcane bioethanol 181; sugarcane industry in 228; territory 218, 229–30
Mozambique case study 408
MTBE *see* methyl *tert*-butyl ether (MTBE)
multi-criteria analysis (MCA) 202

National Administrative Department of Statistics (DANE) 219, 243, 250
National Alcohol Fuel Program 9
National Center for Ecological Analysis and Synthesis (NCEAS) 51
National Interconnected System (NIS) 373
National Natural Parks (PNN) 219, 250
National Renewable Energy Laboratory (NREL) 19
Natural National Parks System (SPNN) 219, 250
NCEAS *see* National Center for Ecological Analysis and Synthesis (NCEAS)
net present value (NPV) 310
NGOs *see* nongovernmental organizations (NGOs)
Nitisols 223
non-ethanol biofuel production: aviation accounts for 103–5; biorefinery for 99; description of 97–8; economic point 106; environmental impacts of 105; gasification-synthesis processes 97; lignocellulose, integration of 98–9
nongovernmental organizations (NGOs) 166
non-mechanizable Areas 219
nonsuitable soil orders 224
NPV *see* net present value (NPV)
NREL *see* National Renewable Energy Laboratory (NREL)

ODP *see* ozone depletion potential (ODP)
OEMs *see* original equipment manufacturers (OEMs)
Optima Program 72
original equipment manufacturers (OEMs) 129
Otto cycle engines 72, 73
ozone depletion potential (ODP) 304

Paris Climate Accord 52
particulate matter formation (PMF) 304
pasture-based agriculture 48
pastureland: bioenergy, intensification for 50–5; evaluation of 48; global intensification potential of 48; GSB geospatial project 48, 55–6
PC&C *see* Public Consultation and Communication (PC&C)
physical restrictions, agricultural expansion 217–18
PMF *see* particulate matter formation (PMF)
PNN *see* National Natural Parks (PNN)
pretreatment-hydrolysis-fermentation methods 100
Proálcool Program 122, 270, 367
production model: mechanization 347–9; scale 350–1; smallholder farmers 351–3; sugarcane ethanol 353–5
"productive uses of energy" 274
Public Consultation and Communication (PC&C) 199, 206

Raízen Project 14
ReCiPe Midpoint (H) 304
RED *see* Renewable Energy Directive (RED)
REFIT policy 275
Renewable Energy Directive (RED) 220
Renewable Fuel Standard (RFS) 114, 122
Restricted soil 217
RFS *see* Renewable Fuel Standard (RFS)
Roundtable on Sustainable Biomaterials (RSB) 65, 138, 352

SAM *see* Social Account Matrix (SAM)
SASA *see* South Africa Sugar Association (SASA)
scale 350–1
Schizosaccharomyces pombe 335
"scorecard for sustainable biofuels" 289

SDGs *see* Sustainable Development Goals (SDGs)
SD models *see* system dynamics models
SE4ALL *see* Sustainable Energy for All (SE4ALL)
SE4ALL High Impact Opportunities 289
second-generation (2G) biofuels: commercial deployment of 117; configuring technology 116–18; ethanol plants 114; ethanol research 12; need and potential benefits of 110–13; production technology 113–16; technology development 14, 21
SIByl-LACAf approach: addressing complexity 201–4; applying indicators 204; data, complexity of 201; decision makers, guidelines for 208; feasibility and acceptability 204–8; objectives, definition of 201; taking decision 208
SimaPro® 304
S-LCA *see* social life cycle assessment (S-LCA)
smallholder farmers 351–3
small-scale biofuel initiatives 318
SNA *see* social network analysis (SNA)
Social Account Matrix (SAM) 137, 182, 186
social infrastructure 215, 241
social life cycle assessment (S-LCA) 304
social network analysis (SNA) 202
socioeconomic development 407
South African Sugar Research Institute (SASRI) 396
South Africa Sugar Association (SASA) 64
Southern Africa (SA) *see* Latin America and the Caribbean (LAC)
SPNN *see* Natural National Parks System (SPNN)
stand-alone conversion processes 103, 104
Strengths, Weaknesses, Opportunities, and Threats (SWOT) 199, 206
Sub-Saharan Africa (SSA), bioenergy in 173, 178–9, 273; agriculture and processing coefficients 276–7; baseline scenario 275–6; challenges and drawbacks 279–80; energy efficiency and yield coefficients 277; equivalence ratio 277; ethanol plus scenario 276; ethanol scenario 276; power scenario 276; results and discussion 277–9; sugarcane-based energy 275; traditional biomass 274–5
subsistence agriculture 220

sugarcane bioeconomy: Africa and Latin America, experiences in 65–8; bagasse-based electricity 62; energy access, implications for 61–2; health and rural development 62–3; poverty reduction 62–3; socioeconomic impacts and livelihoods 63–5; sustainability indicators and certification 65
sugarcane bioethanol 9–12, 181
sugarcane ethanol production 162, 353–5
sugarcane ethanol sector 13
sugarcane lignocelluloses: biofuel production from 105; bio-jet production 103; butanol production from 99–100; gasification of 100; integration of 98–9; methanol-synfuels production 101; stand-alone processes 101
sugarcane potential production: climatic restrictions 221–2; high carbon stocks 219–20; legal restrictions 218–19; non-mechanizable Areas 219; physical restrictions 217–18; potential land use conflicts 220–1; soil group description 223–4
sugarcane production systems: agricultural phase 300–2; economic assessment 309–11; environmental assessment 303–4, 311–12; ethical basis of 137–9; impact of 136; industrial phase 302–3; principles of 139; production costs of 306–8; risk assessment 305; social assessment 304–5, 312–14; techno-economic analysis 303; technologies 318–19
Sugarcane Research Center (CTC) 14, 217, 221–3
Sugar Milling Research Institute (SMRI) 396
survey approach method 203
sustainability assessment methods 321
Sustainable Bioenergy Roadmap 110
Sustainable Development Goals (SDGs) 61, 173, 274
Sustainable Energy for All (SE4ALL) 172

Sybil-LACAf approach 137
system dynamics (SD) models 203

thermally accelerated short-time evaporation (TASTE) 337
thermochemical pretreatment-fungal cellulase paradigm 114, 115
traditional bioenergy (TB): characteristics 284–5; comparative analysis 293–5; concept of 286; as cultural activity 290; in developing countries 285–6; good practices 289; production 286; science and technology development 295
"traditional biomass" 27, 274–5
transportation markets: bioenergy, role of 16–18; fuel consumption 16, 17; global technology 15; passenger light-duty vehicles 15–16

United Nations Industrial Development Organization (UNIDO) 385
United Nations Sustainable Development Goals (UN SDGs) 263–4
United States Geological Survey (USGS) 89

vinasse: anaerobic biodigestion of 339–40; BIOFOM process 340–1; Biostil process 335–6; Brazilian distilleries case 332–4; Colombian distilleries case 338–9; combustion of 341–2; composting of 337–8; in fermentation process 335; fertirrigation 332–4; physical and chemical properties 331–2; reduction of 334; thermal concentration of 336–7
virtual biorefinery tool 408–9
Virtual Sugarcane Biorefinery (VSB) 300, 303, 318, 321

world energy consumption 296n2
World Reference Base for Soil Resources 222, 233, 252

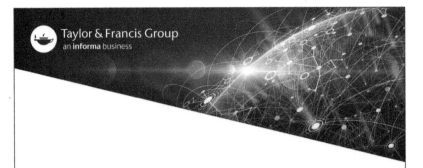

Taylor & Francis eBooks

www.taylorfrancis.com

A single destination for eBooks from Taylor & Francis with increased functionality and an improved user experience to meet the needs of our customers.

90,000+ eBooks of award-winning academic content in Humanities, Social Science, Science, Technology, Engineering, and Medical written by a global network of editors and authors.

TAYLOR & FRANCIS EBOOKS OFFERS:

- A streamlined experience for our library customers
- A single point of discovery for all of our eBook content
- Improved search and discovery of content at both book and chapter level

REQUEST A FREE TRIAL
support@taylorfrancis.com